D0906212

ENZYMATIC BASIS
OF DETOXICATION
Volume I

BIOCHEMICAL PHARMACOLOGY AND TOXICOLOGY

A Series of Monographs

WILLIAM B. JAKOBY, Editor

Section on Enzymes and Cellular Biochemistry
National Institute of Arthritis, Metabolism, and Digestive Diseases
National Institutes of Health
Bethesda, Maryland

William B. Jakoby (editor). ENZYMATIC BASIS OF DETOXICATION,
Volumes I and II, 1980

ENZYMATIC BASIS OF DETOXICATION

Volume I

Edited by

William B. Jakoby

Section on Enzymes and Cellular Biochemistry
National Institute of Arthritis, Metabolism, and Digestive Diseases
National Institutes of Health
Bethesda, Maryland

1980

ACADEMIC PRESS
A Subsidiary of Harcourt Brace Jovanovich, Publishers
New York   London   Toronto   Sydney   San Francisco

ACADEMIC PRESS, INC.
111 Fifth Avenue, New York, New York 10003

United Kingdom Edition published by
ACADEMIC PRESS, INC. (LONDON) LTD.
24/28 Oval Road, London NW1 7DX

Library of Congress Cataloging in Publication Data
Main entry under title:

Enzymatic basis of detoxication.

   (Biochemical pharmacology and toxicology series)
   Includes bibliographical references and index.
   1. Enzymes. 2. Xenobiotic metabolism. 3. Meta-
bolism. I. Jakoby, William B. , Date II. Series.
[DNLM: 1. Metabolic detoxication, Drug. 2. Enzymes
--Metabolism. QU120 E61]
QP601.E515   615.9   80-17350
ISBN 0-12-380001-3 (v. 1)

PRINTED IN THE UNITED STATES OF AMERICA

80 81 82 83   9 8 7 6 5 4 3 2 1

Contents

Chapter 1   Detoxication Enzymes

WILLIAM B. JAKOBY

Part I.   Physiological Aspects

Chapter 2   Kinetic Aspects of Metabolism and Elimination of Foreign Compounds in Animals

JAMES R. GILLETTE

Chapter 3   Human Genetic Variation in the Enzymes of Detoxication

DANIEL W. NEBERT

Chapter 11   Aldehyde Reductase

JEAN-PIERRE VON WARTBURG AND BENDICHT WERMUTH

Chapter 12   Aldehyde Oxidizing Enzymes

HENRY WEINER

Chapter 13   Ketone Reductases

RONALD L. FELSTED AND NICHOLAS R. BACHUR

Chapter 14   Xanthine Oxidase and Aldehyde Oxidase

K. V. RAJAGOPALAN

Chapter 15 Superoxide Dismutases: Detoxication of a Free Radical

H. MOUSTAFA HASSAN AND IRWIN FRIDOVICH

Chapter 16 Glutathione Peroxidase

ALBRECHT WENDEL

Chapter 17 Monoamine Oxidase

KEITH F. TIPTON

List of Contributors

Numbers in parentheses indicate the pages on which the authors' contributions begin.

Nicholas R. Bachur (281), Laboratory of Clinical Biochemistry, Baltimore Cancer Research Program, Division of Cancer Treatment, National Cancer Institute, Baltimore, Maryland 21205

William F. Bosron (231), Departments of Medicine and Biochemistry, Indiana University School of Medicine, Indianapolis, Indiana 46223

Edward Bresnick (69), Department of Biochemistry, University of Vermont College of Medicine, Burlington, Vermont 05405

John Caldwell (85), Department of Biochemical and Experimental Pharmacology, St. Mary's Hospital Medical School, London W2 1PG, England

Minor J. Coon (117), Department of Biological Chemistry, Medical School, The University of Michigan, Ann Arbor, Michigan 48109

Ronald L. Felsted (281), Laboratory of Clinical Biochemistry, Baltimore Cancer Research Program, Division of Cancer Treatment, National Cancer Institute, Baltimore, Maryland 21205

Irwin Fridovich (311), Department of Biochemistry, Duke University Medical Center, Durham, North Carolina 27710

James R. Gillette (9), Laboratory of Chemical Pharmacology, National Heart, Lung, and Blood Institute, National Institutes of Health, Bethesda, Maryland 20205

H. Moustafa Hassan* (311), Department of Biochemistry, Duke University Medical Center, Durham, North Carolina 27710

William B. Jakoby (1), Section on Enzymes and Cellular Biochemistry, National Institute of Arthritis, Metabolism, and Digestive Diseases, National Institutes of Health, Bethesda, Maryland 20205

Ting-Kai Li (231), Departments of Medicine and Biochemistry, Indiana University School of Medicine, Indianapolis, Indiana 46223

* *Present Address:* Department of Microbiology and Immunology, McGill University, Montreal, Quebec H3A 2B4, Canada.

Anthony Y. H. Lu (135), Department of Animal Drug Metabolism and Radiochemistry, Merck Sharp & Dohme Research Laboratories, Rahway, New Jersey 07065

Bettie Sue Siler Masters (183), Department of Biochemistry, The University of Texas Health Science Center at Dallas, Dallas, Texas 75235

Gerald T. Miwa (135), Department of Animal Drug Metabolism and Radiochemistry, Merck Sharp & Dohme Research Laboratories, Rahway, New Jersey 07065

Daniel W. Nebert (25), Developmental Pharmacology Branch, National Institute of Child Health and Human Development, National Institutes of Health, Bethesda, Maryland 20205

Anders V. Persson (117), Department of Biological Chemistry, Medical School, The University of Michigan, Ann Arbor, Michigan 48109

K. V. Rajagopalan (295), Department of Biochemistry, Duke University Medical Center, Durham, North Carolina 27710

Keith F. Tipton (355), Department of Biochemistry, Trinity College, Dublin 2, Ireland

Jean-Pierre von Wartburg (249), Medizinische-Chemisches Institut der Universität, CH-3000 Berne, Switzerland

Henry Weiner (261), Department of Biochemistry, Purdue University, West Lafayette, Indiana 47907

Albrecht Wendel (333), Physiologisch-Chemisches Institut, University of Tubingen, Tubingen, Federal Republic of Germany

Bendicht Wermuth (249), Medizinische-Chemisches Institut der Universität, CH-3000 Berne, Switzerland

Peter G. Wislocki (135), Department of Animal Drug Metabolism and Radiochemistry, Merck Sharp & Dohme Research Laboratories, Rahway, New Jersey 07065

Daniel M. Ziegler (201), Clayton Foundation Biochemical Institute and Department of Chemistry, University of Texas at Austin, Austin, Texas 78712

Preface

During the last decade, the enzymes that appear to act primarily in preparing foreign compounds for excretion, those referred to here as the enzymes of detoxication, have received at last an appropriate degree of attention at the level of the purified catalytic protein. The impressive work with the cytochrome P-450 monooxygenases, which followed isolation of these enzymes in homogeneous form, has been balanced by a similar analysis of enzymes catalyzing other oxidations, as well as of conjugation and hydrolytic reactions. Undoubtedly, additional enzymes that function in detoxication remain to be discovered or correctly assigned. Certainly, a great deal of important information concerning the catalytic mechanisms and physiological expression of the enzymes recorded here is unknown; yet, the main outlines seem to be in place. That much of this has been accomplished recently has prompted the present summary.

Major interest in this field stems from the provocative and challenging problems offered to investigators from a variety of disciplines. Therefore, in editing these two volumes of the new series, Biochemical Pharmacology and Toxicology, I have attempted and have asked the active investigators who are the authors to write in a manner that would provide the pharmacologists and toxicologists with the biochemical view of detoxication, and the biochemists with the corresponding pharmacological and toxicological aspects. This approach will also apply to a third volume in this series: "Metabolic Basis of Detoxication," edited with J. Bend and J. Caldwell, which is designed to evaluate the routes of metabolism of foreign compounds with emphasis on their functional groups.

Laboratory procedures are not included. Rather, many of the applicable techniques for the oxidative enzymes have been described in S. P. Colowick and N. O. Kaplan, Methods in Enzymology, Volume LII "Biomembranes" (S. Fleischer and L. Packer, volume editors), and for conjugation in a volume that I am editing for Methods in Enzymology entitled "Detoxication and Drug Metabolism: Conjugation and Related Systems."

"Enzymatic Basis of Detoxication," Volumes I and II, presents the current state of our knowledge of foreign compound metabolism at the level of what specific enzymes can do. This is not to suggest that the description of enzymes from a few species and organs will be definitive, nor is it intended as a compendium of pH optima and kinetic constants. Rather, we are concerned here with a holistic view of the information gleaned from work with specific, purified enzymes encompassing as many mammalian sources as have been studied. As such, both our knowledge and the content of these volumes will be incomplete, but as Mark Twain emphasized, "Even if you are on the right track, you'll get run over if you just sit there."

William B. Jakoby

Contents of Volume II

Chapter 1

Detoxication Enzymes

WILLIAM B. JAKOBY

It is remarkable that our universal experience with the products of Nature and those of man's synthetic ingenuity and waste does not lead more frequently to physiological catastrophe. Despite the variety and quantity of substances which we absorb, inhale, ingest, and imbibe, not to speak of toxic materials that we produce endogenously, most of us achieve about five decades or so of healthy existence before degeneration and chronic toxic effects become evident. Our normal intake of foreign substances undoubtedly is a source of phenomena that eventually lead to pathological changes; it is safe to assume that they testify to our mortality.

Governmental decrees to the contrary, the major means of protection are within ourselves. Our defense is provided by a group of mechanisms that have evolved as a consequence of their obvious selective advantage. Whatever the contribution of the modern international chemical industry, the problem is not new; toxic materials similar to those we inhale from cigarettes or from an automobile's exhaust were present after lightning struck primeval forests and caused them to burn. Fortunately, the evolutionary process has allowed the accumulation of a number of anatomic, physiological, and enzymatic barriers that provide a means of resistance.

From the viewpoint of biochemistry, mammals are able to avail themselves of a number of enzymes that catalyze the oxidation, reduction, cleavage, rearrangement, and conjugation of the enormous number of

1

ENZYMATIC BASIS OF DETOXICATION, VOL. I

foreign compounds with which they come in contact, and thereby prepare them for rapid elimination. Although the emphasis here is on the enzymatic aspects of detoxication, catalytic conversion of foreign compounds (xenobiotics) is only one of several mechanisms participating in the processes for the removal of foreign substances. Noncatalytic reactions, for example, occur between nucleophiles and electrophiles in which one is of exogenous and the other of intracellular origin; glutathione, present in most tissues at concentrations of about 5 mM, is a major resource of nucleophilic capacity and is capable of reacting even without the intervention of protein catalysts. The detoxication role of enzymes is made possible by the availability of means for partitioning of substrates between phases, e.g., between the lipophilic environment of the endoplasmic reticulum and the more hydrophilic cytosol; by the transport of xenobiotics from their point of entrance to the initial site of conversion and of subsequent shuttling of intermediates from one organ to another; by the activity of binding proteins that serve as a vehicle for transport both in the circulation and through permeability barriers; and by the exquisitely tuned systems for excretion.

I. DETOXICATION

Despite the choice of *detoxication** to refer to the subject matter, rather than "metabolism of foreign compounds," "metabolism of xenobiotics," or "biotransformation," the term is less than ideal.† In the first edition of his classic monograph, "Detoxication Mechanisms," R. T. Williams defined detoxication as, ". . . those chemical changes which foreign organic compounds undergo in the animal body."[1] It is in this sense that we discuss the enzymes catalyzing the reactions by which the numerous compounds that are without physiological value, whether of exogenous or endogenous origin, are prepared for elimination from the animal. The emphasis is on the mechanism of conversion rather than on the toxicological aspects.

Whatever the label applied to them, most, if not all, of the enzymes under discussion are capable of forming products that are pharmacologi-

---

* The proper term is detoxication and not detoxification. The latter implies the correction of a state of toxicity as in the efforts made to produce sobriety in one who is inebriated.

† The need for definition points to the lack of an adequately precise alternative term. Biotransformation, for example, is too broad in its coverage without a modifying phrase, whereas undue emphasis on the *foreign* nature of the compounds involved eliminates consideration of substrates synthesized endogenously which require the same enzyme systems for removal.

cally and toxicologically more potent than the parent compound. A major contributor is the cytochrome P-450 oxidation system that provides intensely reactive and, therefore, potentially toxic substances to the cell (this volume, Chapter 7). It is not, however, sophistry to argue that these enzymes represent only an initial step in preparation for excretion after subsequent conjugation. Thus, detoxication is frequently not the act of a single enzyme but the result of sequential action by several of them. Benzo[a]pyrene, for example, is sequentially subjected to oxidation by the cytochrome P-450 oxidases to form an epoxide, hydration catalyzed by epoxide hydratase to form a diol, and a second stage of oxidation to yield a diol epoxide before the transformation into a powerful electrophile and carcinogen is complete.[2,3] The subsequent detoxication of the epoxide could be catalyzed by glutathione transferase to form the corresponding thioether conjugate with glutathione (Volume II, Chapter 4), or by epoxide hydratase which would hydrolyze the epoxide ring (Volume II, Chapter 15).

II. CHARACTERIZATION OF THE ENZYMES
OF DETOXICATION

It is pertinent to question whether these enzymes form a cohesive group based on factors other than the assumption of a convenient label. Although exceptions will be evident, two interrelated general properties of the enzymes considered as active in detoxication appear to apply: the general lipophilic nature of the substrates that are converted to more hydrophilic products and the loose specificity of the enzymes for substrates.

One view of the function of these enzymes proposes that they serve to prepare lipophilic substances for excretion by catalyzing reactions yielding more water-soluble and readily excretable products. Charged compounds are often poor substrates. It may be significant that such physiological thiols as cysteine and glutathione are not substrates for thiol S-methyltransferase despite the variety of lipophilic thiols that are successful methyl group acceptors (Volume II, Chapter 7). Similarly, maleic and other dicarboxylic acids are not substrates for the glutathione transferases, whereas the diethyl ester of maleic acid is readily converted to the corresponding thioether (Volume II, Chapter 4).

Within our concept of enzymes as highly selective catalysts, both with respect to the substrate and the reaction catalyzed, it is noteworthy to find a group of proteins, possessing broad substrate specificity and capable of catalyzing a range of reactions. The enzymes of detoxication appear to have surfaces adapted for interaction with the lipophilic compounds, thereby providing acceptance for a large number of substrates. Figuratively, each of the enzymes may be viewed as a sea of oil. However, within that sea is an island in the form of a binding site for the specific substrate of the enzyme: for NAD with pyridine nucleotide enzymes; for S-adenosylmethionine with the methyltransferases; for glutathione with glutathione peroxidase; and, in general, for the specific moiety to be conjugated by the enzymes catalyzing those reactions. The sea of oil, of course, will accept most lipophilic compounds. If one wishes to modify the terrain by substituting an aqueous lake covered by oil, additional compounds may be accommodated; large planar molecules bearing a polar group would have the polar group within the aqueous phase exposing a point of attack to the island.

Similar considerations apply to the types of reactions that are catalyzed. The cytochrome P-450 proteins are capable of an enormous range in that they serve to oxidize at aliphatic and aromatic carbon, nitrogen and sulfur (this volume, Chapter 7). The term glutathione transferases is really a misnomer; these enzymes should not be viewed as participating in the transfer of glutathione, which they do, but as proteins that catalyze any reaction in which the glutathione thiolate anion participates and that includes, among others, conjugation with glutathione, isomerization, disulfide interchange, and organic nitrate reduction (Volume II, Chapter 4).

If one considers the evolutionary processes as the transient resolutions of a design problem that is achieved by trial and error over a period of millennia, it is possible to cavil at the individual resolutions of the problem but not with the general scheme. The pattern which has evolved is an economical one for the cell with a spectrum of design solutions which emphasize enzymes of multiple function. At one extreme are the specific enzymes, those with affinity for a single substrate that allow rapid removal from the cell of the smallest quantities and provide the possibility of delicate regulation. Superoxide dismutase may represent this class as a catalyst specific for the superoxide radical (this volume, Chapter 15); it is probably not incidental that this specificity is for a product common to many oxidation systems. Considering the variety of xenobiotics with which the animal contends, however, specific proteins for each compound must be ruled out as a general solution on the basis of both cell economy and ineffectiveness in the face of the continual supply of unfamiliar products.

At the other end of the spectrum are systems prepared to act upon the realistically large number of materials, whatever their size and shape. The merit of this approach is that a single enzyme can accommodate many compounds and interrelated reactions, thereby saving expensive capacity in the machinery for information storage and protein synthesis. The price for such simplicity is inefficiency: low turnover rates and generally poor substrate affinity. Almost all of the enzymes covered in these volumes fall into this category. It is noteworthy that in many instances, the low efficiency of these catalysts is compensated for by their presence in relatively high concentration.

That this design solution may be only one stage in the development of the detoxicating enzymes is emphasized by contrast with bacterial systems. Bacteria, with generation times measured in minutes rather than years, are capable of metabolizing the gamut of natural and synthetic compounds but usually do so with greater specificity. From garden soil or estuary mud it is possible to obtain a microorganism by the enrichment culture technique[4] that can utilize almost any organic compound as the sole source of carbon. The cytochrome P-450 system from a pseudomonad, for example, was isolated by reason of the bacterium's ability to grow on camphor as its source of carbon.[5] The enzyme exhibits a reaction mechanism entirely analogous to that of the mammalian cytochrome P-450 but with radically restricted specificity. Similar examples of high specificity for the products and waste of industry are common. Bacteria able to grow on acetylenedicarboxylic acid for their supply of carbon use an enzyme specific in the hydration of the acetylenic bond of the dicarboxylic acid but are unable to act on its esters or on acetylenemonocarboxylic acid[6]; conversely, an organism isolated for growth on acetylenemonocarboxylic acid is specific for that compound.[7] This diversity and specificity of function must be viewed in the context of the number of mutational events possible in large populations. The bacteria under discussion were isolated by *enrichment* for a particular function and represent only a minute fraction of the total bacterial population.

III. THE QUESTION OF NATURAL SUBSTRATE

The enzymes of detoxication may form a coherent group of catalysts in much the same manner that other groups function in fatty acid oxidation and synthesis, in glycolysis, and in catabolic reactions such as those in lysozomes. Despite their occasional participation in anabolic reactions, the evolutionary development of these enzymes suggests xenobiotics as their natural substrate. Thus, higher organisms appear to have developed a class of enzymes designed to meet the need for detoxication—a class that is characterized by simplicity rather than efficiency in the removal of the multitude of xenobiotics to which the organism is repeatedly exposed.

REFERENCES

1. Williams, R. T. (1947). "Detoxication Mechanisms," p. 2. Wiley, New York.
2. Sims, P., Grover, P. L., Swaisland, A., Pal, K., and Hewer, A. (1974). Metabolic activation of benzo[*a*]pyrene proceeds by a diol-epoxide. *Nature (London)* **252**, 326.
3. Jerina, D. M., Lehr, R., Schaefer-Ridder, M., Yagi, H., Karle, J. M., Thakker, D. R., Wood, A. W., Lu, A. Y. H., Ryan, D., West, S., Levin, W., and Conney, A. H. (1977). Bay region epoxides of dihydrodiols. *In* "Origins of Human Cancer" (H. Hiatt, J. D. Watson, and I. Winstein, eds.), pp. 639–658. Cold Spring Harbor Lab., Cold Spring Harbor, New York.
4. Kluyver, A. J., and van Niel, C. B. (1956). "The Microbe's Contribution to Biology." Harvard Univ. Press, Cambridge, Massachusetts.
5. Gunsalus, I. C., and Wagner, G. C. (1978). Bacterial P-450$_{cam}$ methylene monoogygenase components: Cytochrome m, putidaredoxin, and putidaredoxin reductase. *In* "Methods in Enzymology" (S. Fleisher and L. Packer, eds.), Vol. 52, pp. 166–188. Academic Press, New York.
6. Yamada, E. W., and Jakoby, W. B. (1958). Enzymatic utilization of acetylenic compounds. I. An enzyme converting acetylenedicarboxylic acid to pyruvate. *J. Biol. Chem.* **233**, 706–711.
7. Yamada, E. W., and Jakoby, W. B. (1956). Enzymatic utilization of acetylenic compounds. II. Acetylenemonocarboxylic acid hydrase. *J. Biol. Chem.* **234**, 941–945.

Part I

Physiological Aspects

Chapter 2

Kinetic Aspects of Metabolism and Elimination of Foreign Compounds in Animals

JAMES R. GILLETTE

I. INTRODUCTION

It is evident that the various metabolites formed from a specific foreign compound result from the action of several enzymes acting in concert. Studies of the mechanisms and the kinetic constants of the enzymes with the various substrates therefore represent only the first step toward our understanding of the roles each enzyme plays in the elimination and detoxication of foreign compounds. Moreover, it is evident that the role of metabolism in limiting the actions of foreign compounds cannot be completely understood solely from information obtained from *in vitro* studies of the various drug metabolizing enzymes. Indeed, the concentrations of a parent foreign compound and its various metabolites at sites of action depend on a number of interrelated processes including the rates of absorption of the parent compound from the sites of administration into blood; the transport of the substance to the various organs; the distribution of the parent compound into tissues; the conversion of the compound to various metabolites; and the excretion of the parent drug and its

9

ENZYMATIC BASIS OF DETOXICATION, VOL. I
Copyright © 1980 by Academic Press, Inc.
All rights of reproduction in any form reserved.
ISBN 0-12-380001-3

metabolites into urine, bile, and air. The relative amounts of the parent compound and its metabolites at sites of action also depend on the relative activities of the enzymes that catalyze their metabolism in various organs, the diffusion of these substances between the immediate environment of the enzymes and blood within the organs of elimination, and the blood flow rates through the organs.

The interrelationships among these factors are obviously complex, but in many instances the concepts can be greatly simplified because the contribution of a given factor is so small that it can be ignored. For example, when the elimination of a foreign compound is very slow compared with the rates at which it enters and leaves the various tissues, the entire animal may be viewed as a single compartment. In these cases, the blood flow rates through the organs are irrelevant, and the intracellular and even the intertissue distribution of the enzymes that catalyze the metabolism of the foreign compound may be ignored; it is the total activity of the enzyme within the body that is important. By contrast, when the elimination of a compound is very rapid, the rate of its elimination may depend mainly on the rates of blood flow through the organs containing the enzymes that catalyze the metabolism of the foreign compound. In these cases, the intertissue localization of the enzymes become important, but the activity of the enzymes within a given tissue may be largely irrelevant. In order to recognize the conditions under which a given factor has a dominant influence on the elimination of foreign compounds, pharmacokineticists have developed an idealized model that may be modified to include more complicated situations.[1-4] The model also provides insights into the strengths and limitations of studies obtained with various preparations ranging from purified enzymes to living animals and points out the type of data needed for plausible extrapolations from one type of preparation to another. The model also points to areas where our knowledge is insufficient to make completely accurate extrapolations. Because the enzyme kineticist and the pharmacokineticist approach their problems differently, an extensive discussion of the pharmacokinetic approach may be useful in order to bridge the gap between them.

II. PHARMACOKINETIC MODEL

Central to the pharmacokinetic approach used in describing the idealized model is the concept of "intrinsic clearance"[5-7] of an enzyme, $Cl_{int_{(S-M1;\ cell)}}$ which is defined as the rate, $V_{(S-M1;\ cell)}$, of conversion of a substance to metabolite (M1) divided by the unbound concentration of the substance [S], in the immediate environment of the enzyme within a cell.

That is

$$Cl_{int_{(S-M1;cell)}} = V_{(S-M1;cell)}/[S] \tag{1}$$

The dimensions of $Cl_{int_{(S-M1;\ cell)}}$ are volume per time. For complex enzyme mechanisms, one simply writes the equation for the rate of metabolism of substance under steady-state conditions and substitutes it into Eq. (1). For example, for the two-substrate random equilibrium mechanism

$$Cl_{int_{(S-M1;cell)}} = Enz_{(S-M1;cell)}k \, [C]_{(cell)}/(K_c + [C]_{(cell)}(K_s + [S]_{(cell)}) \tag{2}$$

where $Enz_{(S-M1;\ cell)}$ is the molar amount enzyme within the cell, k is the rate constant for the transformation of the ternary complex to the products of the reaction, $[C]_{(cell)}$ is the concentration of the cosubstrate, K_c is the dissociation constant of the enzyme–cosubstrate complex, $[S]_{(cell)}$ is the unbound concentration of the foreign compound within the cell, and K_s is the dissociation constant of the enzyme–foreign compound complex.

Clearly, $Cl_{int_{(S-M1;\ cell)}}$ is not a constant, but the pharmacokineticist would prefer that it were a constant and frequently points out that at high concentrations of the cosubstrate and low concentrations of the foreign compound, it approaches a constant. Under these conditions, Eq. (2) degenerates to

$$Cl_{int_{(S-M1;cell)}} = \frac{Enz_{(S-M1;cell)}k}{K_s} = \frac{V_{m(S-M1;cell)}}{K_m} \tag{3}$$

Thus, in assessing the importance of an enzyme in the pharmacokinetics of non-steady-state systems, it is generally assumed that the concentration of the foreign compound is smaller than K_s and that the concentration of the cosubstrate is either much greater than K_c or remains constant as the concentration of the foreign compound decreases to negligible values during its metabolism.

When a foreign compound is metabolized by several enzymes within the same cell, the total intrinsic clearance of the foreign compound within the cell becomes the sum of the enzyme intrinsic clearances. Thus

$$Cl_{int_{(cell)}} = \Sigma Cl_{int_{(S-i)}} \tag{4}$$

In relating the intracellular and the extracellular concentrations of the foreign compound, one must also take into account the rate of diffusion of the compound into and out of the cell. Under steady-state conditions the ratio of the unbound concentration of the foreign compound within the cell, $[S]_{(cell)}$, to the total concentration in blood $[S]_B$, that is outside of the cell, may be expressed as

$$\frac{[S]_{cell}}{[S]_B} = \frac{f_B \, D_{in_{(cell)}}}{Cl_{int_{(cell)}} + D_{out_{(cell)}}} \tag{5}$$

in which f_B is the unbound fraction of the foreign compound in the blood and D_{in} and D_{out} are, respectively, the diffusional clearances of the foreign compound into and out of the cell. Accordingly, the cellular intrinsic clearance expressed in terms of the external concentration of the foreign compound would be

$$f_{B_{(S)}}Cl_{int_{(cell)}} = \frac{f_{B_{(S)}} D_{in_{(cell)}} Cl_{int_{(cell)}}}{Cl_{int_{(cell)}} + D_{out_{(cell)}}} \tag{6}$$

In a preparation of isolated hepatocytes containing cells having different amounts of enzyme, the total intrinsic clearance of the enzyme within the preparation is the sum of each of the intrinsic clearances of the enzyme within individual cells. Thus

$$f_{B_{(S)}}Cl_{int_{(S-M1)}} = \sum f_{B_{(S)}}Cl_{int_{(S-M1;cell)}} \tag{7}$$

The total intrinsic activity in the preparation would be

$$f_{B_{(S)}}Cl_{int_{(S)}} = \sum f_{B_{(S)}}Cl_{int_{(cell)}} \tag{8}$$

In the living animal and in perfused organs, the clearance of a foreign compound by a given organ is dependent on the blood flow (Q) through the organ. Regardless of the model, the clearance may be expressed as

$$Cl_{(S)} = Q \left[1 - \frac{[S]_{out}}{[S]_{in}}\right] = Q\,(1 - F_{(S)}) = Q\,E_{(S)} \tag{9}$$

in which $[S]_{out}$ and $[S]_{in}$ represent, respectively, the concentrations of the foreign compound in blood leaving and entering the organ under steady-state conditions. $F_{(S)}$ is the ratio, $[S]_{out}/[S]_{in}$, and is referred to as the available fraction of the foreign compound. $E_{(S)}$ is $(1 - F_{(S)})$ and is called the extraction ratio of the foreign compound by the organ. Equation (9) is thus a restatement of Fick's principle. Several models have been developed to relate the organ clearance to the cellular intrinsic clearances. In the simplest of them, the unbound concentration of the drug in the extracellular fluid within the organ is assumed to be the same throughout the organ and the same as that in the blood leaving the organ.[5,6] In this, the "well-stirred" model,[6] cells possessing different enzyme activities act as though they were randomly distributed within the organ despite their known heterogenous histological distribution. Equation (10) relates organ blood flow to the intrinsic activity

$$Cl_{(S)} = \frac{Q\,f_{B_{(S)}}D_{in_{(S)}}Cl_{int_{(S)}}}{Q\,(Cl_{int_{(S)}} + D_{out_{(S)}}) + f_{B_{(S)}}D_{in_{(S)}}Cl_{int_{(S)}}} \tag{10}$$

Ordinarily, the diffusivities of foreign compounds into and out of cells are about the same and are rapid relative to the cellular intrinsic clearances. Therefore, Eqs. (5), (6), and (10) may usually be simplified. For example, the usual form of Eq. (10) appearing in the literature[5,6] is

$$Cl_{(S)} = \frac{Q f_{B_{(S)}} Cl_{int_{(S)}}}{Q + f_{B_{(S)}} Cl_{int_{(S)}}} \tag{11}$$

Equation (11) may also be written as

$$Cl_{(S)} = Q E_{(S)} \tag{12}$$

where

$$E_{(S)} = \frac{f_{B_{(S)}} Cl_{int_{(S)}}}{Q + f_{B_{(S)}} Cl_{int_{(S)}}}$$

or

$$Cl_{(S)} = F_{(S)} f_{B_{(S)}} Cl_{int_{(S)}} \tag{13}$$

in which

$$F_{(S)} = Q/(Q + f_{B_{(S)}} Cl_{int_{(S)}})$$

The relationship between the blood flow rate and the intrinsic activity of the enzymes may be easily visualized. When the value of $f_{B_{(S)}} Cl_{int_{(S)}}$ is very large relative to the blood flow rate, the organ clearance nearly equals the blood flow rate; when it is very small relative to the blood flow through the organ, clearance nearly equals the value of $f_{B_{(S)}} Cl_{int_{(S)}}$.

Another model that relates the intrinsic activity and the blood flow rate through the organ is called the "parallel tube" or "sinusoid" model.[8-10] Here the cells, which line a collection of tubes of equal length, contain exactly the same amount of enzyme. The concentration of the foreign compound within the lumen of the tubules decline exponentially as the blood passes from the proximal to the distal ends of the tubules. Under these conditions, the equations for $F_{(S)}$, $E_{(S)}$, and $Cl_{(S)}$ would be[8]

$$F_{(S)} = \exp(- f_{B_{(S)}} Cl_{int_{(S)}}/Q) \tag{14}$$

$$E_{(S)} = 1 - \exp(- f_{B_{(S)}} Cl_{int_{(S)}}/Q) \tag{15}$$

$$Cl_{(S)} = Q [1 - \exp(- f_{B_{(S)}} Cl_{int_{(S)}}/Q)] \tag{16}$$

When investigators have compared the models, they have found that the "well-stirred" model frequently provides better predictions of the function of the organ.[11,12]

Most pharmacokinetic models are based on the assumption that the foreign compound is eliminated from the body by only one organ, namely, the liver. On this assumption, many of the concepts of the relationship between the organ clearance and the concentration of the foreign compound may be illustrated by a model of a liver perfusion experiment in which the blood recirculates (Fig. 1). If a foreign compound were continuously infused into the reservoir and samples from the reservoir assayed, the model would be equivalent to an intravenous infusion of the foreign compound when the compound is eliminated solely by the liver or the kidney. The concentration of the foreign compound within the reservoir would be the concentration of the compound in the medium entering the organ. If the infusion is maintained until a steady state is reached, the steady-state concentration $[S]_{ss,i.v.}$ depends on Eq. (17)

$$[S]_{ss,iv} = k_{0(S,i.v.)}/Cl_{(S)} \tag{17}$$

in which $k_{0(S,i.v.)}$ is the rate of infusion into the reservoir and $[S]_{ss,i.v.}$ is the steady-state concentration of the foreign compound within the reservoir.

By contrast, if the foreign compound were continuously infused at a point after the blood leaves the reservoir but before it enters the organ, and samples from the reservoir were assayed, the model would be equivalent to an intraportal infusion of a foreign compound that is eliminated solely by the liver. If the infusion were maintained until steady-state conditions were reached, the concentration of the foreign compound within the reservoir would be the concentration in the medium leaving the organ, rather than entering it. The effective rate of infusion of the drug

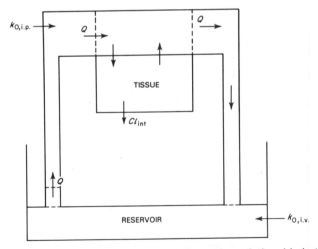

Fig. 1.   Model of liver perfusion experiment illustrating relationship between organ clearance and concentration of foreign compound.

into the reservoir would thus be the rate of infusion multiplied by the fraction of the amount of drug that escapes the organ during its first passage through the organ, i.e., $k_{0(S,i.p.)} F_{(S)}$. Thus, the steady-state concentration of the foreign compound within the reservoir would be

$$[S]_{ss,ip} = \frac{k_{0(S,i.p.)} F_{(S)}}{Cl_{(S)}} \tag{18}$$

in which $k_{0(S,i.p.)}$ is the rate of infusion at a point after the reservoir, and $[S]_{ss,i.p.}$ is the steady-state concentration of the foreign compound within the reservoir.

Equations (17) and (18) are valid for any model we may visualize for the mechanism of elimination of the foreign compound, because they are based solely on mass balance considerations. Nevertheless, if we assume the validity of the "well-stirred" model, we may substitute Eq. (13) into Eq. (18) and obtain

$$[S]_{ss,ip} = \frac{k_{0(S,i.p.)} F_{(S)}}{F_{(S)} f_{B_{(S)}} Cl_{int_{(S)}}} \tag{19}$$

By Eq. (19), the steady-state concentration of the foreign compound would depend solely on the unbound concentration of the foreign compound and the intrinsic clearance.

The model is also useful in visualizing the formation of metabolites by the liver under conditions in which the foreign compound is not appreciably eliminated by other organs. When a foreign compound is constantly infused into the reservoir (or infused intravenously) until a steady state is reached, the steady-state concentration of the metabolite in the reservoir may be predicted from[13]

$$(M1_{(S)})_{ss} = \frac{k_{0(S,\ i.v.\ or\ i.p.)} f_{B_{(S)}} Cl_{int_{(S-M1)}} F_{(M1_{(S)})}}{f_{B_{(S)}} Cl_{int_{(S)}} Cl_{(M1)}} \tag{20}$$

or

$$(M1_{(S)})_{ss} = \frac{k_{0(S,\ i.v.\ or\ i.p.)} f_{(S-M1)} F_{(M1_{(S)})}}{Cl_{(M1)}} \tag{21}$$

in which $f_{B_{(S)}}$ is the unbound fraction of the foreign compound, $Cl_{int_{(S-M1)}}$ is the intrinsic clearance of the enzyme that catalyzes the formation of the metabolite, $F_{(M1_{(S)})}$ is the fraction of the amount of metabolite leaving the organ as it is formed, $Cl_{int_{(S)}}$ is the sum of the intrinsic clearances of all the enzymes in liver that catalyze the metabolism of the foreign compound, $Cl_{(M1)}$ is the clearance of the metabolite, and $f_{(S-M1)}$ is the fraction of the metabolized foreign compound that is converted to metabolite.[1] Note that

$f_{B_{(S)}}$ appears in both the numerator and the denominator of Eq. (20) and that the steady-state concentration of M1 is independent of the degree of binding of the parent compound to components in the perfusion medium.

In this system the steady-state concentration of metabolite (M1), is independent of the route of administration. The same steady state concentration of the metabolite would be obtained whether the parent foreign compound is infused into the reservoir or at some point after the reservoir. An equation analogous to Eq. (13) shows that if the organ acts as though it were "well-stirred," $(M1)_{ss}$ may be expressed as $F_{(M1)} f_{B_{(M1)}} Cl_{int_{(M1)}}$, when M1 is infused into the reservoir. $F_{(M1)}$ would equal $F_{(M1_{(S)})}$ and, under these conditions, Eqs. (20) and (21) may be simplified to

$$(M1_{(S)}) = \frac{k_{0_{(S,\ i.v.\ or\ i.p.)}} f_{B_{(S)}} Cl_{int_{(S-M1)}}}{f_{B_{(S)}} Cl_{int_{(S)}} f_{B_{(M1)}} Cl_{int_{(M1)}}} \tag{22}$$

or

$$(M1_{(S)})_{ss} = \frac{k_{0_{(S,\ i.v.\ or\ i.p.)}} f_{(S-M1)}}{f_{B_{(M1)}} Cl_{int_{(M1)}}} \tag{23}$$

or

$$f_{B_{(M1)}} (M1_{(S)})_{ss} = \frac{k_{0_{(S,\ i.v.\ or\ i.p.)}} f_{(S-M1)}}{Cl_{int_{(M1)}}} \tag{24}$$

Under these conditions the steady-state unbound concentration of metabolite (M1) depends on the fraction of the parent compound that is converted to metabolite (M1) (that is, $f_{(S-M1)}$) and the sum of the intrinsic clearances of the enzymes that metabolize metabolite (M1) (that is, $Cl_{int_{(M1)}}$).

The equation showing the ratio of the steady-state concentration of metabolite (M1) in the reservoir to that of the parent foreign compound during constant infusion into the reservoir (or intravenously), may be obtained by dividing Eq. (23) by Eq. (17).

$$\left[\frac{(M1_{(S)})}{[S]} \right]_{ss,\ i.v.} = \frac{Cl_{(S)} f_{(S-M1)} F_{(M1_{(S)})}}{Cl_{(M1)}} \tag{25}$$

or, in a "well-stirred" model,

$$\left[\frac{(M1_{(S)})}{[S]} \right]_{ss,\ i.v.} = \frac{F_{(S)} f_{B_{(S)}} Cl_{int_{(S-M1)}}}{f_{B_{(M1)}} Cl_{int_{(M1)}}} \tag{26}$$

which may be rearranged to

$$\left[\frac{f_{B_{(M1)}} (M1_{(S)})}{f_{B_{(S)}} [S]} \right]_{ss, \, i.v.} = \frac{F_{(S)} \, Cl_{int_{(S-M1)}}}{Cl_{int_{(M1)}}} \tag{27}$$

Thus, the ratio of the steady-state unbound concentrations of metabolite (M1) and the parent compound depends on the availability of the parent compound, $F_{(S)}$ as well as the intrinsic clearance of the enzyme that catalyzes the formation of metabolite M1 and the sum of the intrinsic clearances of the enzymes that catalyze the elimination of the metabolite.

By contrast, the equation for the ratio of the steady-state concentration of metabolite (M1) to that of the parent foreign compound during the constant infusion of the parent compound at some point after the reservoir (or intraportally) may be obtained by dividing Eq. (2) by Eq. (18)

$$\left[\frac{(M1_{(S)})}{[S]} \right]_{ss, \, i.p.} = \frac{Cl_{(S)} \, f_{(S-M1)} \, F_{(M1_{(S)})}}{F_{(S)} \, Cl_{(M1)}} \tag{28}$$

or in a "well-stirred" model, by dividing Eq. (22) by Eq. (19)

$$\left[\frac{(M1_{(S)})}{[S]} \right]_{ss, \, i.p.} = \frac{f_{B_{(S)}} \, Cl_{int_{(S-M1)}}}{f_{B_{(S)}} \, Cl_{int_{(M1)}}} \tag{29}$$

which may be rearranged to

$$\left[\frac{f_{B_{(M1)}} (M1_{(S)})}{f_{B_{(S)}}[S]} \right]_{ss, \, i.p.} = \frac{Cl_{int_{(S-M1)}}}{Cl_{int_{(M1)}}} \tag{30}$$

The "well-stirred" model predicts that the ratio of the unbound concentrations of metabolite (M1), and the parent foreign compound depends only on the intrinsic clearance of the enzyme that catalyzes the formation of metabolite M1 and the sum of the intrinsic clearances of the enzymes that catalyze its elimination when the parent compound is infused intraportally. In other words, the ratio should be independent of the other reactions by which the parent compound is eliminated.

After constant infusion of metabolite M1 into the reservoir until a steady state is reached, the concentration of the metabolite should be

$$(M1)_{ss, \, i.v.} = k_{0_{(M1, \, iv)}}/Cl_{(M1)} \tag{31}$$

However, after constant infusion of metabolite M1 at a point after the reservoir, the equation should be

$$(M1)_{ss, \, i.p.} = \frac{k_{0_{(M1, \, i.p.)}} \, F_{(M1)}}{Cl_{(M1)}} \tag{32}$$

The constant infusion of the parent foreign compound labeled with one isotope ($k_{0_{(S)}, \text{ i.v.}}$) along with the constant infusion of metabolite M1 labeled with another isotope ($k_{0_{(M1)}, \text{ i.v.}}$) may be used to evaluate other parameters of the system. For example, when the labeled metabolite is constantly infused into the reservoir (or intravenously), the ratio of the steady-state concentration of metabolite (M1) formed from the parent foreign compound (M1$_{(S)}$) to the steady-state concentration of the pre-formed metabolite (M1) may be calculated from Eqs. (21) and (31)

$$\left[\frac{(M1_{(S)})/k_{0_{(S)}}}{(M1)/k_{0_{(M1)}}} \right]_{\text{ss, i.v.}} = f_{(S\text{-}M1)}\, F_{(M1_{(S)})} \tag{33}$$

The ratio depends not only on the fraction of the foreign compound converted to metabolite (M1) but also on the availability of the metabolite formed from the foreign compound.

By contrast, when the preformed metabolite is constantly infused at a point after the reservoir (or intraportally), the ratio of the steady-state concentration of the metabolite formed from the parent foreign compound to the steady-state concentration of the preformed metabolite may be calculated from Eqs. (21) and (32)

$$\left[\frac{(M1_{(S)})/k_{0_{(S)}}}{(M1)/k_{0_{(M1)}}} \right]_{\text{ss, i.p.}} = \frac{f_{(S\text{-}M1)}\, F_{(M1_{(S)})}}{F_{(M1)}} \tag{34}$$

which for a "well-stirred" model becomes

$$\left[\frac{(M1_{(S)})/k_{0_{(S)}}}{(M1)/k_{0_{(M1)}}} \right]_{\text{ss, i.p.}} = f_{(S\text{-}M1)} \tag{35}$$

The "well-stirred" model predicts that the ratio of the two concentrations of metabolite M1, when the preformed metabolite is infused intraperitoneally, may be used to estimate the fraction of the parent foreign compound that is converted to metabolite M1. If the metabolism of the foreign compound is sufficiently slow, however, the value of $F_{(M1_{(S)})}$ may approach 1.0 and the value of $f_{(S\text{-}M1)}$ may be estimated from Eq. (33) as well as Eq. (35).

It is possible to predict the activity of a given enzyme in liver from *in vivo* data. If both the parent foreign compound and the preformed metabolite are continuously infused at a point after the reservoir (or intraportally), the following equation may be obtained from Eqs. (18) and (34)

$$\left[\frac{(M1_{(S)})/[S]}{(M1)/k_{0_{(M1)}}} \right]_{\text{ss, i.p.}} = \frac{Cl_{(S)}\, f_{(S\text{-}M1)}\, F_{(M1_{(S)})}}{F_{(M1)} F_{(S)}} \tag{36}$$

If the organ acts as though it were a "well-stirred" compartment, substitution of Eq. (13) into Eq. (36) gives

$$\left[\frac{(M1_{(S)})/[S]}{(M1)/k_{o_{(M1)}}} \right]_{ss,\ i.p.} = f_{B_{(S)}}\ Cl_{int_{(S-M1)}} \tag{37}$$

which may be rearranged to

$$\left[\frac{(M1_{(S)})/f_{B_{(S)}}\ [S]}{(M1)/k_{o_{(M1)}}} \right]_{ss,\ i.p.} = Cl_{int_{(S-M1)}} \tag{38}$$

The "well-stirred" model thus predicts that the intrinsic activity of the enzyme that converts a foreign compound to metabolite (M1) may be estimated from the steady-state concentrations of $M1_{(S)}$ and M1, and the steady-state unbound concentration of the parent foreign compound.

Because "well-stirred" models are relatively easy to interpret one would like to assess their validity in evaluating the results of perfusion experiments. In most studies designed to compare the "well-stirred" model with the "parallel tube" model, the effects of changes in blood flow rate through the organ on changes in the availability of the foreign compound in blood, have been measured [Eqs. (13) and (14)]. Such studies have usually indicated that the "well-stirred" model was better in predicting changes in the availability of many foreign compounds passing through the liver than was the "parallel tube" model. By contrast, Pang and Gillette[13,14] have assessed the two models by studying the availability of acetaminophen derived from phenacetin. In this system the "parallel tube" model is difficult to assess because it predicts that the availability of the metabolite will depend on the relative localization of the enzymes in liver that catalyze the formation and elimination of the metabolite. Clearly, the availability of the metabolite would be greater when the enzyme catalyzing the elimination of the metabolite are located in the proximal end of the tubules than when they are located in the distal end. The finding of Pang and Gillette[13] that the availability of acetaminophen was invariably less than that predicted by the "homogenous parallel tube" model, in which the relative activities of these enzymes is assumed to be the same in all hepatocytes, was therefore surprising because cytochrome P-450 enzymes, including the one that catalyzes the conversion of phenacetin to acetaminophen, are now known to be preferentially localized in the distal portion (centrilobular) of the sinusoids.[15] Nevertheless, the availability of acetaminophen was invariably greater than that predicted by the "well-stirred" pool model. While neither model is completely accurate, the "well-stirred" model is probably the better of the two for most substrates.

The steady-state concentrations of a foreign compound and its metabolites in blood are even more difficult to assess when all the metabolites are eliminated from the body by more than one organ. Consider the problem when the foreign compound and its metabolites are metabolized by enzymes in intestinal bacteria and intestinal mucosa and are secreted into bile or diffuse from intestinal blood into the intestinal contents.[16] Some of the difficulties can be illustrated by models for the elimination of the foreign compound and its metabolites by way of the liver and excretion by the kidneys. For example, enzymes in liver may catalyze the formation of a foreign compound to metabolite (M1) and, in turn, convert metabolite (M1) to metabolite (N1). In addition, the drug and both metabolites may be excreted into urine. When the compound is continuously infused intraportally or intravenously the steady-state concentrations of the foreign compound and both metabolites in systemic blood would be predicted by the following equations.

After intraportal infusion of the foreign compound,
Foreign compound (S):

$$[S]_{ss,\ i.p.} = \frac{k_{0(S,\ ip)}\ F_{H_{(S)}}}{Cl_{H_{(S)}} + Cl_{R_{(S)}}} \tag{39}$$

First generation metabolite (M1):

$$(M1_{(S)})_{ss,\ i.p.} = \frac{k_{0(S,ip)}\ A_H\ [1 + (Cl_{R_{(S)}}/Q_H)]\ F_{H_{(M1_{(S)})}}}{Cl_{H_{(M1)}} + Cl_{R_{(M1)}}} \tag{40}$$

Second generation metabolite (N1):

$$(N1_{S,M1})_{ss,i.p.} = \frac{k_{0(S,ip)}\ A_H\ [1 + (Cl_{R_{(S)}}/Q_H)]\ B_H\ F_{H_{(N1_{(M1)})}}}{Cl_{H_{(N1)}} + Cl_{R_{(N1)}}} \tag{41}$$

After intravenous infusion of the foreign compound:
Foreign compound (S):

$$[S]_{ss,i.v.} = \frac{k_{0(S,iv)}}{Cl_{H_{(S)}} + Cl_{R_{(S)}}} \tag{42}$$

First generation metabolite (M1):

$$(M1_{(S)})_{ss,\ i.v.} = \frac{k_{0(S,iv)}\ A_H\ F_{H_{(M1_{(S)})}}}{Cl_{H_{(M1)}} + Cl_{R_{(M1)}}} \tag{43}$$

Second generation metabolite (N1):

$$(N1_{(S,M1)})_{ss,\ i.v.} = \frac{k_{0(S,iv)}\ A_H\ B_H\ F_{H_{(N1_{(M1)})}}}{Cl_{H_{(N1)}} + Cl_{R_{(N1)}}} \tag{44}$$

where

$$A_H = \frac{f_{(S\text{-}M1)} \, Cl_{H_{(S)}}}{Cl_{H_{(S)}} + Cl_{R_{(S)}}} \tag{45}$$

$$B_H = \frac{f_{(M1\text{-}N1)} \, Cl_{H_{(M1)}} [1 + (Cl_{R_{(M1)}}/Q_H)]}{Cl_{H_{(M1)}} + Cl_{R_{(M1)}}} \tag{46}$$

in which, $Cl_{H_{(S)}}$ and $Cl_{R_{(S)}}$ are the clearances of the foreign compound by the liver and the kidney, respectively, and $f_{(S\text{-}M1)}$ is the fraction of the metabolized foreign compound that is converted to metabolite (M1). $Cl_{H_{(M1)}}$, $Cl_{R_{(M1)}}$ and $f_{(M1\text{-}N1)}$ have analogous meanings in respect to the clearances and metabolism of metabolite (M1) and $Cl_{H_{(N1)}}$ and $Cl_{R_{(N1)}}$ are the clearances of metabolite (N1). Q_H is the blood flow rate through the liver, $F_{H_{(S)}}$, $F_{H_{(M1_{(S)})}}$ and $F_{H_{(N1_{(M1)})}}$ are the hepatic availabilities of the foreign compound and metabolites (M1) and (N1), respectively.

Inspection of these equations reveals that the steady-state concentration of the metabolites may depend on the site of administration of the foreign compound, on the relative values of the kidney clearance and the hepatic blood flow (Cl_R/Q_H), on the relative clearances of the foreign compound and metabolites by the kidney and liver, and on the effective intrinsic clearances of the enzymes in liver.

As with the simple hepatic model, in which these foreign compounds and their metabolites are eliminated solely by the liver, much valuable information may be obtained by comparing the relative steady-state concentrations of the metabolites during simultaneous infusions of the labeled foreign compound intraportally and intravenously. For example,

$$\frac{\left[(M1_{(S)})/k_{0_{(S)}}\right]_{ss, \, i.p.}}{\left[(M1_{(S)})/k_{0_{(S)}}\right]_{ss, i.v.}} = 1 + (Cl_{R_{(S)}}/Q_H) \tag{47}$$

The equations for the ratio indicates that the steady-state concentration of the metabolite will depend on the route of administration when the parent foreign compound is readily cleared by the kidney. It may also be shown, that the ratio of the steady-state concentrations of metabolite (N1) after intravenous and intraportal administration of S will also be $1 + (Cl_{R_{(S)}}/Q_H)$.

Moreover, the kidney clearance of the parent drug will also affect the estimate of the fraction of the dose of the foreign compound that becomes converted to a metabolite. When labeled foreign compound and labeled metabolite (M1) are infused intraportally, the ratio of the steady-state concentrations of the two labels associated with metabolite (M1) would follow Eq. (48)

$$\frac{\left[(M1_{(S)})/k_{0_{(S)}}\right]_{ss,i.p.}}{\left[(M1)/k_{0_{(M1)}}\right]_{ss,i.p.}} = \frac{A_H \left[1 + (Cl_{R_{(S)}}/Q_H)\right] F_{H_{(M1_{(S)})}}}{F_{H_{(M1)}}} \qquad (48)$$

where $F_{H_{(M1)}}$ is the hepatic availability of preformed metabolite (M1). If the liver acted as though it were a "well-stirred" compartment, the value of $F_{H_{(M1_{(S)})}}/F_{H_{(M1)}}$ would be unity and, therefore, both factors might be removed from the equation. Nevertheless, the ratio of the steady-state concentration of the generated and preformed species of metabolite (M1) would overestimate the fraction of the dose of foreign compound that is converted to metabolite (M1), which in this case is A_H.

The equations derived for the steady-state concentrations of the foreign compound and its metabolites during constant infusion into the animal should be valid even when one or more of the enzymes or excretory processes approach saturation with respect to the substrate concentration. They are not valid for non-steady-state situations unless all of the reactions follow first-order kinetics.

When all processes approach first-order conditions, however, other dosage schedules may be used.[17] For example, a foreign compound may be administered repetitively until a quasi-steady-state is achieved after which time the entire dose is eliminated from the body during the dosage interval. The area under the blood concentration curve $(AUC)_{ss}$ during the dosage interval may then be used to estimate the average steady-state concentration. Alternatively, a single dose may be administered and the area under the curve (AUC) may be measured until all of the foreign compound and its metabolites are eliminated from the body. These approaches are valid because under first-order conditions, the following terms are mathematically equivalent

$$C_{ss} \Delta t = (AUC)_{ss} = AUC = \frac{\text{dose}}{\text{apparent clearance}} \qquad (49)$$

| Constant | Multidose | Single dose |
| infusion | injection | injection |

in which the "apparent clearance" is dependent not only on the true total body clearance but also on the route of administration and other terms. Thus, the equation for the steady-state concentrations of the foreign compound and its metabolites derived for constant infusion models may also be used when the foreign compound is administered as a single dose or in multiple doses.

Although biologists frequently refer to the half-life of a foreign compound in the body, it should be emphasized that the term has meaning only during the terminal phase of elimination of the foreign compound (usually referred to as the β-phase) at which time first-order kinetics occur and the body acts as though it were a single compartment into which the foreign compound is distributed. The apparent size of the compartment at this time is frequently referred to as $V_{d\beta}$. The half-life of the foreign compound depends on the following relationship

$$(\ln 2/t_{1/2}) = \beta = Cl_{TB}/V_{d\beta} \qquad (50)$$

in which Cl_{TB} is the total body clearance. However, Cl_{TB} is not always the apparent clearance estimated from the area under the curve of the steady-state concentration of the foreign compound in blood by the relationship shown in Eq. (49). Instead it may be calculated from the following relationship

$$Cl_{TB} = \frac{k_{0(S, \text{ left ventricle})}}{[S]_{ss}} \qquad (51)$$

in which $k_{0(S, \text{ left ventricle})}$ is the constant infusion rate of the foreign compound directly into the left ventricle of the heart. In the liver–kidney model discussed above, the total body clearance would be the sum of the clearances of the foreign compound by the liver and kidney (that is $Cl_{H_{(S)}}$ + $Cl_{R_{(S)}}$). But when a foreign compound is eliminated by organs that are connected in sequence by the blood vessels, the total body clearance is not the sum of the individual organ clearances. For example, when a compound is eliminated by the liver, lung and kidney, the total body clearance would be expressed as

$$Cl_{TB} = Q_H (1 - F_H F_L) + Cl_R + Q_X E_L \qquad (52)$$

in which F_H and F_L are availabilities of the foreign compound as it passes through the liver, and lung, respectively. Q_H is the sum of the portal venous and hepatic arterial blood flows. Q_X is the cardiac output minus the portal venous, hepatic arterial and kidney blood flows and E_L is the extraction ratio of the foreign compound by the lung.

III. COMMENTS

The extrapolation from the kinetic parameters of individual enzymes within subcellular organelles to the effects of these enzymes on the concentrations of the foreign compound and its metabolites in blood and thence into other organs, depends on a number of factors which are still incompletely understood. The equations described in this overview may be useful in developing a framework on which to build our future knowledge of the relationships among various enzyme activities and the biological effects of foreign compounds.

REFERENCES

1. Teorell, T. (1937). Kinetics of distribution of substances administered to the body. I. The extravascular modes of administration. *Arch. Intern. Pharmacodyn.* **57**, 205–225.
2. Teorell, T. (1937). Kinetics of distribution of substances administered to the body. II. The intravascular modes of administration. *Arch. Intern. Pharmacodyn.* **57**, 226–240.
3. Riggs, R. S. (1963). The mathematical approach to physiological problems. Williams and Wilkens, Baltimore.
4. Various authors (1974). *In* "Pharmacology and Pharmacokinetics" (T. Teorell, R. L. Dedrick, P. G. Condliffe, ed). Plenum Press, New York.
5. Gillette, J. R. (1971). Factors affecting drug metabolism. *Ann. N.Y. Acad. Sci.* **179**, 43–66.
6. Rowland, M., Benet, L. Z., and Graham, G. G. (1973). Clearance concepts in pharmacokinetics. *J. Pharmacokinet. Biopharm.* **1**, 123–136.
7. Wilkinson, G. R., and Shand, D. G. (1975). Commentary. A physiological approach to hepatic drug clearance. *Clin. Pharmacol. Ther.* **18**, 377–390.
8. Pang, K. S., and Rowland, M. (1977). Hepatic clearance of drugs. I. Theoretical considerations of a "well-stirred" model and a "parallel tube" model. Influence of hepatic blood flow, plasma and blood cell binding and the hepatocellular enzymatic activity on hepatic drug clearance. *J. Pharmacokinet. Biopharm.* **5**, 625–653.
9. Winkler, K., Keiding, S., and Tygstrup, N. (1973). Clearance as a quantitative measure of liver function. *In* "The Liver: Quantitative Aspects of Structure and Function" (P. Paugartner and R. Presig, eds.), pp. 144–155. Karger, Basel.
10. Bass, L., and Bracken, A. J. (1977). Time-dependent elimination of substrates flowing through the liver or kidney. *J. Theor. Biol.* **67**, 637–652.
11. Pang, K. S., and Rowland, M. (1977). Hepatic clearance of drugs. II. Experimental evidence for acceptance of the "well-stirred" model over the "parallel tube" model using lidocaine in the perfused rat liver *in situ* preparation. *J. Pharmacokinet. Biopharm.* **5**, 655–680.
12. Shand, D. G., Kornhauser, D. M., and Wilkinson, G. R. (1975). Effects of route of administration and blood flow on hepatic drug elimination. *J. Pharmacol. Exp. Ther.* **195**, 424–432.
13. Pang, K. S., and Gillette, J. R. (1978). Kinetics of metabolite formation and elimination in the perfused rat liver preparation. Difference between the elimination of preformed acetaminophen and acetaminophen formed from phenacetin. *J. Pharmacol. Exp. Ther.* **207**, 178–194.
14. Pang, K. S., and Gillette, J. R. (1979). Sequential first-pass elimination of a metabolite derived from a precursor. *J. Pharmacokinet. Biopharm.* **7**, 275–290.
15. Baron, J., Redick, J. A., and Guengerich, F. P. (1978). Immunohistochemical localizations of cytochrome *P*-450 in rat liver. *Life Sci.* **23**, 2627–2631.
16. Gillette, J. R., and Pang, K. S. (1977). Theoretical aspects of pharmacokinetic drug interactions. *Clin. Pharmacol. Ther.* **22**, 623–639.
17. Gibaldi, M., and Perrier, D. (1975). "Pharmacokinetics: Drugs and the Pharmaceutical Sciences," Vol. 1. Dekker, New York.

Chapter 3

Human Genetic Variation in the Enzymes of Detoxication

DANIEL W. NEBERT

I. INTRODUCTORY REMARKS

A. Genetics and Detoxication

Among the mechanisms responsible for toxicity and detoxication, there is the growing appreciation that a delicate balance exists in each tissue between enzymes that form toxic intermediates and those that detoxify these highly reactive intermediates. This balance may vary, depending on different tissues, strains, and species. Other factors contributing to an individual's susceptibility to drug toxicity may include genetic "predis-

25

ENZYMATIC BASIS OF DETOXICATION, VOL. I
Copyright © 1980 by Academic Press, Inc.
All rights of reproduction in any form reserved.
ISBN 0-12-380001-3

position," age, nutrition, hormone levels, diurnal rhythm, compartmentalization of the enzyme (local pH, saturating versus nonsaturating substrate concentrations, K_m, V_{max}), efficiency of DNA repair and RNA and protein *de novo* synthesis, and immunological competence. The scope of this chapter will emphasize pharmacogenetics, which may be defined simply as an attempt at understanding why two different individuals (with the possible exception of monozygotic twins), given the same dose of any foreign chemical, may respond differently.

If one looks at any pharmacology textbook, almost every xenobiotic is found to be metabolized by a myriad of pathways (Fig. 1). Observing such a complexity of interacting pathways, one might expect that pharmacogenetic differences would typically be expressed as polygenic (two or more genes) multifactorial traits such as the ear length in corn (Fig. 2A). This expectation has often been shown *not* to be the case. Why not? If, for example (Fig. 1), compound A causes toxicity, any factor decreasing enzyme 1 or 5 would increase the steady-state level of compound A, thereby enhancing its duration and toxic effects. If compound A requires metabolism to toxic intermediate B*, any factor decreasing enzyme 2, 3, or 5 or increasing enzyme 1 would enhance the steady-state level of the reactive intermediate, thereby causing greater toxicity. On the other hand, if any of the other, more distant enzymes such as 4, 6, or 7 were rate-limiting for the overall pathway, any factor changing the level of such an enzyme could be most important in affecting the steady-state level of compound A or B*. If one accepts the "one gene, one enzyme" hypothesis and if the enzyme responsible for the rate-limiting step can vary by a large amount (by 2-fold, or especially 10- or 100-fold), allelic differences in the gene encoding for that enzyme would predominate over any subtle changes in the remainder of the enzymes involved.

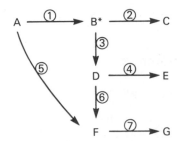

Fig. 1. Schematic representation of cascading pathways by which most xenobiotics are metabolized. A, parent drug; B* through G, various metabolites; circled numbers depict enzymes.

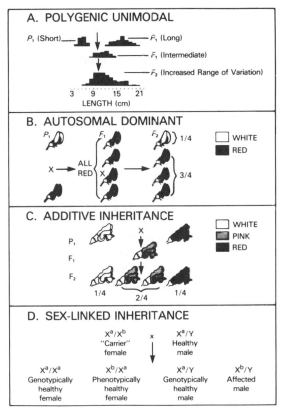

Fig. 2. Classical examples of genetic expression. (A) Multifactorial, or polygenic, inheritance, of the trait ear length in corn. (B) Mendelian autosomal dominant, as shown by the red dominant trait and the white recessive trait in the garden pea. (C) Additive inheritance, as shown by the red and white traits in the snapdragon. (D) Sex-linked, or X-linked, inheritance in which the allele X^b on the X chromosome denotes phenotypic expression of a disorder when paired with a Y chromosome but not when paired with a "normal" allele (X^a) on an X chromosome.

B. "Introduction" to Genetics

Most pharmacologists and biochemists have not worked daily with genetic concepts and experiments since having taken an undergraduate course in the subject. To familiarize the reader with genetic terminology, Fig. 2 illustrates the four most common modes of inheritance. As stated above, two or more genes expressed concomitantly result in polygenic, multifactorial inheritance (Fig. 2A), which is generally unimodal (no obvious distinct groupings) and which can be distributed in Gaussian fashion or skewed to the left or right. A gene, or a genetic locus, represents two alleles at a particular chromosomal site—one allele on each member of a chromosome pair in diploid organisms (eukaryotes). All chromosomes other than the sex chromosome pair are called autosomes. The allele for red color in garden peas (Fig. 2B) dominates over white color; when the red homozygote (CC) is crossed with white (cc), all F_1 progeny (Cc) are red. When F_1 heterozygotes are intercrossed (Cc × Cc), the F_2 distribution is genotypically (composition of the gene) 1:2:1 CC:Cc:cc and is phenotypically (what can be seen or measured) 3:1 red:white. The allele for red color is therefore inherited as a Mendelian autosomal dominant trait, white color as a recessive trait.

Red does not dominate over white color in the snapdragon (Fig. 2C); these traits represent additive inheritance (also codominant, midparent, gene-dose). Autosomal autonomous also indicates a gene-dose effect; according to strict Mendelian terminology, this condition would be called autosomal dominant because the heterozygote is affected clinically. However, the term "autonomous" provides a useful distinction for situations in which the severity of phenotypic response can be predicted on the basis of whether an individual is heterozygous or homozygous for the variant gene. When a trait is said to be expressed with partial penetrance, most likely modifier genes (or other unknown factors) contribute to the genetic expression in a manner not yet understood by the geneticist.

Lastly, the allele X^b (Fig. 2D) represents sex-linked inheritance. If an apparently healthy "carrier" female is crossed with a healthy male, the expected distribution of offspring is one-fourth genotypically healthy females, one-fourth phenotypically healthy females (i.e., carrier females), one-fourth genotypically healthy males, and one-fourth affected males. Half of all males born to a carrier female will thus express the sex-linked trait under consideration.

The Hardy–Weinberg distribution of alleles p and q follows the binomial expansion denoted by the equation $(p + q)^2 = p^2 + 2pq + q^2$. Hence, if the frequency of $p = 0.70$ and $q = 0.30$ in a given population, the expected occurrence of patients homozygous for p is 0.49, of heterozygotes is 0.42, and of patients homozygous for q is 0.09.

An important distinction has been made[1] among different types of genes: regulatory, structural, temporal, and processing. Regulatory genes control enzyme activity, synthesis, etc.; because such control is usually "on" or "off," alleles of a regulatory gene are usually dominant or recessive. A receptor may be viewed, for example, as the product of a regulatory allele. Structural genes encode for the enzyme or other protein that is being regulated. Because many proteins are composed of subunits and both alleles may contribute different subunits, allelic differences in structural genes are often expressed additively. Temporal genes govern when, during development, enzymes undergo characteristic changes in activity. Processing genes modify the enzyme or other protein posttranslationally by glycosylation, phosphorylation, adenylylation, subcellular localization and degradation.

C. Identification of Pharmacogenetic Subgroups

Genetic variation in any cellular process involving drug detoxication would conceivably lead to atypical drug responses: (i) transport (absorption, plasma protein binding); (ii) transducer mechanisms (receptors, enzyme induction, or inhibition); (iii) biotransformation; and (iv) excretory mechanisms (renal and biliary transport). The only published genetic differences under transport comprise two recent reports. There exist significant interindividual differences of two- to threefold in the displacing effect of drugs such as sulfisoxazole, salicylic acid, and salicyluric acid on bilirubin plasma binding.[2] Genetically inherited structural differences in the albumin molecule (one binding site versus two binding sites) may affect the transport of certain drugs such as warfarin.[3] Several examples of transducer mechanisms will be discussed in this chapter. Of course, the bulk of understood pharmacogenetic disorders reflects alterations in biotransformation. To my knowledge, there are no published genetic differences involving drug excretion; certain syndromes with gross anomalies such as biliary or renal atresia are excluded.

There are two ways in which identification of pharmacogenetic subgroups becomes possible. First, if one specific aspect of drug disposition assumes paramount importance in the overall host response to that drug, genetic variation in that process may be identified as a discontinuous response in the population. An example is atypical pseudocholinesterase, first detected as a prolonged apneic response to succinylcholine.[4] If the patient does not breathe for a while, this can be dramatic. Second, one can select a drug or an enzyme activity to test, rather than looking for the more obvious atypical drug response. Recent examples include clinical surveys for polymorphisms of dopamine β-hydroxylase activity,[5] catechol O-methyltransferase activity,[6] and barbiturate metabolites in the urine.[7-9] The drug acetylation polymorphism[10] was uncovered following documentation that genetic variability exists in the rate of isoniazid metabolism in man.[11-14]

Even if enzyme activities are determined directly, it is possible to miss an existing polymorphism. The best example is acid phosphatase in erythrocytes. Five possible human phenotypes, each having a different electrophoretic pattern of enzyme, are controlled by three alleles at a single genetic locus.[15] Yet the distribution of red cell acid phosphatase activities in parents and offspring from 50 British families was found[16] to be unimodal and Gaussian. It was only after the population was subdivided on the basis of electrophoretic phenotype that the influence of polymorphism on enzymatic activity could be discerned. Hence, even though net enzyme "activity" *in vivo* may be the crucial factor in polymorphisms of drug metabolism, determinations of *in vitro* "activity" may not be sufficiently discriminating for detecting enzyme variants. Rather, qualitative variations in the charge, stability, kinetics of product formation, or inhibitor specificity may be more precise in reflecting genetically mediated alterations in protein structure or conformation. Isoelectric focusing techniques are now able to detect substitutions of a single neutral amino acid. With this technique, for example, it was shown[17] that catalytically "normal" glucose-6-phosphate dehydrogenase activity in the red blood cells of several hundred children is genetically heterogeneous and that the structural *Gd* alleles concerned are polymorphic in the Nigerian population.

D. Investigative Approaches to Pharmacogenetics

At least four general investigative approaches might be used for uncovering new atypical drug responses which have a genetic basis.

1. Clinical Observations

Such an approach may be hampered by the frequent occurrence of multifactorial variation, as mentioned above.

2. Family or Twin Studies

With the use of twins, several investigators have established that interindividual differences in drug metabolism are largely under genetic control for isoniazid,[11] phenylbutazone, antipyrine, dicumarol,[18] nortryptyline,[19] ethanol,[20] and halothane.[21] Family studies have suggested that metabolism is under polygenic control for dicumarol,[22] phenylbutazone,[23] and nortriptyline.[24] Evidence also exists[19,25] that differences in response to inducers of drug metabolism, such as phenobarbital, are genetically controlled. For certain drugs, determination of plasma half-life,[18] steady-state plasma concentration,[14,19] or rates of urinary metabolite excretion[10] can be quite specific estimates of metabolism. In instances in which drug response is easily measured and directly reflects tissue levels, e.g., in the initiation and maintenance of anesthesia,[26] clinical observation may also be a fair estimate of drug metabolism. It is anticipated that, as genetic variability of specific drug-metabolizing enzymes is studied directly, new polymorphisms will be discovered.

3. Protein Polymorphisms

Investigating polymorphism of potential pharmacological significance represents another even more promising approach. Harris[27] estimated that 30% of randomly selected gene products may exhibit polymorphic variation, i.e., frequency greater than 0.01 for variant alleles. Many such structural variants are functionally normal *in vitro*, i.e., with respect to enzymatic activities, or in the case of hemoglobin variants, oxygen-carrying capacities. It is quite possible, however, that such "isoallelic" variants may be responsible for significant functional derangement *in vivo*. As is evident from the erythrocyte acid phosphatase and glucose-6-phosphate dehydrogenase polymorphisms mentioned above, a suitable qualitative test, such as electrophoretic pattern, is often required just to detect enzyme variants with appreciable differences in activity, especially in heterozygotes of autosomal traits.

4. Animal Models

Animal systems may provide examples of mutants which are likely to be stable in mammalian gene pools. Several pharmacologically important genetic models have been described, including warfarin resistance in rats,[28] susceptibility to chemical carcinogenesis in mice,[29] and polymorphic isoniazid acetylation in rabbits.[30] The obvious advantages of animal studies are that a pharmacogenetic disorder can be explored in great detail in numerous tissues, strains, and species, and that sophisticated predictions can be made for the human based on the detailed understanding of the disorder in other mammals. An important disadvantage is that the pharmacogenetic disorder characterized in an animal model system need not be identical to that found in the human (e.g., cf. Weber *et al.* 30, 31). Pharmacogenetic differences may also be strikingly similar, however, between man and laboratory animals. For example, the caffeine sensitive type of human malignant hyperthermia may reflect two or more genes,[32] and the same phenomenon is probably also true in the pig.[33] "Low" catechol *O*-methyltransferase activity is inherited as an autosomal recessive trait in the rat[34] and in the human.[35] Although many interesting animal models exist for pharmacogenetics, the emphasis in this chapter is on human pharmacogenetics and atypical drug responses.

II. Classification of Human Pharmacogenetic Disorders

During the two decades since Motulsky[36] first promulgated the basic concepts of pharmacogenetics and Vogel[37] proposed the term, large advances have been made in our understanding of basic pharmacological mechanisms. As is evident from past comprehensive reviews,[18, 38-42] any attempt to classify these disorders into subgroups is arbitrary. The attempt at classification of these disorders that is presented here (Table I) is admittedly arbitrary as well.

A. Disorders with Increased Sensitivity to Drugs

1. Enzyme Deficiencies Resulting in Decreased Detoxication

Conditions under this classification are caused by decreased or total absence of enzymatic activity. If clinical symptoms exist, they probably result from increased tissue levels of parent compounds that are not cleared or detoxified as rapidly as in normal individuals.

    a. Succinylcholine Apnea. This atypical drug response is associated with the autosomal recessive inheritance[43] of an atypical pseudo-cholinesterase,[44] the serum enzyme which hydrolyzes acylcholines such as succinylcholine and is responsible for its short duration of action in normal subjects. The incidence of this condition is 1:2500. At least three variant alleles at a single locus (designated E_1) have been described which result in structural and functional changes in the enzyme molecule.[45] Since cocaine hydrolysis is mediated by human plasma cholinesterase,[46] it would be of interest if differences in response to this drug could also be explained on a genetic basis.

TABLE I

Arbitrary Classification of Human Pharmacogenetics and Atypical Drug Response

---

Disorders with increased sensitivity to drugs
  Enzyme deficiencies resulting in decreased detoxication
  Disorders of unknown etiology
Disorders with increased resistance to drugs
  Defective [?]"receptor"
  Defective absorption
  Increased metabolism
Disorders exacerbated by enzyme-inducing drugs
Disorders in which presence or absence of metabolic potentiation may play a role

---

b. Serum Paraoxonase Polymorphism. This arylesterase (EC 3.1.1.2) cleaves *p*-nitrophenol from paraoxon and shows a polymorphism consistent with a simple two-allele model.[47] The chemical structure of paraoxon is illustrated in Fig. 3. Although paraoxon is claimed to be the "natural" substrate for this enzyme, one wonders if a "true" endogenous substrate does not exist. Because the organophosphate paraoxon is a potent cholinesterase inhibitor, an accidental exposure of this insecticide to industrial workers, for example, would affect more severely those persons having the homozygous autosomal recessive trait for "low" paraoxonase activity.

c. Sulfite Oxidase Deficiency. This autosomal recessive disorder in sulfur metabolism[48] results in large amounts of abnormal sulfur-containing metabolites (sulfite, thiosulfate, and *S*-sulfocysteine) in the urine due to enzyme deficiency. A diet low in sulfur amino acids may benefit the individual. An atypical response to sulfur-containing drugs might be anticipated.

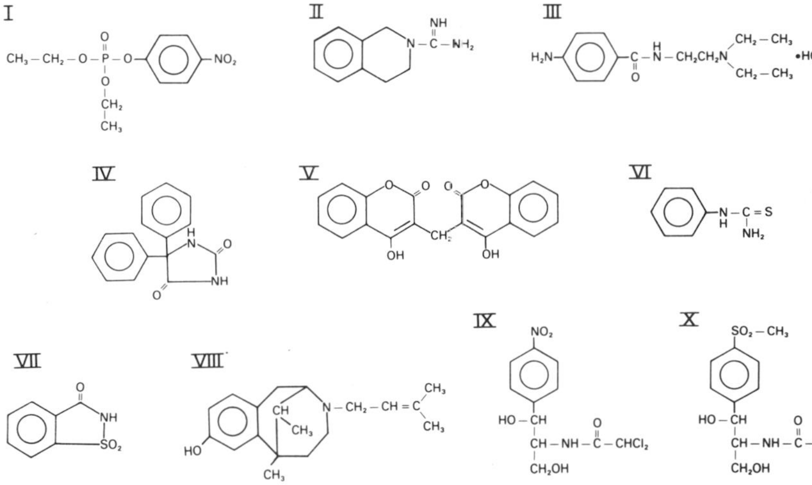

Fig. 3. Chemical structures of various xenobiotics. I, Paraoxon, diethyl-*p*-nitrophenylphosphate; II, Debrisoquine, 2-amidino-1,2,3,4-tetrahydroisoquinoline; III, procainamide, *p*-amino-*N*-(2-diethylaminoethyl)benzamide hydrochloride; IV, phenytoin, 5,5'-diphenyl-2,4-imidazolidinedione; V, dicumarol, 3,3'-methylenebis(4-hydroxy-1,2-benzopyrone); VI, phenylthiocarbamide; VII, saccharin, 2,3-dihydro-3-oxo-benzisosulfonazole; VIII, pentazocine, 1,2,3,4,5,6-hexahydro-6,11-dimethyl-3-(3-methyl-2-butenyl)-2,6-methano-3-benzazocin-8-ol); IX, chloramphenicol, D-(-)-*threo*-2,2-dichloro-*N*-[β-hydroxy-α(hydroxymethyl)-*p*-nitrophenethyl]acetamide; X, thiamphenicol, D-*threo*-2,2-dichloro-*N*-[β-hydroxy-α(hydroxymethyl)-*p*-(methylsulfonyl)-phenethyl]acetamide.

d. Debrisoquine Hydroxylation. Humans have been phenotyped for their ability to metabolize the antihypertensive drug debrisoquine.[49] About 9% of the British population tested have the autosomal recessive trait for little or no metabolism; these patients, with a given dosage, exhibit a significant orthostatic hypotensive effect, compared with "extensive metabolizers." Of further interest, these individuals with poor 4-hydroxylation of debrisoquine exhibit relatively low metabolism of other drugs such as phenytoin.[50] The individuals may represent genetic deficiencies in one or more forms of cytochrome *P*-450; further studies are clearly indicated.

e. Slow Acetylators. The slow inactivation of isoniazid is an autosomal recessive trait[13,14] which reflects decreased activity of a hepatic soluble *N*-acetyltransferase requiring acetyl-CoA.[10] The acetylation of isoniazid and sulfadiazine appear to be under similar genetic control in the rabbit, whereas acetylation of *p*-aminobenzoic acid is not.[31] Monosubstituted hydrazines, such as isoniazid, phenelzine, hydralazine, and sulfamethazine,[51] are acetylated with a bimodal distribution in the population; approximately half the European population are phenotypically "slow" acetylators and half phenotypically "rapid" acetylators.[22] The population distributions of the acetylation rates of *p*-aminosalicylic acid and sulfanilamide, on the other hand, are unimodal, and these drugs are therefore probably substrates for another enzyme.[51] Before understanding of this polymorphism was reached, it was known that slow isoniazid inactivators are more prone to develop a toxic peripheral neuropathy during isoniazid therapy[52,53] and that the simultaneous administration of pyridoxine prevents the disorder.[54] Moreover, among individuals receiving both isoniazid and phenytoin, slow isoniazid inactivators are much more likely to develop signs and symptoms of phenytoin toxicity,[55] apparently because isoniazid is a potent inhibitor of phenytoin p-hydroxylation.[56]

Slow acetylators are also more likely to develop phenelzine toxicity[57] and the hydralazine lupus syndrome.[58,59] More recently it was shown[60] that procainamide (administered to patients with cardiac arrhythmias) will cause a positive anti-nuclear antibody test in slow acetylators much more rapidly than in fast acetylators. Moreover, if one examines the phenotype of patients with "idiopathic" systemic lupus erythematosis,[61] the preponderance of slow acetylators is statistically significant ($P < 0.01$ by chi-square analysis). These data suggest that unknown agents in the environment, possibly azo dyes in foods and cosmetics that require acetylation for detoxication, in combination with a "predisposition" in the host for developing the autoimmune disorder, result in this life-shortening disease.

f. Phenytoin Toxicity. Kutt and co-workers[62] described a family in which the mother and two of four male offspring developed symptoms of phenytoin toxicity while taking relatively low doses of the drug. Extremely high blood levels of the drug and abnormally low urinary levels of 5-phenyl-5'-p-hydroxyphenylhydantoin, either free or as the glucuronide, were found; in the human this hydroxylated product is known to be the principal metabolite. The presumed decrease in p-hydroxylation of diphenylhydantoin represented the first known example of a human genetic defect in a cytochrome P-450-dependent monooxygenase.

g. Dicumarol Sensitivity. Solomon[63] reported extreme sensitivity to the anticoagulant bishydroxycoumarin (dicumarol) in a patient whose plasma half-life for the drug was three times longer than normal. Although the family refused to cooperate in further studies, the patient's mother had earlier suffered a spinal cord hematoma while being treated with low doses of warfarin.

h. Acatalasia (Acatalasemia). The disorder is a decreased level, or absence, of the enzyme catalase, which is transmitted as an autosomal recessive trait.[64] The only established clinical consequence is an abnormal response to topical hydrogen peroxide. A total of 54 affected families have been reported from Japan, Korea, Switzerland, and Israel; a recent comprehensive review[65] considers the genetic heterogeneity of the condition and its possible implications.

i. Crigler–Najjar Syndrome. This severe form of congenital nonhemolytic jaundice associated with kernicterus,[66] is inherited as an autosomal recessive trait.[67] In liver samples from these patients, *in vitro* formation of glucuronide conjugates is normal with p-nitrophenol,[68] markedly decreased with 4-methylumbelliferone and o-aminophenol, and totally undetectable with bilirubin[69,70] as substrate. Impairment in the capacity for glucuronidation of acetaminophen,[69] tetrahydrocortisol,[71] chloral hydrate, trichloroethanol, and salicylate[67] has been described in affected individuals. Although no cases of drug sensitivity have been reported, these individuals would be expected to exhibit a greater susceptibility to adverse reactions from drugs which are conjugated appreciably by the particular microsomal glucuronyltransferase that is defective.

2. *Pharmacogenetic Disorders of Unknown Etiology*

Several other disorders resulting from an increased sensitivity to drugs by as yet unknown mechanisms have either a proven or suspected genetic basis (Table II).[72-84] About 5% of the United States population is

TABLE II

Pharmacogenetic Disorders of Unknown Etiology

Condition	Inheritance	References
Steroid-induced glaucoma (dexamethasone)	Autosomal autonomous	72
Malignant hyperthermia caused by anesthetics (halothane, succinylcholine, methoxyfluorane, ether, and cyclopropane)	Autosomal dominant	73,74
Norepinephrine sensitivity and absence of flare in response to intradermal histamine, in patients with familial dysautonomia (Riley–Day syndrome)	Autosomal recessive	75–77
Phenytoin-induced protein synthesis and collagen production enhanced in cultured fibroblasts of patients with phenytoin-induced gingival hyperplasia	Unknown	78
Thromboembolic complications caused by anovulatory agents	(Increased incidence in A,B, and AB blood groups)	79,80
Atropine sensitivity in patients with Down's syndrome	[?]Associated with trisomy 21	81,82
Meperidine-induced myopathy	[?](mother and daughter)	83
Norethisterone-induced jaundice	[?](two sisters)	84

homozygous for the recessive allele causing glaucoma induced by corticosteroids.[72] The incidence of malignant hyperthermia (usually associated with muscular rigidity) in response to anesthetics is about 1:20,000.[39] A promising insight into the etiology of familial dysautonomia was recently described[85]: a qualitative abnormality in the β subunit of nerve growth factor. In view of the common clinical observation that gingival hyperplasia occurs in some, but not other, patients receiving phenytoin and that this disorder appears to run in families, the tissue culture study cited[78] represents the first attempt at trying to understand this pharmacogenetic lesion of unknown etiology.

B. Genetic Disorders Resulting from Increased Resistance to Drugs

1. Possibility of Defective Receptor

a. **Coumarin Resistance.** This autosomal dominant trait has been described in two large pedigrees; the affected individual requires a dose of anticoagulant 5 to 20 times greater than normal.[86] Hypoprothrombinemic responses to large doses of oral anticoagulants in such individuals are reversed by lower doses of vitamin K than are required in normal subjects; yet, paradoxically, following vitamin K deprivation, greater doses of the vitamin are needed to restore prothrombin activity to normal levels.[87] Thus, O'Reilly postulated the existence in affected individuals of an altered receptor site with moderately decreased affinity for vitamin K and a markedly decreased affinity for anticoagulants.

b. Inability to Taste Phenylthiourea or Phenylthiocarbamide. This pharmacogenetic difference is transmitted as an autosomal recessive trait; about 30% of Caucasians are homozygous for the recessive allele.[88,89] The mechanism for the deficiency is unknown but could represent an altered receptor for these phenylthio analogues. An interesting association between this trait and the ability to perceive saccharin as a bitter taste was recently reported.[90]

c. Familial Hypercholesterolemia. Although cholesterol may be considered as an endogenous compound synthesized by man, dietary cholesterol can be regarded as a foreign chemical, which must be avoided by patients with familial hypercholesterolemia. Genetic differences in the low-density lipoprotein (LDL) receptor system,[91] therefore, can be regarded as a pharmacogenetic disorder. Two-thirds of the cholesterol in human plasma is contained within LDL. A sequence of events occurs by which LDL binds to a cell surface receptor and becomes internalized. The cholesterol, derived from internalized LDL, suppresses a microsomal enzyme, 3-hydroxy-3-methylglutaryl-CoA reductase, the rate-controlling enzyme in cholesterol biosynthesis, thereby turning off cholesterol synthesis in the cell.[92,93] The newly arrived cholesterol also activates a microsomal cholesterol esterifying enzyme, acyl-CoA:cholesterol acyltransferase, so that excess cholesterol can be reesterified and stored as cholesteryl esters.[94,95] Lastly, the newly arrived cholesterol turns off the synthesis of the LDL receptor, thereby preventing further entry of LDL and protecting cells against an overaccumulation of cholesterol.[96] Genetic defects in this system have been described: (i) receptor-negative; (ii) receptor-defective; and (iii) internalization-defective.[91] Inheritance is autosomal autonomous, i.e., heterozygotes display a two- to threefold elevation in plasma LDL levels and often develop myocardial infarcts as early as 35 to 45 years of age; homozygotes exhibit plasma LDL levels six times above normal and typically have a myocardial infarction before the age of 15.

2. Defective Absorption

Patients with juvenile pernicious anemia are resistant to massive, i.e., pharmacological, oral doses of vitamin B_{12} because of defective ileal absorption. Several causes of the disorder have been described, including congenital absence of intrinsic factor[97] and secretion of a functionally defective intrinsic factor with normal immunological properties.[98] Although the mode of transmission is not known, genetic control seems probable from the high incidence among siblings.[97] The disorder can be effectively treated by administration of vitamin B_{12} intramuscularly.

3. Increased Metabolism

a. **Succinylcholine Resistance.** Patients with this atypical drug response have been described in only one family.[99] The disorder represents the Cynthiana variant of serum cholinesterase, in which condition succinylcholine and other acylcholines are hydrolyzed about three times more rapidly than normal. The defect presumably reflects an increased number of molecules rather than any structural change in the enzyme.[100]

b. **Atypical Alcohol Dehydrogenase.** The frequency of individuals with this pharmacogenetic difference was found to be 20% in 59 liver specimens from Switzerland and 4% in 50 liver samples from England.[101] The atypical cytosol enzyme displays a five- to sixfold increase in ethanol oxidation *in vitro*, whereas ethanol metabolism *in vivo* is increased only 40 or 50%; reasons for the discrepancy are unclear. Increased ethanol metabolism might result in higher acetaldehyde levels, which are known to be elevated in alcoholics[102] and in patients whose primary relatives suffer from chronic alcoholism.[103] The inheritance of atypical alcohol dehydrogenase and its possibly direct relationship to alcoholism appear not to have been studied.

c. **Pentazocine Resistance.** The analgesic pentazocine is used to supplement nitrous oxide relaxant anesthesia. A larger pentazocine dosage was reported[26] to be required during surgery on patients who smoked cigarettes or resided in urban areas, as compared with nonsmokers and rural dwellers, respectively. Although the excess drug requirement for maintenance analgesia probably reflects increased metabolism due to environmental factors that induce enzyme activities, differences in the capacity of certain forms of cytochrome *P*-450 to be induced clearly exist in the human population. Further clinical studies on anesthesia and analgesia pharmacogenetics would be bf great interest.

C. Genetic Disorders Exacerbated by Enzyme-Inducing Drugs

1. Hepatic Porphyrias

Barbiturates and steroids cause profound exacerbations of clinical symptoms in porphyria patients, the explanation of which remains obscure. These chemicals induce δ-aminolevulinic acid synthetase, the rate-limiting step in heme synthesis. A recent proposed explanation for the neuropsychiatric symptoms includes the fact that δ-aminolevulinic acid is a potent agonist for γ-aminobutyric acid autoreceptors.[104] Each of the types of porphyria appears to be autosomal dominant. Although the precise nature of each defect has not been established,[105] enzyme deficiencies appear likely for acute intermittent porphyria (uroporphyrinogen I synthetase) and hereditary coproporphyria (coproporphyrinogen oxidase); variegate porphyria may result from an exaggerated responsiveness of hepatic δ-aminolevulinic acid synthetase. Because these drugs induce enzymes that are proximal to defective enzymes in a pathway, the result is increased urinary excretion of certain precursors.

2. Pentosuria

In patients with this autosomal recessive trait, increased urinary excretion of L-xylulose appears to result from a marked decrease in NADP-dependent xylitol dehydrogenase activity,[106] an enzyme in the glucuronic acid oxidation pathway. Aminopyrine increases the excretion of L-xylulose in affected individuals[107]; the increase probably occurs by induction of a preceding enzyme in the pathway, since the inducers, Chloretone and barbital, stimulate free glucuronic acid excretion in normal mammals.[108] No clinical abnormalities are known in individuals with pentosuria.

3. Environmental Carcinogenesis

Polycyclic hydrocarbons and phenobarbital are inducers of various forms of cytochrome P-450, as well as other enzyme activities. Many polycyclic hydrocarbons are well known environmental contaminants and potent carcinogens in laboratory animals. Phenobarbital has been shown[109,110] to enhance tumor promotion in laboratory animals. In view of known human genetic differences in enzyme inducibility by both polycyclic hydrocarbons[111] and phenobarbital,[19,25] the predisposition toward environmentally caused cancer might be regarded as an important pharmacogenetic disorder. Further clinical studies are necessary. The subject is considered in greater detail toward the end of this chapter.

D. Disorders in Which the Presence or Absence of Metabolic Potentiation May Play a Role

The conditions listed in Table III involve the possibility of metabolic activation at several different levels. In the first two disorders, the enzyme deficiency responsible for the disease state also results in an increased resistance to drugs commonly used for treating the primary disorder. The next three conditions represent recently recognized human enzyme abnormalities in which atypical drug responses may be suspected but have not been reported. Although the next condition represents interesting enzyme abnormalities that are well characterized only in laboratory animals, the possibility for human pharmacogenetic disorders of these types could be very high.

1. Vitamin D-Dependent Rickets

Several forms of vitamin D-refractory syndromes exist and have in common a partial or complete inability to respond to pharmacological doses of the vitamin. The basic and clinical aspects of vitamin D, the prohormone, and the mechanism of action of $1\alpha,25$-dihydroxy-D_3, regarded as the active hormone (Fig. 4), have been extensively reviewed.[112,113] Vitamin D_3 (cholecalciferol) is hydroxylated in the 25-position by a microsomal enzyme that is predominantly in the liver of mammals but exists in the intestine and kidney of the chicken. This enzyme is inhibited by its product, 25-hydroxyvitamin D_3 (25-HO-D_3).

TABLE III

Genetic Disorders in Which Presence or Absence of Metabolic Potentiation May Play a Role in Human Disease

Hereditary vitamin D-dependent rickets

Hypoxanthine-guanine phosphoribosyltransferase-deficient gout

[?]Consequences of the cytidine deaminase polymorphism

[?]Consequences of polymorphisms in catecholamine metabolic pathway

[?]Consequences of steroid hydroxylase deficiency

[?]Consequences of aberrant metabolism of halogenated anesthetics

Chloramphenicol-induced aplastic anemia

Isoniazid-induced hepatitis in fast acetylators

Increased susceptibility to drug-induced hemolysis (glucose-6-phosphate dehydrogenase deficiency, other defects in glutathione formation or utilization, hemoglobinopathies)

Hereditary methemoglobinemia (NADH methemoglobin reductase deficiency, hemoglobinopathies, acetophenetidin O-dealkylase deficiency)

[?]Consequences of polymorphism in rate of chemical carcinogen metabolism

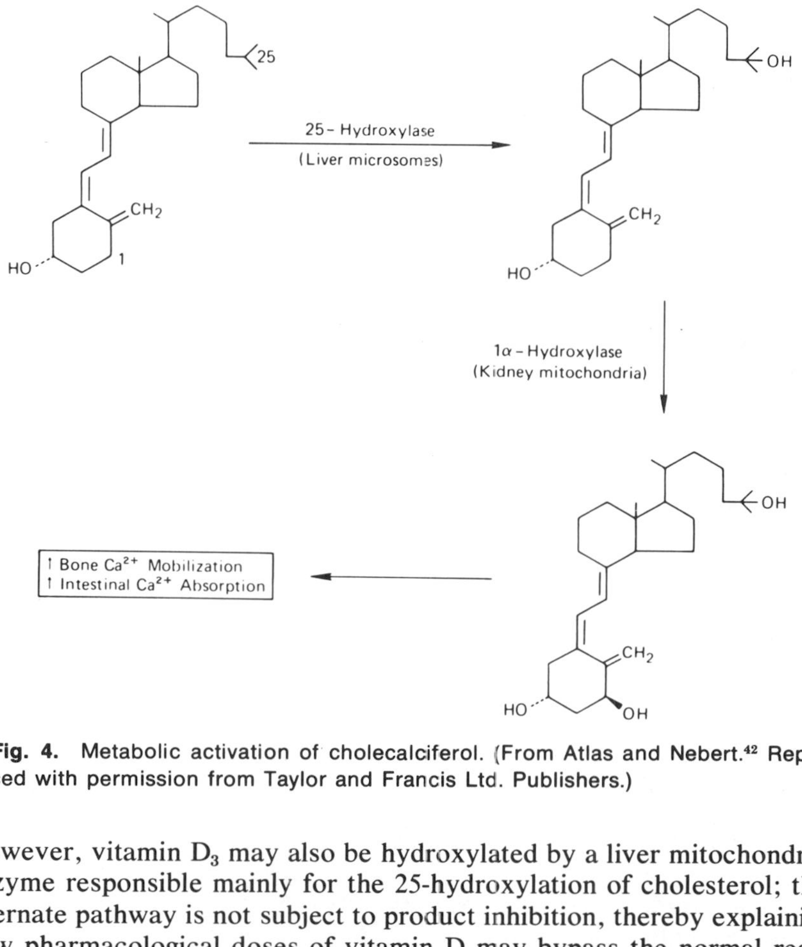

Fig. 4. Metabolic activation of cholecalciferol. (From Atlas and Nebert.[42] Reproduced with permission from Taylor and Francis Ltd. Publishers.)

However, vitamin D_3 may also be hydroxylated by a liver mitochondrial enzyme responsible mainly for the 25-hydroxylation of cholesterol; this alternate pathway is not subject to product inhibition, thereby explaining how pharmacological doses of vitamin D may bypass the normal regulatory step. 25-Hydroxyvitamin D_3 is subsequently hydroxylated to 1α,25-dihydroxyvitamin D_3 (1α,25-diHO-D_3) by a kidney mitochondrial enzyme analogous to the adrenal P-450-mediated steroid hydroxylases.[114,115] The 1α,25-diHO-D_3 promotes intestinal calcium absorption and, together with parathormone, bone recalcification and is considered to be the active form of vitamin D. Low dietary phosphorus stimulates 1α-hydroxylase activity and may increase accumulation of 1α,25-diHO-D_3 in both intestine and plasma.[116] There exists, especially in kidney mitochondria, an enzyme which hydroxylates 25-HO-D_3 and 1α,25-diHO-D_3 in the 24-position; the purpose of this enzyme remains unknown. There must be a hydroxy group in the 25-position, however, before either the 1α- or the 24-hydroxylase will act on the substrate.[117]

In patients with vitamin D-dependent rickets, the defect is inherited as an autosomal recessive trait. This disorder represents either a complete absence of the renal 1α-hydroxylase or a congenital defect in its regulation.[118] Patients with vitamin-D-dependent rickets thus respond completely to physiological doses of $1\alpha,25$-diHO-D_3 whereas abnormally large doses of vitamin D and 25-HO-D_3 are required to maintain normocalcemia. In contrast, patients with hypophosphatemic vitamin D-resistant rickets do not respond as favorably to $1\alpha,25$-diHO-D_3; there is some increase in intestinal calcium absorption but no improvement in their profound hypophosphatemia.[119] This disorder, which is inherited in both X-linked and autosomal dominant forms, is generally considered to be due to a primary defect in renal tubular transport of phosphorus.[119,120]

Patients with parathyroid dysfunction have lesser deficiencies in the 1α-hydroxylation of 25-HO-D_3,[121,122] which agrees with the observation[123] that parathormone regulates renal levels of 1α-hydroxylase activity, perhaps by changes in inorganic phosphate concentration. Oral $1\alpha,25$-diHO-D_3 therapy can reverse bone disease and increase growth in uremic children with renal osteodystrophy.[124] The defect in sarcoidosis was recently shown[125] to be associated with impaired regulation of the production and(or) degradation of $1\alpha,25$-diHO-D_3.

A recent exciting finding involves discovery of the high-affinity receptor for the hormone $1\alpha,25$-diHO-D_3 in bone[126] and parathyroid gland,[127] separable from a high-affinity serum transport binding protein for 25-HO-D_3.[128] The 25-HO-D_3 binding protein inhibits the 1α-hydroxylase activity.[129] Could it be that patients with one or more forms of vitamin D-resistant rickets lack the normal receptor or have abnormally high levels of the serum transport binding protein for 25-HO-D_3?

2. Hypoxanthine-Guanine Phosphoribosyltransferase (HGPRT) Deficient Gout

Complete deficiency of HGPRT is responsible for the Lesch-Nyhan syndrome, an X-linked inborn error of metabolism in which neurological abnormalities, mental retardation, and compulsive self-mutilation are associated with marked hyperuricemia.[130] About 1% of normal HGPRT activity is found in certain patients with gout; this disorder is also inherited as an X-linked recessive trait. The phosphoribosyltransferase is the enzyme of principal importance in the purine salvage pathway (Fig. 5). When deficient, conversion of guanine and hypoxanthine to their respective nucleotides is diminished. Since these nucleotides are feedback inhibitors of the rate-limiting initial step in *de novo* purine biosynthesis, a decrease in the steady-state concentration of guanine and hypoxanthine nucleotides results in an increased (unregulated) rate of purine synthesis

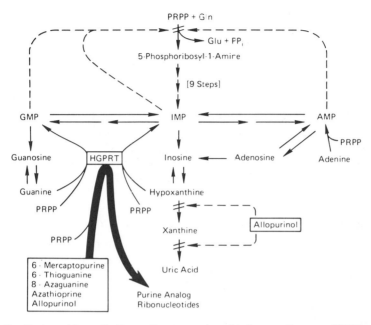

Fig. 5. Purine biosynthetic, salvage and oxidative pathways. PRPP, phosphoribosyl pyrophosphate; PP_i, inorganic pyrophosphate. (From Atlas and Nebert.[42] Reproduced with permission from Taylor and Francis Ltd. Publishers.)

and is therefore the major cause of hyperuricemia in affected individuals.[131] At least 10 to 12 different mutant forms of the enzyme are known.[130]

Because purine base analogues used in cancer chemotherapy and immunosuppression must be converted to their respective nucleotides by the transferase in order to inhibit purine biosynthesis, the antimetabolites shown in Fig. 5 would be expected to be ineffective in patients with a deficiency in the enzyme. This possibility has been documented for 6-mercaptopurine in a patient with leukemia[132] and for the immunosuppressive agent azathioprine.[133]

Allopurinol, one of the major drugs used in the treatment of gout, inhibits uric acid production principally by blocking xanthine oxidase (Fig. 5). In normal individuals the resulting accumulation of hypoxanthine leads, via HGPRT, to increased levels of the nucleotide IMP and thus to feedback inhibition of purine biosynthesis. Further, allopurinol is a substrate for the transferase and, as the ribonucleotide form, inhibits *de novo* purine synthesis. These latter two effects of allopurinol are absent in patients with HGPRT-deficient gout, thereby accounting for an atypical drug response. Hence, whereas normal individuals taking allopurinol exhibit a decrease in both uric acid and total purine urinary excretion,

HGPRT patients taking allopurinol show a decrease in uric acid excretion only; total purine urinary output, accounted for primarily by the oxypurines hypoxanthine and xanthine, is unchanged or slightly increased.[131] Although allopurinol may still be effective in treating gout and preventing urinary calculi formation in enzyme-deficient individuals, such patients may develop xanthine stones while taking allopurinol. At least three genetic variants of phosphoribosyl pyrophosphate (PRPP) synthetase (also X-linked) are known[130] to increase production of PRPP and therefore to enhance the rate of purine biosynthesis *de novo*. This enzyme defect in purine metabolism will respond to allopurinol therapy in the usual manner, however, with urinary decreases in both uric acid and total purine.

3. Polymorphism of Leukocyte Cytidine Deaminase

Three electrophoretic forms of cytidine deaminase occur in human leukocytes, and family studies are consistent with autosomal inheritance of two allelic genes. The data strongly suggest that cytidine deaminase is a tetramer and that heterozygotes at this locus produce two homotetramers plus the three possible heterotetramers.[134]

To my knowledge there has been no quantitation of enzymatic activity from the three phenotypes characterized by gel electrophoresis. Besides cytidine, deoxycytidine and the nucleoside analogues 5-fluorocytidine, 5-azacytidine, and cytosine arabinoside are all substrates for this deaminase.[135] The latter two analogues are used in cancer chemotherapy and become inactive as chemotherapeutic agents following deamination. Thus, a relative increase or decrease in cytidine deaminase activity might result in individual differences in resistance to chemotherapy and in toxic side effects of these antimetabolites. Levels of cytidine deaminase activity in leukemic cells are correlated with resistance to cytosine arabinoside[136]; whether possible interindividual differences reflect genetic polymorphism or the degree of neoplastic dedifferentiation in leukemic cells is not known.

4. Polymorphisms in Catecholamine Metabolic Pathway

The catecholamines, an important class of neutrotransmitters, are thought to play a role in the pathophysiology of diseases such as mental depression, schizophrenia, Parkinson's disease, and hypertension. Polymorphisms of the enzymes of catecholamine biosynthesis and degradation are presently being uncovered, although it is not clear whether such genetic information will enhance our understanding of these diseases. Pharmacogenetic differences should be found, however, since various drugs, e.g., monoamine oxidase inhibitors and adrenergic agonists and antagonists, will competitively or noncompetitively inhibit one or more of these enzymes. Of the major enzymes involved in catecholamine biosynthesis, no human biochemical genetic data are available on tyrosine hydroxylase (believed to be the rate-limiting step)[137] and phenylethanolamine N-methyltransferase (which converts norepinephrine to epinephrine).[138] Among inbred strains of mice, autosomal additive inheritance of both these enzymes is believed to be due to differences in rates of degradation,[139] i.e., coordinate control by a processing gene. The lack of an easily obtained tissue has hindered human genetic studies of L-aromatic amino acid decarboxylase, tyrosine hydroxylase, and phenylethanolamine N-methyltransferase.

 a. Dopamine β-Hydroxylase. This enzyme converts dopamine to norepinephrine[140] and is associated with the catecholamine-containing vesicles in sympathetic nerve terminals and the adrenal medulla. Though most of the enzyme is believed to be bound to the vesicle membrane, some of the enzyme is released with catecholamines in response to nerve stimulation and is found circulating in the blood. Abnormalities in plasma dopamine β-hydroxylase have been demonstrated in patients with Down's syndrome, torsion dystonia, and familial dysautonomia.[141] In a normal population surveyed,[5] an autosomal recessive trait for "very low" plasma dopamine β-hydroxylase activity was characterized; the gene frequency for this allele was about 0.2. The possibility of a temporal gene defect is being considered,[141,142] but the biochemical mechanism is unknown. The significance of low plasma dopamine β-hydroxylase is equally unclear since it cannot be assumed that these levels reflect dopamine β-hydroxylase activity in neural tissue.

 b. Catechol O-Methyltransferase. This methyltransferase catalyzes the O-methylation of endogenous catecholamines and such catechol drugs as isoproterenol.[143] Data from family studies are compatible with an allele for low human erythrocyte catechol O-methyltransferase that is inherited in an autosomal recessive fashion.[6] The significance of this polymorphism is not known, since the regulation of the enzyme in experimental animals is known to vary from tissue to tissue.[141]

 c. Monoamine Oxidase. Inhibitors of monoamine oxidase are used clinically in the treatment of depression, hypertension, and orthostatic hypotension. Although some studies have hinted at a bimodal distribution of platelet monoamine oxidase activity and heritability of this enzyme is clear from twin studies, the mode of inheritance and the biochemical basis of the genetic effects is unknown.[141]

5. Steroid Hydroxylase deficiencies

Many human genetic defects exist in enzymes involved in sex steroid[144] or glucocorticoid[145] biosynthesis and degradation. Since many drugs act as agonists or antagonists of one or another of these enzymes, genetic differences in drug response may be expected. For example, metyrapone [2-methyl-1,2-di(3-pyridyl)-1-propanone] inhibits steroid hydroxylations at 11β-, 18-, and 19-positions but has no effect on 20α-, 21-, or 22-hydroxylations[146] and is used clinically to test adrenal function; dicumarol inhibits steroid 11β-hydroxylation.[147] A mutant strain of rats that responds to high NaCl intake with a marked increase in blood pressure, apparently lacks a form of adrenal mitochondrial cytochrome P-450 catalyzing both the 11β- and 18-hydroxylations.[148] For reasons that are not clear, cadmium induces hypertension in this sensitive strain but not in the normal strain of rats.[149] Cholesterol 7α-hydroxylase activity is induced sixfold in Wistar rats by phenobarbital but not at all in Sprague-Dawley derived Charles River rats.[150] Although not characterized in man, genetic differences, e.g., hypertension, in response to dietary NaCl, heavy metals, or steroids, if found, would constitute a pharmacogenetic disorder.

6. Metabolism of Halogenated Anesthetics

The metabolism and toxicity of various halogenated anesthetics appear to be associated in rat liver with metabolites (so-called "suicide substrates") that covalently bind to nucleic acids and protein, especially to the form of cytochrome P-450 involved in the metabolism.[151-155] Although no genetic studies of anesthesia toxicity appear to have been performed in any species, fruitful pharmacogenetic data in both laboratory animals and in man should soon be reported. This subject is especially important since, from the clinical standpoint alone, it is necessary to be able to make sophisticated predictions about anesthetic dose, maintenance dosage, and avoidance of liver toxicity.

7. Chloramphenicol-Induced Aplastic Anemia

Chloramphenicol remains the leading *known* cause of drug-induced aplastic anemia. Compelling evidence indicates that chloramphenicol produces two distinct forms of bone marrow toxicity: (i) the common dose-related reversible suppression involving primarily the erythroid series; and (ii) the rare complication of bone marrow aplasia that is not dose-related. The latter form can be regarded as a pharmacogenetic disorder. Whereas concentrations of chloramphenicol greater than 100 μg/ml of growth medium are necessary to inhibit significantly total gross DNA or RNA synthesis in cultures of marrow taken from normal subjects, significant inhibition can be demonstrated at chloramphenicol concentrations of 25 to 50 μg/ml with marrow from patients who have recovered from chloramphenicol-induced aplastic anemia.[156] In each of two instances in which parents of affected patients were tested, the significant inhibition of DNA synthesis by chloramphenicol at 25 to 50 μg/ml was seen in cultures from the father's marrow.[157]

The metabolism of chloramphenicol has become a recent subject of interest in several laboratories.[158,159] The introduction into the European market in recent years of a chloramphenicol analogue, thiamphenicol (Fig. 3), has reopened the question of relationship between structure and toxicity. No cases of aplastic anemia appear to have been associated with this analogue; this is of particular interest because thiamphenicol produces reversible dose-dependent bone marrow suppression as readily as chloramphenicol.[157] It appears possible, therefore, that the nitrobenzene moiety of chloramphenicol is essential for the idiosyncratic drug response.

8. Isoniazid-Induced Hepatitis

As mentioned above, the slow acetylator phenotype is associated with a toxic response to several hydrazine derivatives. The population frequency of the slow acetylator allele exhibits striking geographical variation, ranging from 0.2 to 0.4 in Eskimos and Orientals but from 0.7 to 0.9 in European and Middle Eastern populations.[22] Although the rapid inactivator is better protected against isoniazid-induced neuropathy and hydralazine-induced lupus, he is more susceptible to isoniazid-induced hepatitis than the slow inactivator. By comparing profiles of urinary isoniazid metabolites in rapid and slow acetylators, Mitchell and associates[160] showed that rapid acetylators excrete smaller amounts of isoniazid and significantly greater amounts of isonicotinic acid than do slow acetylators. Yet, following administration of acetylisoniazid, rapid and slow acetylators excrete identical amounts of isonicotinic acid. The proposed scheme for the activation of isoniazid[160] to toxic intermediates (Fig. 6) involves the hydrolysis of acetylisoniazid, liberating stoichiometric amounts of isonicotinic acid and acetylhydrazine. The latter is known to be converted in rats to potent acylating agents capable of causing liver necrosis,[161] and this N-hydroxylation probably involves a cytochrome P-450-mediated monooxygenase.[162] Thus, acetylation of isoniazid appears to be the rate-limiting step in the formation of acetylhydrazine, and a higher steady-state level of this intermediate in rapid acetylators may be the basis for increased susceptibility to hepatic injury.

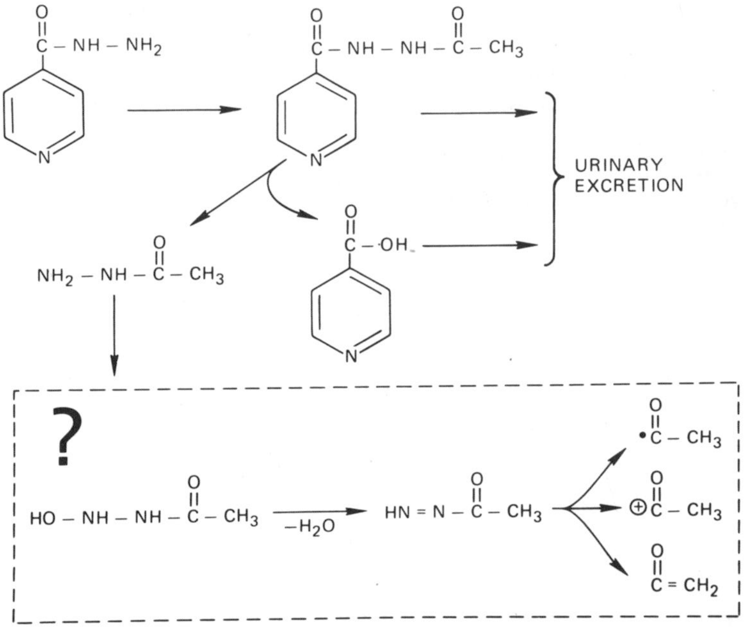

Fig. 6. Proposed pathway for the activation of isoniazid to reactive acylating agents. The rectangle drawn by the dashed line represents postulated, rather than experimentally proven, reactive intermediates capable of binding covalently with cellular macromolecules. (From Atlas and Nebert.[42] Reproduced with permission from Taylor and Francis Ltd. Publishers.)

9. Drug-Induced Hemolysis

a. **Glucose-6-Phosphate Dehydrogenase (G6PD) Deficiency.** The antimalarial primaquine causes acute hemolysis in approximately 10% of Black Americans.[163] The finding that reduced glutathione concentration is diminished in primaquine-sensitive erythrocytes[164] led to the discovery of red cell G6PD deficiency in affected individuals.[165] This enzyme, part of the hexose monophosphate pathway, is one of the principal sources of NADPH in the normal red cell. NADPH is, in turn, the cofactor for glutathione reductase, which reduces oxidized glutathione. More than 80 G6PD variants, all inherited as X-linked traits, are now known, and approximately 100 million individuals are affected.[42] The two principal variants occur in Black Americans (A-type) and in Mediterranean populations. Hemolysis associated with the A-type enzyme is usually mild and self-limited, affecting mainly the older erythrocytes. In the Mediterranean type, red-cell G6PD is generally lower in younger cells and is associated with more severe symptoms[166]; this condition is also known as "favism" because G6PD-deficient Mediterraneans are especially sensitive to an unknown hemolytic agent in the fava bean.

Numerous drugs (Table IV) are known to precipitate hemolytic crises in G6PD-deficient subjects. The precise mechanisms whereby G6PD deficiency leads to hemolysis are not fully understood. It is commonly felt that oxidation of GSH plays a central role, since GSH is essential for maintaining protein sulfhydryl groups in the reduced state, thereby preventing denaturation of enzymes and hemoglobin (i.e., Heinz body formation) and perhaps even preserving the integrity of the erythrocyte membrane. The reader is referred to several comprehensive reviews[42,166-168] for a detailed discussion of these mechanisms. The view that decreased red cell GSH is primary in the pathogenesis of hemolysis might be supported by the fact that hemolysis can be initiated, in the absence of drug ingestion, by a number of infections and metabolic conditions.[166] There is, however, ample evidence that many of the agents listed in Table IV require metabolic activation before they can produce hemolysis.[169] It is therefore quite likely that toxic intermediates of such agents may directly cause red-cell damage.

As outlined in Fig. 7, reactive metabolites may be involved in the pathogenesis of hemolysis in two ways. First, they may cause either oxidation or depletion of erythrocyte GSH; the decrease in GSH may itself be the ultimate event leading to hemolysis. More likely, such metabolites, e.g., epoxides and other electrophiles, may themselves be

TABLE IV

Known or Suspected Agents Causing Hemolysis in Glucose-6-Phosphate Dehydrogenase-Deficient Subjects

Aniline derivatives
    N-Acetylarylamines (acetanilide, phenacetin, [?]acetaminophen)
    Phenylhydrazine
    Sulfonamides (N^4-acetylsulfanilamide, sulfanilamide, sulfamethoxypyridazine, sulfapyridine, salicylazosulfapyridine [Azulfidine])
    Sulfones (diphenyl sulfone, thiazolesulfone)
Chloramphenicol
[?]Menadione derivatives
Naphthalene
Nitrofuran derivatives (furazolidine, nitrofurantoin, nitrofurazone)
Quinoline derivatives
    8-Aminoquinoline antimalarials (pentaquine, primaquine, quinocide)
    [?]Quinine, [?]quinidine
[?]Salicylates (acetylsalicylic acid, p-aminosalicylic acid)

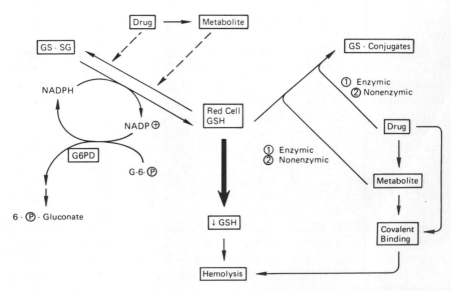

Fig. 7. Mechanisms by which erythrocyte glutathione may become depleted. A drug or metabolite may, because of differences in redox potential, oxidize GSH to GS-SG or may become conjugated enzymically or nonenzymically with GSH to form GS conjugates. Regeneration of GSH from GS-SG requires glucose-6-phosphate dehydrogenase (G6PD). Depletion of GSH and perhaps covalent binding of drug or metabolite causes hemolysis in some manner. (From Atlas and Nebert.[42] Reproduced with permission from Taylor and Francis Ltd. Publishers.)

responsible for cytotoxicity, resulting in oxidative or peroxidative damage to cellular constituents or in covalent modification of cellular macromolecules. In this sense, GSH may serve as a protective mechanism against xenobiotic toxicity, in a manner somewhat analogous to the proposed model for acetaminophen-produced hepatotoxicity in rats.[170] Such protection would be insufficient in G6PD-deficient erythrocytes.

Heritable variation in any of the steps in metabolic activation of hemolytic agents may lead, of course, to heterogeneity of the G6PD-deficiency syndrome. Because it is a relatively common clinically recognizable disorder, G6PD-deficiency may provide a worthwhile model for investigating the genetic control of drug metabolism in man.

b. Allied Conditions. Inherited deficiencies in several other enzymes related to GSH production and utilization have been associated with both spontaneous and drug-induced hemolysis (Table V).[166, 171-177] The two enzymes in the biosynthesis of GSH from amino acid precursors may also be responsible for transport of amino acids into cells via the γ-glutamyl

TABLE V

Other Enzyme Deficiencies Possibly Associated with Increased Susceptibility to Drug-Induced Hemolysis

Enzyme	Inheritance	Hemolysis	References
Defects in GSH synthesis			
γ-Glutamylcysteine synthetase	[?]Two cases	Yes	171
GSH synthetase			
with 5-oxoprolinuria	Autosomal recessive	Yes	172,173
without 5-oxoprolinuria	Autosomal recessive	Yes	174,175
Defects in maintenance of reduced GSH			
GSH reductase	[?]Autosomal dominant	Yes	176
6-Phosphogluconate dehydrogenase	Autosomal autonomous	[?]	166
Defects in GSH utilization			
GSH peroxidase	Autosomal autonomous	Yes	177

cycle.[178] The second enzyme of the hexose monophosphate shunt, 6-phosphogluconate dehydrogenase, also supplies NADPH to the erythrocyte, and deficiency in its activity might also be expected to affect levels of GSH; however, there is no unequivocal evidence that inherited deficiency of this enzyme increases susceptibility to drug-induced hemolysis.[166]

c. **Unstable Hemoglobins.** In several abnormal hemoglobins, amino acid substitutions occur at or near the histidine residues that serve as the sixth ligand for heme iron (α^{58} and β^{63}), the site for reversible oxygen binding. Such hemoglobins are "unstable" and are easily denatured to form Heinz bodies.[166] Of the five known α-chain substitutions and the 22 known β-chain substitutions, only three variant hemoglobins are documented to be associated with drug-induced hemolysis[179]: Torino ($\alpha^{43 \text{ Phe}\rightarrow\text{Val}}$), Zürich ($\beta^{63 \text{ His}\rightarrow\text{Arg}}$) and Shepherds Bush ($\beta^{74 \text{ Gly}\rightarrow\text{Asp}}$). Phenotypically, all three are inherited as autosomal dominant traits and are associated with susceptibility to sulfonamide-induced hemolysis. In addition, hemolysis due to primaquine and other quinoline derivatives have been reported in individuals with hemoglobin Zürich.[166]

10. Hereditary Methemoglobinemia

It is difficult to separate entirely discussions of drug-induced hemolysis and drug-induced methemoglobinemia, since methemoglobin formation may play some role in hemolysis in G6PD-deficient individuals.[166] The first two groups of disorders discussed below are due to protein polymorphisms other than abnormalities in drug metabolism. However, many of the same considerations in metabolic activation of causative agents that have been discussed in the preceding section apply to methemoglobin formation as well. Nitrites, chlorates, and quinones can produce methemoglobin *in vitro*, suggesting that direct chemical oxidation rather than metabolic potentiation is involved[180]; nitrates are converted to nitrites by intestinal bacteria before they are effective. Kiese[181] has extensively reviewed the evidence that metabolism of aromatic amines, e.g., aniline derivatives, is necessary before such agents may cause methemoglobinemia. N-Hydroxylation appears to be the most important activation step, although formation of aminophenols and *N*-oxides may also play a role. Nitrobenzenes may be activated by reduction to the corresponding arylhydroxylamine.[181]

a.   **NADH Methemoglobin Reductase Deficiency.**   This is the most important oxidoreductase involved in the reduction of ferric heme iron. Its complete absence from erythrocytes is inherited as an autosomal recessive trait[182] and is associated with congenital cyanosis as well as drug-induced exacerbations.

b.   **Abnormal Hemoglobins.**   In several abnormal hemoglobins, collectively known as hemoglobin M (for "methemoglobinemia"), at least two of the four heme irons are especially prone to oxidation by drugs. These amino acid substitutions also occur near the heme binding sites on the α- and β-chains.[183] Inheritance is autosomal dominant. Hemoglobin H, a β^4 tetramer found in certain patients with recessively inherited α-thalassemia, is also associated with drug-induced methemoglobinemia.[184]

c.   **Deficient O-Deethylation of Phenacetin.**   The usual pathway for phenacetin metabolism in man (Fig. 8) involves predominantly cytochrome *P*-450-mediated O-deethylation and excretion of large amounts of the phenol acetaminophen (Fig. 8, upper left), mainly as the glucuronide. GSH conjugation with acetaminophen also occurs, with resultant small amounts of excreted mercapturic acid derivative (Fig. 8, middle left). Arene oxide formation by a cytochrome *P*-450-mediated monooxygenase also presumably occurs and normally results in the excretion of small amounts of 2-hydroxyphenacetin glucuronide (Fig. 8, middle) and 2-hydroxyphenetidin sulfuric acid ester (Fig. 8, lower right). Two sisters in Switzerland were found[185] to excrete only 30% of the total dose as acetaminophen (compared with normal values of 62% or higher)

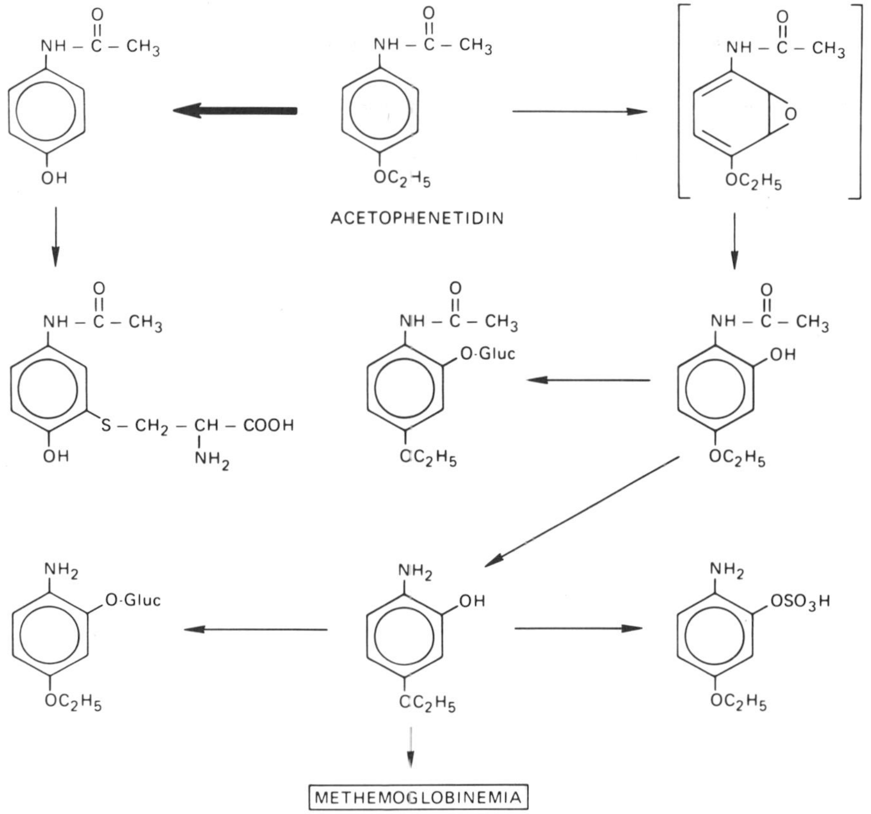

Fig. 8. Major and minor pathways of phenacetin metabolism in man. The arene oxide (in the brackets at upper right) is postulated but was not isolated or characterized. (From Atlas and Nebert.[42] Reproduced with permission from Taylor and Francis Ltd. Publishers.)

and to excrete large amounts of 2-hydroxyphenacetin glucuronide, 2-hydroxyphenetidin sulfuric acid ester, and 2-hydroxyphenetidin glucuronide (Fig. 8, lower left). This last product, in the presence of β-glucuronidase, and 2-hydroxyphenetidin (Fig. 8, lower middle), in the absence of β-glucuronidase, were shown[185] to cause marked methemoglobinemia with red cells *in vitro*. The mechanism is presumed to be the direct chemical oxidation of oxyferrohemoglobin by the *o*-aminophenol.

It was felt that phenobarbital might aid the disorder by inducing the *O*-deethylase; on the contrary, phenobarbital in these sisters appeared to exacerbate the condition by inducing the monooxygenase responsible for arene oxide formation.[185] In sum, this disorder with resultant methemoglobinemia appears to involve a genetic decrease in a cyto-chrome P-450-mediated O-deethylase. The urinary profile of a large population following a challenge dose of oral phenacetin should be studied in an attempt to establish the scope of this pharmacogenetic disorder.

11. Polymorphism of Aryl Hydrocarbon Hydroxylase (AHH) (EC 1.14.14.2) Inducibility

For about a decade[186,187] it has been clear that certain forms of cyto-chrome P-450 and associated aromatic hydrocarbon hydroxylase (AHH) activity differ among inbred strains of mice and that this trait follows expected Mendelian inheritance. This genetic locus is now called the "Ah complex," or the "Ah cluster."[188] More than 20 cytochrome P-450-mediated monooxygenase activities are usually induced in concert in aromatic hydrocarbon-"responsive" mice treated with various polycyclic aromatic inducers, many of which are ubiquitous environmental contaminants. A cytosolic receptor is believed to be the major regulatory gene product of the Ah complex. "Nonresponsive" mutants have at least 50-fold less hepatic receptor than responsive C57BL/6N mice.[189] The several enzymes induced are believed to be the structural gene products of the Ah locus. Temporal genes are postulated also to be part of the Ah cluster.[188] Practically all tissues of the mouse exhibit Ah locus-associated responsiveness or nonresponsiveness, i.e., systemic genetic expression. The metabolism of many drugs and other environmental pollutants by cytochrome P-450 represents an important rate-limiting step. It is no doubt for these reasons that striking differences in chemical car-cinogenesis, mutagenesis, binding of metabolites to DNA, drug toxicity, and teratogenesis have all been demonstrated.[29,188]

Evidence for the Ah locus exists in man,[42,190] and many of the tumorigenic and toxicologic data shown so clearly in the inbred mouse have been at least partially demonstrable in the (very outbred) human (Table VI).[29,188,191-216] Experimental difficulties in the AHH assay with cultured lymphocytes or monocytes account for the equivocal data listed in the table. An alternative assay (such as measurement of cytosolic receptor,[189] a radioimmunoassay for specific forms of cytochrome P-450,[217] or detection of an appropriate ^{13}C-labeled metabolite exhaled[218] or in the urine) is highly desirable.

III. AVENUES FOR FUTURE EXPLORATION

Several pharmacogenetic disorders are well characterized in laboratory animals and have not yet been studied in man. In some cases the difficulty

TABLE VI

Human Disorders That Appear to Be Associated with the *Ah* Locus

Disorder	Associated with high or low AHH inducibility	References
Malignancy		
Bronchogenic carcinoma	High[a]	191–199
	No association found	200–205
Laryngeal carcinoma	High[b]	206
Cancer of oral cavity	High[b]	207,208
Cancer of renal pelvis or ureter	No association found	209
Cancer of urinary bladder	No association found	210,211
Acute leukemia of childhood	Low[c]	212
Toxicity		
Zoxazolamine-induced fatal hepatic necrosis	Unknown[c]	213
Earlier menopause among cigarette smokers	Unknown[c]	214
Acetaminophen-induced diffuse bilateral cataracts	Unknown[c]	215

[a] Consistent with genetic data from inbred strains of mice.[28,188]

[b] Studies of these disorders in mice have not been specifically carried out, but the human data are consistent with what is known[216] about environmental carcinogens and their effect on local and distant tissue sites in genetically "*Ah*-responsive" and "*Ah*-nonresponsive" mice.

[c] Genetically responsive mice are at increased risk for these disorders.[188] In retrospect (or in studies to be designed in the future), it would have been (or would be) of interest to know the *Ah* phenotype of afflicted clinical patients.

is in obtaining adequate tissue or in designing a simple and ethical experiment. As indicated, important differences in atypical drug responses are anticipated among patients with several drug-induced hemolytic anemias and the various steroid hydroxylase deficiencies and in patients receiving halogenated anesthetics.

The concept of ecology and genetics ("ecogenetics") will become increasingly important in the future. The interaction of environmental carcinogens and mutagens with the human population, specifically with the *Ah* locus polymorphism, demands careful study and widespread interest because of a growing concern with increasing pollution of this planet. Because differences in susceptibility to toxicity from isoniazid, hydralazine, and procainamide are associated with different acetylator phenotypes, the possibility has been raised[219] that these phenotypes may differ in susceptibility to chemical carcinogenesis from exposure to arylamines. A very interesting report from South Africa[220] recently showed a correlation between increased incidence of mutagens in the feces and the population at highest risk for colon cancer.

In the mouse, resistance to cadmium toxicity is expressed as an autosomal recessive trait.[221] Metallothioneins are cytoplasmic proteins that occur in many tissues, contain a very large number of sulfhydryl groups, bind heavy metal cations such as cadmium, and are induced by these same heavy metals.[222] Human HeLa cells in culture possess one or more of these inducible proteins.[223] It would be of interest if humans display pharmacogenetic differences in response to heavy metals which pollute our environment. In numerous industries involved with various toxic chemicals, it would be ideal to identify prospectively those workers that are genetically at lowest risk if exposed to any given chemical.[224]

IV. COMMENTS

Human pharmacogenetic disorders with increased sensitivity to drugs are classified into those enzyme deficiencies resulting in decreased detoxication and into disorders of unknown etiology. Defects associated with increased resistance to drugs include possible deficiencies in some sort of "receptor," defects in absorption, and disorders involving increased metabolism. Several diseases are exacerbated in one way or another by drugs which induce enzyme activities.

There exist a number of pathologic conditions in which genetic differences in the metabolic potentiation of drugs may play an important role in the human. Chloramphenicol-induced aplastic anemia, isoniazid-induced hepatitis, and certain drug-induced hemolytic anemias and methemoglobinemias are offered as examples of drug toxicity. Differences in the *Ah* locus are associated with susceptibility to chemical carcinogenesis and drug toxicity in the mouse, whereas such an association in the human requires further studies. Differences in aromatic hydrocarbon responsiveness in the mouse are also associated with increased *in utero* toxicity and malformations.[225] Demonstration of a human pharmacogenetic disorder correlated with an increased likelihood of drug-induced birth defects is an interesting possibility but is far from being proved experimentally. It is conceivable that certain instances of drug allergy or autoimmune disease may be caused by genetic differences in the metabolic potentiation of drugs, since haptenes include drugs covalently bound to macromolecules.

New pharmacogenetic disorders associated with drug-induced hemolytic anemias, steroid hydroxylase deficiencies, toxicity of halogenated anesthetics, and toxicity to various other environmental contaminants and industrial chemicals are predicted as future discoveries. Drug tolerance and physical dependence similarly may be explained in part on a pharmacogenetic basis. Lastly, conditions in which the basic genetic defect contributes to therapeutic refractoriness have been discussed here. Vitamin D-dependent rickets and HGPRT-deficient gout clearly involve metabolic potentiation; the consequences of the cytidine deaminase polymorphism and catecholamine metabolic pathway polymorphisms must await further clinical pharmacological studies.

ACKNOWLEDGMENT

The expert secretarial assistance of Ms. Ingrid E. Jordan is greatly appreciated.

REFERENCES

1. Lusis, A. J., and Paigen, K. (1977). Mechanisms involved in the intracellular localization of mouse glucuronidase. *Isozymes: Curr. Top. Biol. Med. Res.* **2,** 63–106.
2. Øie, S., and Levy, G. (1977). Interindividual differences in the effect of drugs on bilirubin plasma binding in newborn infants and in adults. *Clin. Pharmacol. Ther.* **21,** 627–632.
3. Wilding, G., Paigen, B., and Vesell, E. S. (1977). Genetic control of interindividual variations in racemic warfarin binding to plasma and albumin of twins. *Clin. Pharmacol. Ther.* **22,** 831–842.
4. Evans, F. T., Gray, P. W. S., Lehmann, H., and Silk, E. (1952). Sensitivity to succinylcholine in relation to serum cholinesterase. *Lancet* **1,** 1229–1230.
5. Weinshilboum, R. M. (1979). Serum dopamine β-hydroxylase. *Pharmacol. Rev.* **30,** 133–166.
6. Weinshilboum, R., and Raymond, F. A. (1977). Inheritance of low erythrocyte catechol *O*-methyltransferase activity in man. *Am. J. Hum. Genet.* **29,** 125–135.
7. Kalow, W., Kadar, D., Inaba, T., and Tang, B. K. (1977). A case of deficiency of N-hydroxylation of amobarbital. *Clin. Pharmacol. Ther.* **21,** 530–535.
8. Tang, B. K., Inaba, T., and Kalow, W. (1978). N-Hydroxylation of barbiturates. *In* "Biological Oxidation of Nitrogen" (J. W. Gorrod, ed.), pp. 151–156. Elsevier/North-Holland Biomedical Press, New York.
9. Tang, B. K., Kalow, W., and Grey, A. A. (1978). Amobarbital metabolism in man: *N*-glucoside formation. *Res. Commun. Chem. Pathol. Pharmacol.* **21,** 45–53.
10. Evans, D. A. P., and White, T. A. (1964). Human acetylation polymorphism. *J. Lab. Clin. Med.* **63,** 394–403.
11. Bönicke, R., and Lisboa, B. P. (1957). Über die Erbbedingtheit der intraindividuellen Konstanz der Isoniazidausscheidung beim Menschen. *Naturwissenschaften* **44,** 314.
12. Harris, H. W., Knight, R. A., and Selin, M. J. (1958). Comparison of isoniazid concentrations in the blood of people of Japanese and European descent—therapeutic and genetic implications. *Am. Rev. Tuberc. Pulm. Dis.* **78,** 944–948.

13. Knight, R. A., Selin, M. J., and Harris, H. W. (1959). Genetic factors influencing isoniazid blood levels in humans. *Trans. Conf. Chemother. Tuberc., 18th Conf., 1958,* pp. 52–58.
14. Evans, D. A. P., Manley, K., and McKusick, V. A. (1960). Genetic control of isoniazid metabolism in man. *Br. Med. J.* **2,** 485–491.
15. Hopkinson, D. A., Spencer, N., and Harris, H. (1963). Red cell acid phosphatase variants: A new human polymorphism. *Nature (London)* **199,** 969–971.
16. Eze, L. C., Tweedie, M. C. K., Bullen, M. F., Wren, P. J. J., and Evans, D. A. P. (1974). Quantitative genetics of human red cell acid phosphatase. *Ann. Hum. Genet.* **37,** 333–340.
17. Modiano, G., Battistuzzi, G., Esan, G. J. F., Testa, U., and Luzzatto, L. (1979). Genetic heterogeneity of "normal" human erythrocyte glucose-6-phosphate dehydrogenase: An isoelectrophoretic polymorphism. *Proc. Natl. Acad. Sci. U.S.A.* **76,** 852–856.
18. Vesell, E. S. (1973). Advances in pharmacogenetics. *Prog. Med. Genet.* **9,** 291–367.
19. Alexanderson, B., Evans, D. A. P., and Sjöqvist, F. (1969). Steady state plasma levels of nortriptyline in twins: Influence of genetic factors and drug therapy. *Br. Med. J.* **4,** 764–768.
20. Vesell, E. S., Page, J. G., and Passananti, G. T. (1971). Genetic and environmental factors affecting ethanol metabolism in man. *Clin. Pharmacol. Ther.* **12,** 192–201.
21. Cascorbi, H. F., Vesell, E. S., Blake, D. A., and Hebrich, M. (1971). Genetic and environmental influence on halothane metabolism in twins. *Clin. Pharmacol. Ther.* **12,** 50–55.
22. Motulsky, A. G. (1964). Pharmacogenetics. *Prog. Med. Genet.* **3,** 49–74.
23. Whittaker, J. A., and Evans, D. A. P. (1970). Genetic control of phenylbutazone metabolism in man. *Br. Med. J.* **4,** 323–328.
24. Åsberg, M., Evans, D. A. P., and Sjöqvist, F. (1971). Genetic control of nortriptyline kinetics in man: A study of relatives of propositi with high plasma concentrations. *J. Med. Genet.* **8,** 129–135.
25. Vesell, E. S., and Page, J. G. (1969). Genetic control of phenobarbital-induced shortening of plasma antipyrine half-lives in man. *J. Clin. Invest.* **48,** 2202–2209.
26. Keeri-Szanto, M., and Pomeroy, J. R. (1971). Atmospheric pollution and pentazocine metabolism. *Lancet* **1,** 947–949.
27. Harris, H. (1975). "The Principles of Human Biochemical Genetics." Elsevier/North-Holland Biomedical Press, New York.
28. Greaves, J. H., and Ayres, P. (1967). Heritable resistance to warfarin in rats. *Nature (London)* **215,** 877–878.
29. Kouri, R. E., and Nebert, D. W. (1977). Genetic regulation of susceptibility to polycyclic hydrocarbon-induced tumors in the mouse. *In* "Origins of Human Cancer" (H. H. Hiatt, J. D. Watson, and J. A. Winsten, eds.), pp. 811–835. Cold Spring Harbor Lab., Cold Spring Harbor, New York.
30. Weber, W. W., Cohen, S. N., and Steinberg, M. S. (1968). Purification and properties of N-acetyltransferase from mammalian liver. *Ann. N.Y. Acad. Sci.* **151,** 734–741.
31. Weber, W. W., Miceli, J. N., Hearse, D. J., and Drummond, G. S. (1976). N-acetylation of drugs. Pharmacogenetic studies in rabbits selected for their acetylator characteristics. *Drug Metab. Dispos.* **4,** 94–101.
32. Kalow, W., Britt, B. A., and Richter, A. (1976). Individuality in human skeletal muscle, as revealed by studies of malignant hyperthermia. *Can. J. Genet. Cytol.* **18,** 565.
33. Britt, B. A., Kalow, W., and Endrenyi, L. (1978). Malignant hyperthermia—pattern of inheritance in swine. *In* "Malignant Hyperthermia" (J. A. Aldrete and B. A. Britt, eds.), pp. 195–211. Grune & Stratton, New York.

34. Weinshilboum, R. M., Raymond, F. A., and Frohnauer, M. (1979). Monogenic inheritance of catechol O-methyltransferase activity in the rat—Biochemical and genetic studies. *Biochem. Pharmacol.* **28**, 1239–1247.

35. Scanlon, P. D., Raymond, F. A., and Weinshilboum, R. M. (1979). Catechol O-methyltransferase: Thermolabile enzyme in erythrocytes of subjects homozygous for allele for low activity. *Science* **203**, 63–65.

36. Motulsky, A. G. (1957). Drug reactions, enzymes and biochemical genetics. *J. Am. Med. Assoc.* **165**, 835–837.

37. Vogel, F. (1959). Moderne Probleme der Humangenetik. *In* "Ergebnisse der Inneren Medizin und Kinderheilkande," (L. Heilmeyer, R. Schoen, and B. deRudder, eds.) Vol. 12, pp. 52–55. Springer-Verlag, Berlin.

38. Evans, D. A. P. (1963). Pharmacogenetics. *Am. J. Med.* **34**, 639–662.

39. Kalow, W. (1971). Topics in pharmacogenetics. *Ann. N.Y. Acad. Sci.* **179**, 654–659.

40. La Du, B. N. (1972). The isoniazid and pseudocholinesterase polymorphisms. *Fed. Proc., Fed. Am. Soc. Exp. Biol.* **31**, 1276–1285.

41. Vesell, E. S. (1975). Pharmacogenetics. *Biochem. Pharmacol.* **24**, 445–450.

42. Atlas, S. A., and Nebert, D. W. (1976). Pharmacogenetics and human disease. *In* "Drug Metabolism from Microbe to Man" (D. V. Parke and R. L. Smith, eds.), pp. 393–430. Taylor & Francis, London.

43. Lehmann, H., and Ryan, E. (1956). The familial incidence of low pseudocholinesterase level. *Lancet* **2**, 124.

44. Kalow, W., and Genest, K. (1957). A method for the detection of atypical forms of human serum cholinesterase. Determination of dibucaine numbers. *Can. J. Biochem. Physiol.* **35**, 339–346.

45. Lehmann, H., and Liddell, J. (1964). Genetical variants of human serum pseudocholinesterase. *Prog. Med. Genet.* **3**, 75–105.

46. Stewart, D. J., Inaba, T., Tang, B. K., and Kalow, W. (1977). Hydrolysis of cocaine in human plasma by cholinesterase. *Life Sci.* **20**, 1557–1564.

47. Geldmacher-von Mallinckrodt, M., Lindorf, H. H., Petenyi, M., Flügel, M., Fischer, T., and Hiller, T. (1973). Genetisch determinierter Polymorphismus der menschlichen Serum-Paraoxonase (EC 3.1.1.2). *Humangenetik* **17**, 331–335.

48. Shih, V. E., Abroms, I. F., Johnson, J. L., Carney, M., Mandell, R., Robb, R. M., Cloherty, J. P., and Rajagopalan, K. V. (1977). Sulfite oxidase deficiency. Biochemical and clinical investigations of a hereditary metabolic disorder in sulfur metabolism. *N. Engl. J. Med.* **297**, 1022–1028.

49. Idle, J. R., Mahgoub, A., Lancaster, R., and Smith, R. L. (1978). Hypotensive response to debrisoquine and hydroxylation phenotype. *Life Sci.* **22**, 979–984.

50. Idle, J. R., and Smith, R. L. (1979). Polymorphisms of oxidation at carbon centers of drugs and their clinical significance. *Drug Metab. Rev.* **9**, 301–317.

51. Evans, D. A. P. (1965). Individual variations of drug metabolism as a factor in drug toxicity. *Ann. N.Y. Acad. Sci.* **123**, 178–187.

52. Hughes, H. B., Biehl, J. P., Jones, A. P., and Schmidt, L. H. (1954). Metabolism of isoniazid in man as related to the occurrence of peripheral neuritis. *Am. Rev. Tuberc.* **70**, 266–273.

53. Devadatta, S., Gangadharam, P. R. J., Andrews, R. H., Fox, W., Ramakrishnan, C. V., Selkon, J. B., and Velu, S. (1960). Peripheral neuritis due to isoniazid. *Bull. W.H.O.* **23**, 587–598.

54. Carlson, H. B., Anthony, E. M., Russel, W. F., and Middlebrook, G. (1956). Prophylaxis of isoniazid neuropathy with pyridoxine. *N. Engl. J. Med.* **255**, 118–122.

55. Kutt, H., Brennan, R., Dehejia, H., and Verebely, K. (1970). Diphenylhydantoin intoxication. *Am. Rev. Respir. Dis.* **101**, 377–384.

56. Kutt, H., Winters, W., and McDowell, F. H. (1966). Depression of parahydroxylation of diphenylhydantoin by antituberculous chemotherapy. *Neurology* **16**, 594–602.
57. Evans, D. A. P., Davison, K., and Pratt, R. T. C. (1965). The influence of acetylator phenotype on the effects of treating depression with phenelzine. *Clin. Pharmacol. Ther.* **6**, 430–435.
58. Perry, H. M., Jr., Sakamoto, A., and Tan, E. M. (1967). Relationship of acetylating enzyme to hydralazine toxicity. *J. Lab. Clin. Med.* **70**, 1020.
59. Perry, H. M., Jr., Tan, E. M., Carmody, S., and Sakamoto, A. (1970). Relationship of acetyltransferase activity to anti-nuclear antibodies and toxic symptoms in hypertensive patients treated with hydralazine. *J. Lab. Clin. Med.* **76**, 114–125.
60. Woolsely, R. L., Drayer, D. E., Reidenberg, M. M., Nies, A. S., Carr, K., and Oates, J. A. (1978). Effect of acetylator phenotype on the rate at which procainamide induces anti-nuclear antibodies and the lupus syndrome. *N. Engl. J. Med.* **298**, 1157–1159.
61. Reidenberg, M. M., Drayer, D. E., and Robbins, W. C. (1979). Polymorphic drug acetylation and systemic lupus erythematosus. *In* "Advances in Pharmacology and Therapeutics: Clinical Pharmacology" (P. Duchene-Marullaz, ed.), Vol. 6, pp. 51–56. Pergamon, Oxford.
62. Kutt, H., Wolk, M., Scherman, R., and McDowell, F. (1964). Insufficient parahydroxylation as a cause of diphenylhydantoin toxicity. *Neurology* **14**, 542–548.
63. Solomon, H. M. (1968). Variations in metabolism of coumarin anticoagulant drugs. *Ann. N.Y. Acad. Sci.* **151**, 932–935.
64. Takahara, S., Sato, H., Doi, M., and Mihara, S. (1952). Acatalasemia. III. On the heredity of acatalasemia. *Proc. Jpn. Acad.* **28**, 585–589.
65. Aebi, H., and Suter, H. (1971). Acatalasemia. *Adv. Hum. Genet.* **2**, 143–199.
66. Crigler, J. F., and Najjar, V. A. (1952). Congenital familial non-hemolytic jaundice with kernicterus. *Pediatrics* **10**, 169–180.
67. Childs, B., Sidbury, J. B., and Migeon, C. J. (1959). Glucuronic acid conjugation by patients with familial non-hemolytic jaundice and their relatives. *Pediatrics* **23**, 903–913.
68. Crigler, J. F., Jr., and Gold, N. I. (1969). Effect of sodium phenobarbital on bilirubin metabolism in an infant with congenital, nonhemolytic, unconjugated hyperbilirubinemia, and kernicterus. *J. Clin. Invest.* **48**, 42–55.
69. Axelrod, J., Schmid, R., and Hammaker, L. (1957). A biochemical lesion in congenital non-obstructive non-hemolytic jaundice. *Nature (London)* **180**, 1426–1427.
70. Arias, I. M., Gartner, L. M., Cohen, M., Ben Ezzer, J., and Levi, A. J. (1969). Chronic nonhemolytic unconjugated hyperbilirubinemia with glucuronyl transferase deficiency; clinical, biochemical, pharmacologic and genetic evidence for heterogeneity. *Am. J. Med.* **47**, 395–409.
71. Peterson, R. E., and Schmid, R. (1957). A clinical syndrome associated with a defect in steroid glucuronide formation. *J. Clin. Endocrinol. Metab.* **17**, 1485–1488.
72. Armaly, M. F. (1968). Genetic factors related to glaucoma. *Ann. N.Y. Acad. Sci.* **151**, 861–875.
73. Denborough, M. A., Forster, J. F. A., Lovell, R. R. H., Maplestone, P. A., and Villers, J. D. (1962). Anaesthetic deaths in a family. *Br. J. Anaesth.* **34**, 395–396.
74. Kalow, W., Britt, B. A., and Richter, A. (1977). The caffeine test of isolated human muscle in relation to malignant hyperthermia. *Can. Anaesth. Soc. J.* **24**, 678–694.
75. Smith, A. A., and Dancis, J. (1964). Exaggerated response to infused norepinephrine in familial dysautonomia. *N. Engl. J. Med.* **270**, 704–707.
76. Dancis, J. (1968). Altered drug response in familial dysautonomia. *Ann. N.Y. Acad. Sci.* **151**, 876–879.
77. Ziegler, M. G., Lake, C. R., and Kopin, I. J. (1976). Deficient sympathetic nervous response in familial dysautonomia. *N. Engl. J. Med.* **294**, 630–633.

78. Hassell, T. M., Page, R. C., Narayanan, A. S., and Cooper, C. G. (1976). Diphenylhydantoin (Dilantin) gingival hyperplasia: Drug-induced abnormality of connective tissue. *Proc. Natl. Acad. Sci. U.S.A.* **73**, 2909–2912.

79. Jick, H., Slone, D., Westerholm, B., Inman, W. H. W., Vessey, M. P., Shapiro, S., Lewis, G. P., and Worcester, J. (1969). Venous thromboembolic disease and ABO blood type. A cooperative study. *Lancet* **1**, 539–542.

80. Lewis, G. P., Jick, H., Slone, D., and Shapiro, S. (1971). The role of genetic factors and serum protein binding in determining drug response as revealed by comprehensive drug surveillance. *Ann. N.Y. Acad. Sci.* **179**, 729–738.

81. Harris, W. S., and Goodman, R. M. (1968). Hyper-reactivity to atropine in Down's syndrome. *New Engl. J. Med.* **279**, 407–410.

82. Berg, J. M., Brandon, M. W. G., and Kirman, B. H. (1959). Atropine in mongolism. *Lancet* **2**, 441–442.

83. Aberfeld, D. C., Bienenstock, H., Shapiro, M. S., Namba, T., and Grob, D. (1968). Diffuse myopathy related to meperidine addiction in a mother and daughter. *Arch. Neurol. (Chicago)* **19**, 384–388.

84. Somayaji, B. N., Paton, A., Price, J. H., Harris, A. W., and Flewett, T. H. (1968). Norethisterone jaundice in two sisters. *Br. Med. J.* **2**, 281–283.

85. Siggers, D. C., Rogers, J. G., Boyer, S. H., Margolet, L., Dorkin, H., Banerjee, S. P., and Shooter, E. M. (1976). Increased nerve-growth-factor β-chain cross-reacting material in familial dysautonomia. *N. Engl. J. Med.* **295**, 629–634.

86. O'Reilly, R. A. (1970). The second reported kindred with hereditary resistance to oral anticoagulant drugs. *N. Engl. J. Med.* **282**, 1448–1451.

87. O'Reilly, R. A. (1971). Vitamin K in hereditary resistance to oral anticoagulant drugs. *Am. J. Physiol.* **221**, 1327–1330.

88. Snyder, L. H. (1932). Studies in human inheritance. IX. The inheritance of taste deficiency in man. *Ohio J. Sci.* **32**, 436–445.

89. Blakeslee, A. F. (1932). Genetics of sensory thresholds: Taste for phenylthiocarbamide. *Proc. Natl. Acad. Sci. U.S.A.* **18**, 120–130.

90. Bartoshuk, L. M. (1979). Bitter taste of saccharin related to the genetic ability to taste the bitter substance 6-*n*-propylthiouracil. *Science* **205**, 934–935.

91. Brown, M. S., and Goldstein, J. L. (1979). Receptor-mediated endocytosis: Insights from the lipoprotein receptor system. *Proc. Natl. Acad. Sci. U.S.A.* **76**, 3330–3337.

92. Brown, M. S., Dana, S. E., and Goldstein, J. L. (1973). Regulation of 3-hydroxy-3-methylglutaryl coenzyme A reductase activity in human fibroblasts. *Proc. Natl. Acad. Sci. U.S.A.* **70**, 2162–2166.

93. Brown, M. S., Dana, S. E., and Goldstein, J. L. (1974). Regulation of 3-hydroxy-3-methylglutaryl coenzyme A reductase activity in cultured human fibroblasts. Comparison of cells from a normal subject and from a patient with homozygous familial hypercholesterolemia. *J. Biol. Chem.* **249**, 789–796.

94. Goldstein, J. L., Dana, S. E., and Brown, M. S. (1974). Esterification of low density lipoprotein cholesterol in human fibroblasts and its absence in homozygous familial hypercholesterolemia. *Proc. Natl. Acad. Sci. U.S.A.* **71**, 4288–4292.

95. Brown, M. S., Dana, S. E., and Goldstein, J. L. (1975). Cholesterol ester formation in cultured human fibroblasts. Stimulation by oxygenated sterols. *J. Biol. Chem.* **250**, 4025–4027.

96. Brown, M. S., and Goldstein, J. L. (1975). Regulation of the activity of the low density lipoprotein receptor in human fibroblasts. *Cell* **6**, 307–316.

97. McIntyre, O. R., Sullivan, L. W., Jeffries, G. H., and Silver, R. H. (1965). Pernicious anemia in childhood. *N. Engl. J. Med.* **272**, 981–986.

98. Katz, M., Lee, S. K., and Cooper, B. A. (1972). Vitamin B_{12} malabsorption due to a biologically inert intrinsic factor. *N. Engl. J. Med.* **287**, 425–429.
99. Neitlich, H. W. (1966). Increased plasma cholinesterase activity and succinylcholine resistance; a genetic variant. *J. Clin. Invest.* **45**, 380–387.
100. Yoshida, A., and Motulsky, A. G. (1969). A pseudocholinesterase variant ($E_{cynthiana}$) with elevated plasma enzyme activity. *Am. J. Hum. Genet.* **21**, 486–498.
101. von Wartburg, J. P., and Schürch, P. M. (1968). Atypical human liver alcohol dehydrogenase. *Ann. N.Y. Acad. Sci,* **151**, 936–946.
102. Korsten, M. A., Matsuzaki, S., Feinman, L., and Lieber, C. S. (1975). High blood acetaldehyde levels after ethanol administration. Difference between alcoholic and nonalcoholic subjects. *N. Engl. J. Med.* **292**, 386–389.
103. Schuckit, M. A., and Rayses, V. (1978). Ethanol ingestion: Differences in blood acetaldehyde concentrations in relatives of alcoholics and controls. *Science* **203**, 54–55.
104. Brennan, M. J. W., and Cantrill, R. C. (1979). δ-Aminolaevulinic acid is a potent agonist for GABA autoreceptors. *Nature (London)* **280**, 514–515.
105. Marver, H. S., and Schmid, R. (1972). The porphyrias. *In* "The Metabolic Basis of Inherited Disease" (J. B. Stanbury, J. B. Wyngaarden, and D. S. Frederickson, eds.), pp. 1087–1140. McGraw-Hill, New York.
106. Wang, Y. M., and van Eys, J. (1970). The enzymatic defect in essential pentosuria. *N. Engl. J. Med.* **282**, 892–896.
107. Margolis, J. L. (1929). Chronic pentosuria and migraine. *Am. J. Med. Sci.* **177**, 348–351.
108. Burns, J. J., Evans, C., Trousof, N., and Kaplan, J. (1957). Stimulatory effect of drugs on excretion of L-ascorbic acid and non-conjugated D-glucuronic acid. *Fed. Proc., Fed. Am. Soc. Exp. Biol.* **16**, 286.
109. Peraino, C., Fry, R. J. M., and Staffeidt, E. (1973). Enhancement of spontaneous hepatic tumorigenesis in C3H mice by dietary phenobarbital. *J. Natl. Cancer Inst.* **51**, 1349–1350.
110. Kitagawa, T., Pitot, H. C., Miller, E. C., and Miller, J. A. (1979). Promotion by dietary phenobarbital of hepatocarcinogenesis by 2-methyl-*N,N*-dimethyl-4-aminoazobenzene in the rat. *Cancer Res.* **39**, 112–115.
111. Nebert, D. W. (1980). The *Ah* locus. A gene with possible importance in cancer predictability. *Arch. Toxicol.* **43**, 195–207.
112. DeLuca, H. F. (1979). Vitamin D metabolism and function. *In* "Monographs on Endocrinology" (F. Gross, A. Labhart, T. Mann, and J. Zander, eds.), pp. 1–78. Springer-Verlag, Berlin and New York.
113. Haussler, M. R., and McCain, T. A. (1977). Basic and clinical concepts related to vitamin D metabolism and action *N. Engl. J. Med.* **297**, 974–983 and 1041–1050.
114. DeLuca, H. F. (1976). Recent advances in our understanding of the metabolism of vitamin D and its regulation. *Clin. Endocrinol.* **5**, 97s–108s.
115. Ghazarian, J. G., Hsu, P.-Y., and Peterson, B. L. (1977). Chick kidney microsomal cytochrome *P*-450 involvement in the metabolism of 25-hydroxyvitamin D_3. *Arch. Biochem. Biophys.* **184**, 596–604.
116. Ribovich, M. L., and DeLuca, H. F. (1978). 1,25-Dihydroxyvitamin D_3 metabolism. The effect of dietary calcium and phosphorus. *Arch. Biochem. Biophys.* **188**, 164–171.
117. Tanaka, Y., Castillo, L., DeLuca, H. F., and Ikekawa, N. (1977). The 24-hydroxylation of 1,25-dihydroxyvitamin D_3. *J. Biol. Chem.* **252**, 1421–1424.
118. Fraser, D., Kooh, S. W., Kind, H. P., Holick, M. F., Tanaka, Y., and DeLuca, H. F. (1973). Pathogenesis of hereditary vitamin-D-dependent rickets. An inborn error of

vitamin D metabolism involving defective conversion of 25-hydroxyvitamin D to 1α,25-dihydroxyvitamin D. *N. Engl. J. Med.* **289**, 817–822.

119. Brickman, A. S., Coburn, J. W., Kurokawa, K., Bethune, J. E., Harrison, H. E., and Norman, A. W. (1973). Actions of 1,25-dihydroxycholecalciferol in patients with hypophosphatemic, vitamin-D-resistant rickets. *N. Engl. J. Med.* **289**, 495–498.

120. Robertson, B. R., Harris, R. C., and McCune, D. J. (1942). Refractory rickets: Mechanism of therapeutic action of calciferol. *Am. J. Dis. Child.* **64**, 948–949.

121. Kooh, S. W., Fraser, D., DeLuca, H. F., Holick, M. F., Belsey, R. E., Clark, M. B., and Murray, T. M. (1975). Treatment of hypoparathyroidism and pseudohypoparathyroidism with metabolites of vitamin D: Evidence for impaired conversion of 25-hydroxy-vitamin D to 1α,25-dihydroxyvitamin D. *N. Engl. J. Med.* **293**, 840–844.

122. Chesney, R. W., Horowitz, S. D., Kream, B. E., Eisman, J. A., Hong, R., and DeLuca, H. F. (1977). Failure of conventional doses of 1α,25-dihydroxycholecalciferol to correct hypocalcemia in a girl with idiopathic hypoparathyroidism. *N. Engl. J. Med.* **297**, 1272–1275.

123. Gray, R. W., Omdahl, J. L., Ghazarian, J. G., and DeLuca, H. F. (1972). 25-Hydroxycholecalciferol-1-hydroxylase: Subcellular location and properties. *J. Biol. Chem.* **247**, 7528–7532.

124. Chesney, R. W., Moorthy, A. V., Eisman, J. A., Jax, D. K., Mazess, R. B., and DeLuca, H. F. (1978). Increased growth after long-term oral 1α,25-vitamin D_3 in childhood renal osteodystrophy. *N. Engl. J. Med.* **298**, 238–242.

125. Bell, N. H., Stern, P. H., Pantzer, E., Sinha, T. K., and DeLuca, H. F. (1979). Evidence that increased circulating 1α,25-dihydroxyvitamin D is the probable cause for abnormal calcium metabolism in sarcoidosis. *J. Clin. Invest.* **64**, 218–225.

126. Kream, B. E., Jose, M., Yamada, S., and DeLuca, H. F. (1977). A specific high-affinity binding macromolecule for 1,25-dihydroxyvitamin D_3 in fetal bone. *Science* **197**, 1086–1088.

127. Hughes, M. R., and Haussler, M. R. (1978). 1,25-Dihydroxyvitamin D_3 receptors in parathyroid glands. Preliminary characterization of cytoplasmic and nuclear binding components. *J. Biol. Chem.* **253**, 1065–1073.

128. Kream, B. E., DeLuca, H. F., Moriarity, D. M., Kendrick, N. C., and Ghazarian, J. G. (1979). Origin of 25-hydroxyvitamin D_3 binding protein from tissue cytosol preparations. *Arch. Biochem. Biophys.* **192**, 318–323.

129. Ghazarian, J. G., Kream, B., Botham, K. M., Nickells, M. W., and DeLuca, H. F. (1978). Rat plasma 25-hydroxyvitamin D_3 binding protein: An inhibitor of the 25-hydroxyvitamin D_3-1α-hydroxylase. *Arch. Biochem. Biophys.* **189**, 212–220.

130. Boss, G. R., and Seegmiller, J. E. (1979). Hyperuricemia and gout. Classification, complications and management. *N. Engl. J. Med.* **300**, 1459–1468.

131. Kelley, W. N., Greene, M. L., Rosenbloom, F. M., Henderson, J. F., and Seegmiller, J. E. (1969). Hypoxanthine-guanine phosphoribosyltransferase deficiency in gout. *Ann. Intern. Med.* **70**, 155–206.

132. Davidson, J. D., and Winter, T. S. (1964). Purine nucleotide pyrophosphorylases in 6-mercaptopurine-sensitive and resistant human leukemias. *Cancer Res.* **24**, 261–267.

133. Brown, R. S., Kelley, W. N., Seegmiller, J. E., and Carbone, P. P. (1968). The action of thiopurines in lymphocytes lacking hypoxanthine guanine phosphoribosyltransferase. *J. Clin. Invest.* **47**, 12A.

134. Teng, Y.-S., Anderson, J. E., and Giblett, E. R. (1975). Cytidine deaminase: A new genetic polymorphism demonstrated in human granulocytes. *Am. J. Hum. Genet.* **27**, 492–497.

135. Chabner, B. A., Johns, D. G., Coleman, C. N., Drake, J. C., and Evans, W. H. (1974). Purification and properties of cytidine deaminase from normal and leukemic granulocytes. *J. Clin. Invest.* **53,** 922–931.

136. Steuart, C. D., and Burke, P. J. (1971). Cytidine deaminase and the development of resistance to arabinosyl cytosine. *Nature (London), New Biol.* **233,** 109–110.

137. Levitt, M., Spector, S., Sjoerdsma, A., and Udenfriend, S. (1965). Elucidation of the rate-limiting step in norepinephrine biosynthesis on the perfused guinea-pig heart. *J. Pharmacol. Exp. Ther.* **148,** 1–8.

138. Axelrod, J. (1962). Purification and properties of phenylethanolamine *N*-methyltransferase. *J. Biol. Chem.* **237,** 1657–1660.

139. Ciaranello, R. D., Hoffman, H. J., Shire, J. G. M., and Axelrod, J. (1974). Genetic regulation of catecholamine biosynthetic enzymes. *J. Biol. Chem.* **249,** 4528–4536.

140. Kaufman, S., and Friedman, S. (1965). Dopamine-β-hydroxylase. *Pharmacol. Rev.* **17,** 71–100.

141. Weinshilboum, R. M. (1979). Catecholamine biochemical genetics in human populations. *In* "Neurogenetics: Genetic Approaches to the Nervous System" (X. O. Breakefield, ed.), pp. 257–282. Elsevier/North-Holland Biomedical Press, New York.

142. Weinshilboum, R. M., Dunnette, J., Raymond, F., and Kleinberg, F. (1978). Erythrocyte catechol *O*-methyltransferase and plasma dopamine-β-hydroxylase in human umbilical cord blood. *Experientia* **34,** 310–311.

143. Axelrod, J., and Tomchick, R. (1958). Enzymatic O-methylation of epinephrine and other catechols. *J. Biol. Chem.* **233,** 702–705.

144. Bercu, B. B., and Schulman, J. D. (1980). Genetics of abnormalities of sexual differentiation and of female reproductive failure. *OB GYN Surv.* **35,** 1–11.

145. Lee, P. A., Plotnick, L. P., Kowarski, A. A., and Migeon, C. J. (1977). "Congenital Adrenal Hyperplasia." Univ. Park Press, Baltimore, Maryland.

146. Satre, M., and Vignais, P. V. (1974). Steroid 11β-hydroxylation in beef adrenal cortex mitochondria. Binding affinity and capacity for specific [^{14}C]steroids and for [^{3}H] metyrapol, an inhibitor of the 11β-hydroxylation reaction. *Biochemistry* **13,** 2201–2209.

147. Williamson, D. G., and O'Donnell, V. J. (1969). The role of cytochrome *P*-450 in the mechanism of inhibition of steroid 11β-hydroxylation by dicumarol. *Biochemistry* **8,** 1300–1306.

148. Rapp, J. P., and Dahl, L. K. (1976). Mutant forms of cytochrome *P*-450 controlling both 18- and 11β-steroid hydroxylation in the rat. *Biochemistry* **15,** 1235–1242.

149. Ohanian, E. V., Iwai, J., Leitl, G., and Tuthill, R. (1978). Genetic influence on cadmium-induced hypertension. *Am. J. Physiol.* **235,** H385–H391.

150. Shefer, S., Hauser, S., and Mosbach, E. H. (1972). Stimulation of cholesterol 7α-hydroxylase by phenobarbital in two strains of rats. *J. Lipid Res.* **13,** 69–70.

151. Munson, E. S., Malagodi, M. H., Shields, R. P., Tham, M. K., Fiserova-Bergerova, V., Holaday, D. A., Perry, J. C., and Embro, W. J. (1975). Fluroxene toxicity induced by phenobarbital. *Clin. Pharmacol. Ther.* **18,** 687–699.

152. Van Dyke, R. A., and Gandolfi, A. J. (1976). Anaerobic release of fluoride from halothane. Relationship to the binding of halothane metabolites to hepatic cellular constituents. *Drug Metab. Dispos.* **4,** 40–44.

153. Karashima, D., Hirokata, Y., Shigematsu, A., and Furukawa, T. (1977). The *in vitro* metabolism of halothane (2-bromo-2-chloro-1,1,1-trifluoroethane) by hepatic microsomal cytochrome *P*-450. *J. Pharmacol. Exp. Ther.* **203,** 409–416.

154. Brown, B. R., and Sipes, I. G. (1977). Biotransformation and hepatotoxicity of halothane. *Biochem. Pharmacol.* **26,** 2091–2094.

155. Marsh, J. A., Bradshaw, J. J., Sapeika, G. A., Lucas, S. A., Kaminsky, L. S., and

Ivanetich, K. M. (1977). Further investigations of the metabolism of fluroxene and the degradation of cytochromes P-450 *in vitro*. *Biochem. Pharmacol.* **26**, 1601–1606.

156. Yunis, A. A., and Harrington, W. J. (1960). Patterns of inhibition by chloramphenicol of nucleic acid synthesis in human bone marrow and leukemic cells. *J. Lab. Clin. Med.* **56**, 831–838.

157. Yunis, A. A. (1973). Chloramphenicol-induced bone marrow suppression. *Sem. Hematol.* **10**, 225–234.

158. Pohl, L. R., and Krishna, G. (1978). Study of the mechanism of metabolic activation of chloramphenicol by rat liver microsomes. *Biochem. Pharmacol.* **27**, 335–341.

159. Siliciano, R. F., Margolis, S., and Lietman, P. S. (1978). Chloramphenicol metabolism in isolated rat hepatocytes. *Biochem. Pharmacol.* **27**, 2759–2762.

160. Mitchell, J. R., Thorgeirsson, U. P., Black, M., Timbrell, J. A., Snodgrass, W. R., Potter, W. Z., Jollow, D. J., and Keiser, H. R. (1975). Increased incidence of isoniazid hepatitis in rapid acetylators: Possible relation to hydrazine metabolites. *Clin. Pharmacol. Ther.* **18**, 70–79.

161. Mitchell, J. R., and Jollow, D. J. (1974). Biochemical basis for drug-induced hepatotoxicity. *Isr. J. Med. Sci.* **10**, 312–318.

162. Nelson, S. D., Snodgrass, W. R., and Mitchell, J. R. (1975). Metabolic activation of hydrazines to reactive intermediates: Mechanistic implications for isoniazid and proniazid hepatitis. *Fed. Proc., Fed. Am. Soc. Exp. Biol.* **34**, 784.

163. Hockwald, R. S., Arnold, J., Clayman, C. B., and Alving, A. S. (1952). Status of primaquine. IV. Toxicity of primaquine in Negroes. *J. Am. Med. Assoc.* **149**, 1568–1570.

164. Beutler, E., Dern, R. J., and Alving, A. S. (1955). The hemolytic effect of primaquine. VII. Biochemical studies of drug-sensitive erythrocytes. *J. Lab. Clin. Med.* **45**, 286–295.

165. Carson, P. E., Flanagan, C. L., Ickes, C. E., and Alving, A. S. (1956). Enzymatic deficiency in primaquine sensitive erythrocytes. *Science* **124**, 484–485.

166. Beutler, E. (1972). Disorders due to enzyme defects in the red blood cell. *Adv. Metab. Disord.* **6**, 131–160.

167. Kirkman, H. N. (1971). Glucose-6-phosphate dehydrogenase. *Adv. Hum. Genet.* **2**, 1–60.

168. Neely, C. L., and Kraus, A. P. (1972). Mechanisms of drug-induced hemolytic anemia. *Adv. Intern. Med.* **18**, 59–76.

169. Fraser, I. M., and Vesell, E. S. (1968). Effects of drugs and drug metabolites on erythrocytes from normal and glucose-6-phosphate dehydrogenase-deficient individuals. *Ann. N.Y. Acad. Sci.* **151**, 777–794.

170. Mitchell, J. R., Thorgeirsson, S. S., Potter, W. Z., Jollow, D. J., and Keiser, H. (1974). Acetaminophen induced injury: Protective role of glutathione in humans and rationale for possible therapy. *Clin. Pharmacol. Ther.* **16**, 676–684.

171. Boivin, P., Galand, C., and Bernard, J. F. (1974). Deficiencies in GSH biosynthesis. *In* "Glutathione" (L. Flohé, H. C. Benöhr, H. Sies, H. D. Waller, and A. Wendel, eds.), pp. 146–157. Academic Press, New York.

172. Jellum, E., Kluge, T., Börresen, H. C., Stokke, O., and Eldjarn, L. (1970). Pyroglutamic aciduria: A new inborn error of metabolism. *Scand. J. Clin. Lab. Invest.* **26**, 327–335.

173. Wellner, V. P., Sekura, R., Meister, A., and Larsson, A. (1974). Glutathione synthetase deficiency, an inborn error of metabolism involving the γ-glutamyl cycle in patients with 5-oxoprolinuria (pyroglutamic aciduria). *Proc. Natl. Acad. Sci. U.S.A.* **71**, 2505–2509.

174. Mohler, D. N., Majerus, P. W., Minnich, V., Hess, C. E., and Garrick, M. D. (1970). Glutathione synthetase deficiency as a cause of hereditary hemolytic disease. *N. Engl. J. Med.* **283**, 1253–1257.

175. Spielberg, S. P., Garrick, M. D., Corash, L. M., Butler, J. D., Tietze, F., Rogers, L.-V., and Schulman, J. D. (1978). Biochemical heterogeneity in glutathione synthetase deficiency. *J. Clin. Invest.* **61**, 1417–1420.

176. Löhr, G. W., Blume, K. G., Rüdiger, H. W., and Arnold, H. (1974). Genetic variability in the enzymatic reduction of oxidized glutathione. *In* "Glutathione" (L. Flohé, H. C. Benöhr, H. Sies, H. D. Waller, and A. Wendel, eds.), pp. 165–173. Academic Press, New York.

177. Necheles, T. F. (1974). The clinical spectrum of glutathione-peroxidase deficiency. *In* "Glutathione" (L. Flohé, H. C. Benöhr, H. Sies, H. D. Waller, and A. Wendel, eds.), pp. 173–180. Academic Press, New York.

178. Meister, A. (1974). Biosynthesis and utilization of glutathione; the γ-glutamyl cycle and its function in amino acid transport. *In* "Glutathione" (L. Flohé, H. C. Benöhr, H. Sies, H. D. Waller, and A. Wendel, eds.), pp. 56–68. Academic Press, New York.

179. Comings, D. E. (1972). Hemoglobinopathies associated with unstable hemoglobin. *In* "Hematology" (W. J. Williams, E. Beutler, A. J. Erslev, and R. W. Rundles, eds.), pp. 440–447. McGraw-Hill, New York.

180. Prankerd, T. A. J. (1961). "The Red Cell." Thomas, Springfield, Illinois.

181. Kiese, M. (1966). The biochemical production of ferrihemoglobin-forming derivatives from aromatic amines, and mechanisms of ferrihemoglobin formation. *Pharmacol. Rev.* **18**, 1091–1161.

182. Scott, E. M. (1960). The relation of diaphorase of human erythrocytes to inheritance of methemoglobinemia. *J. Clin. Invest.* **39**, 1176–1179.

183. Comings, D. E. (1972). Hemoglobinopathies producing cyanosis. *In* "Hematology" (W. J. Williams, E. Beutler, A. J. Erslev, and R. W. Rundles, eds.), pp. 434–440. McGraw-Hill, New York.

184. Huehns, E. R., and Shooter, E. M. (1965). Human haemoglobins. *J. Med. Genet.* **2**, 48–90.

185. Shahidi, N. T. (1968). Acetophenetidin-induced methemoglobinemia. *Ann. N.Y. Acad. Sci.* **151**, 822–832.

186. Nebert, D. W., and Gelboin, H. V. (1969). The *in vivo* and *in vitro* induction of aryl hydrocarbon hydroxylase in mammalian cells of different species, tissues, strains, and developmental and hormonal states. *Arch. Biochem. Biophys.* **134**, 76–89.

187. Nebert, D. W., Goujon, F. M., and Gielen, J. E. (1971). Genetic regulation of monooxygenase activity. *In* "Fonds de la Recherche Scientifique Medicale, Groupes de Contact," pp. 240–270. Fonds de la Recherche Scientifique Medical Publishers, Bruxelles, Belgium.

188. Nebert, D. W., and Jensen, N. M. (1979). The *Ah* locus: Genetic regulation of the metabolism of carcinogens, drugs, and other environmental chemicals by cytochrome *P*-450-mediated monooxygenases. *Crit. Rev. Biochem.* **6**, 401–437.

189. Okey, A. B., Bondy, G. T., Mason, M. E., Kahl, G. F., Eisen, H. J., Guenthner, T. M., and Nebert, D. W. (1979). Regulatory gene product of the *Ah* locus. Characterization of the cytosolic inducer–receptor complex and evidence for its nuclear translocation. *J. Biol. Chem.* **254**, 11636–11648.

190. Atlas, S. A., Vesell, E. S., and Nebert, D. W. (1976). Genetic control of interindividual variations in the inducibility of aryl hydrocarbon hydroxylase in cultured human lymphocytes. *Cancer Res.* **36**, 4619–4630.

191. Kellermann, G., Shaw, C. R., and Luyten-Kellermann, M. (1973). Aryl hydrocarbon hydroxylase inducibility and bronchogenic carcinoma. *N. Engl. J. Med.* **289**, 934–937.

192. Coomes, M. L., Mason, W. A., Muijsson, I. E., Cantrell, E. T., Anderson, D. E., and Busbee, D. L. (1976). Aryl hydrocarbon hydroxylase 16α-hydroxylase in cultured human lymphocytes. *Biochem. Genet.* **14,** 671–685.

193. Guirgis, H. A., Lynch, H. T., Mate, T., Harris, R. E., Wells, I., Caha, L., Anderson, J., Maloney, K., and Rankin, L. (1976). Aryl-hydrocarbon hydroxylase activity in lymphocytes from lung cancer patients and normal controls. *Oncology* **33,** 105–109.

194. Korsgaard, R., and Trell, E. (1978). Aryl hydrocarbon hydroxylase and bronchogenic carcinomas associated with smoking. *Lancet* **1,** 1103–1104.

195. Emery, A. E. H., Danford, N., Anand, R., Duncam, W., and Paton, L. (1978). Aryl-hydrocarbon-hydroxylase inducibility in patients with cancer. *Lancet* **1,** 470–471.

196. Kärki, N. T., and Huhti, E. (1978). Aryl hydrocarbon hydroxylase activity in cultured lymphocytes from lung carcinoma patients and cigarette smokers. *Abstr. Int. Congr. Pharmacol., 7th, 1978* No. 644, p. 254.

197. Arnott, M. S., Yamauchi, T., and Johnston, D. A. (1979). Aryl hydrocarbon hydroxylase in normal and cancer populations. *In* "Carcinogens: Identification and Mechanisms of Action" (A. C. Griffin, and C. R. Shaw, eds), pp. 145–156. Raven, New York.

198. Gahmberg, C. G., Sekki, A., Kosunen, T. U., Holsti, L. R., and Mäkelä, O. (1979). Induction of aryl hydrocarbon hydroxylase activity and pulmonary carcinoma. *Int. J. Cancer* **23,** 302–305.

199. Korsgaard, R., Trell, E., Simonsson, B. G., Stiksa, G., and Hood, B. (1980). Aryl hydrocarbon hydroxylase induction levels in patients with malignant tumours associated with smoking. *Cancer Lett.* (in press).

200. Gurtoo, H. L., Minowada, J., Paigen, B., Parker, N. B., and Hayner, N. T. (1977). Factors influencing the measurement and the reproducibility of aryl hydrocarbon hydroxylase activity in cultured human lymphocytes. *J. Natl. Cancer Inst.* **59,** 787–798.

201. Paigen, B., Gurtoo, H. L., Minowada, J., Houten, L., Vincent, R., Paigen, K., Parker, N. B., Ward, E., and Hayner, N. T. (1977). Questionable relation of aryl hydrocarbon hydroxylase to lung-cancer risk. *N. Engl. J. Med.* **297,** 346–350.

202. Jett, J. R., Moses, H. L., Branum, E. L., Taylor, W. F., and Fontana, R. S. (1978). Benzo(a)pyrene metabolism and blast transformation in peripheral blood mononuclear cells from smoking and nonsmoking populations and lung cancer patients. *Cancer* **41,** 192–200.

203. Paigen, B., Ward, E., Steenland, K., Houten, L., Gurtoo, H. L., and Minowada, J. (1978). Aryl hydrocarbon hydroxylase in cultured lymphocytes of twins. *Am. J. Hum. Genet.* **30,** 561–571.

204. Ward, E., Paigen, B., Steenland, K., Vincent, R., Minowada, J., Gurtoo, H. L., Sartori, P., and Havens, M. B. (1978). Aryl hydrocarbon hydroxylase in persons with lung or laryngeal cancer. *Int. J. Cancer* **22,** 384–389.

205. Lieberman, J. (1978). Aryl hydrocarbon hydroxylase in bronchogenic carcinoma. *N. Engl. J. Med.* **298,** 686–687.

206. Trell, E., Korsgaard, R., Hood, B., Kitzing, P., Nordén, G., and Simonsson, B. G. (1976). Aryl hydrocarbon hydroxylase inducibility and laryngeal carcinomas. *Lancet* **2,** 140.

207. Trell, E., and Korsgaard, R. (1978). Smoking and oral carcinoma. *Lancet* **1,** 671.

208. Trell, E., Korsgaard, R., Kitzing, P., Lundgren, K., and Mattiasson, I. (1978). Aryl hydrocarbon hydroxylase inducibility and carcinoma of oral cavity. *Lancet* **1,** 109.

209. Trell, E., Oldbring, J., Korsgaard, R., and Mattiasson, I. (1977). Aryl hydrocarbon hydroxylase inducibility in carcinoma of renal pelvis and ureter. *Lancet* **2,** 612.

210. Trell, E., Oldbring, J., Korsgaard, R., Hellsten, S., Mattiasson, I., and Telhammar, E.

(1978). Aryl hydrocarbon hydroxylase inducibility and carcinoma of the urinary bladder. *IRCS J. Med. Sci.* **6**, 138.

211. Paigen, B., Ward, E., Steenland, K., Havens, M., and Sartori, P. (1979). Aryl hydrocarbon hydroxylase inducibility is not altered in bladder cancer patients or their progeny. *Int. J. Cancer* **23**, 312–315.

212. Blumer, J. L., Dunn, R., and Gross, S. (1979). Lymphocyte aryl hydrocarbon hydroxylase (AHH) inducibility in acute leukemia of childhood (AL). *Proc. Am. Assoc. Cancer Res.* **20**, 310.

213. Lubell, D. L. (1962). Fatal hepatic necrosis associated with zoxazolamine therapy. *N.Y. State J. Med.* **62**, 3807–3810.

214. Jick, H., Porter, J., and Morrison, A. S. (1977). Relation between smoking and age of menopause. *Lancet* **1**, 1354–1355.

215. Cohen, S. B., and Burk, R. F. (1978). Acetaminophen overdoses at a county hospital: A year's experience. *South. Med. J.* **71**, 1359–1364.

216. Nebert, D. W. (1980). The *Ah* locus: Genetic differences in toxic response to foreign compounds. *In* "Microsomes, Drug Oxidations, and Chemical Carcinogenesis" (M. J. Coon, A. H. Conney, R. W. Estabrook, H. V. Gelboin, J. R. Gillette, B. N. La Du, and P. J. O'Brien, eds.). Academic Press, New York (in press).

217. Negishi, M., and Nebert, D. W. (1979). Structural gene products of the *Ah* locus. Genetic and immunochemical evidence for two forms of mouse liver cytochrome P-450 induced by 3-methylcholanthrene. *J. Biol. Chem.* **254**, 11015–11023.

218. Hepner, G. W., and Piken, E. P. (1979). Detection of carcinogen-induced stimulation of cytochrome *P*-448-associated enzymes by $^{14}CO_2$ breath analysis studies using dimethylaminoazobenzene. *Gastroenterology* **76**, 267–271.

219. Glowinski, I. B., Radtke, H. E., and Weber, W. W. (1978). Genetic variation in N-acetylation of carcinogenic arylamines by human and rabbit liver. *Mol. Pharmacol.* **14**, 940–949.

220. Ehrich, M., Aswell, J. E., van Tassell, R. L., and Wilkins, T. D. (1979). Mutagens in the feces of 3 South-African populations at different levels of risk for colon cancer. *Mutat. Res.* **64**, 231–240.

221. Taylor, B. A., Heiniger, H. J., and Meier, H. (1973). Genetic analysis of resistance to cadmium-induced testicular damage in mice. *Proc. Soc. Exp. Biol. Med.* **143**, 629–633.

222. Panemangalore, M., and Brady, F. O. (1978). Induction and synthesis of metallothionein in isolated perfused rat liver. *J. Biol. Chem.* **253**, 7898–7904.

223. Rudd, C. J., and Herschman, H. R. (1979). Metallothionein in a human cell line: The response of HeLa cells to cadmium and zinc. *Toxicol. Appl. Pharmacol.* **47**, 273–278.

224. Stokinger, H. E., Mountain, J. T., and Scheel, L. D. (1968). Pharmacogenetics in the detection of the hypersusceptible worker. *Ann. N.Y. Acad. Sci.* **151**, 968–976.

225. Shum, S., Jensen, N. M., and Nebert, D. W. (1979). The *Ah* locus: *In utero* toxicity and teratogenesis associated with genetic differences in benzo[*a*]pyrene metabolism. *Teratology* **20**, 365–376.

Chapter 4

Induction of the Enzymes of Detoxication

EDWARD BRESNICK

I. INTRODUCTION

Control of metabolic activity within prokaryotes is largely accomplished by processes of induction or repression and by end-product inhibition. In induction or repression, RNA synthesis or transcription is either turned on or off, respectively, as a result of interaction at the level of the promoter region of the genome. In prokaryotes, the process of transcription and translation or protein synthesis are linked. Indeed, translation may be initiated while the newly formed messenger RNA is being transcribed.

With eukaryotes, however, the problem of regulation is considerably

69

ENZYMATIC BASIS OF DETOXICATION, VOL. I

more complex. A good deal of the complexity is associated with sophisticated compartmentalization which occurs in the cells of these organisms. Within mammalian cells, for example, may be found a discrete nucleus surrounded by a unit membrane in which pores or annuli are observable; membraneless nucleoli, masses of ribonucleoprotein, and specific chromatin containing the ribosomal RNA genes are located within the nuclei; mitochondria, each of which is also enveloped by a discrete unit membrane with invaginations of the inner membrane into the cell, i.e., cristae; rough endoplasmic reticulum, a canal-like structure which permeates throughout the cell; smooth endoplasmic reticulum occurring in vesicular form; lysosomes, structures that are bounded by single membranes; and the Golgi apparatus.

The complexity at the level of morphology is compounded by changes in structure which occur during the cell cycle. Nucleoli will "disappear" and the nuclear membrane will stretch and ultimately open.

The extreme of compartmentalization, although increasing the apparent complexity, also simplifies the mechanics of regulation. Thus, the formation of the posttranscriptional modifications of RNA takes place within the nucleus. The process of transcription is then followed by transport of the RNA into the cytoplasm and its accommodation at the site of the rough endoplasmic reticulum. Translation also involves the multiple interactions of subcellular organelles. After translation of a specific protein species at the site of the rough endoplasmic reticulum, further modification may occur within the Golgi apparatus prior to the final translocation of the finished protein to other portions of the cell. It has been recognized that the synthesis of proteins for secretion or for intracellular utilization may proceed differently. Recently, the presence of a signal peptide at the N-terminal portion of the protein has been suggested as the mechanism for threading the newly synthesized protein into the endoplasmic reticulum canal for either incorporation into membrane or secretion from the cell; ultimately the signal peptide is proteolytically digested.

A more detailed view of some of the potential sites for regulation is presented in Fig. 1 and is discussed in detail elsewhere.[1,2] Since a large portion of the genetic material present within mammalian cells is unavailable for transcription as a result of the interaction of the DNA with nuclear proteins, a first step in regulation of phenotypic expression relates to the activation of chromatin by the redistribution of nuclear proteins. This redistribution then allows for the transcription of the exposed region of the genome.

The transcription of RNA molecules is not only dependent upon the presence of an activated chromatin template but is also determined to a large extent by the amount of intracellular nuclear RNA polymerase.

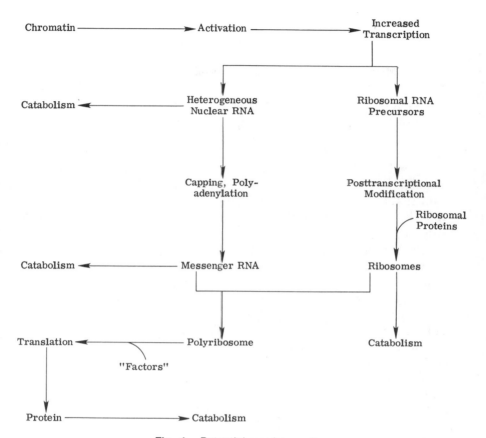

Fig. 1. Potential regulatory sites.

Thus, the synthesis of ribosomal RNA precursor molecules may in large measure be regulated by the amount of RNA polymerase I while the elaboration of messenger RNA precursor substances is under the control of the intracellular concentration of RNA polymerase II. These aspects will also be dealt with subsequently.

Both ribosomal and messenger precursor RNA molecules undergo a series of posttranscriptional modifications prior to the formation of fully completed macromolecules. The rates of methylation, of other processing reactions, and the release of ribosomal RNA as ribosomal subunits may comprise important determinants of the steady-state level of ribosomes. In addition, the posttranscriptional modifications of heterogeneous nuclear RNA, the first transcriptional product ultimately destined for messenger RNA function, will include such reactions as polyadenylation at the 3'-end and the addition of a methyl guanylate cap at the 5'-terminus.

More recently, the presence of inserted sequences within certain genes and the sorting out of these sequences in the elaboration of a transcriptional product are thought to comprise important regulatory mechanisms about which information is scant. The messenger RNA resulting after the modifications at the 3'- and 5'-ends, also possesses a defined turnover rate which is determined in part by a variety of physiological and, perhaps, environmental factors. The latter substances, by either increasing or decreasing the degradation of messenger RNA, may significantly affect the steady-state level of this macromolecule (Fig. 1).

Messenger RNA is only one component of the protein synthetic machinery and functions only in collaboration with ribosomes. The latter organelles are often rate limiting in number, and, where increased protein synthesis is required, the enhanced elaboration of the ribosomes must take precedence.

A number of protein factors play significant roles in regulating the extent of protein synthesis. Specific initiation, elongation, and termination proteins contribute substantially to the protein synthetic rate. Hormones also appear to affect the level of initiation factor and, in this manner, can increase protein synthesis.

A final aspect of regulation, presented in Fig. 1, relates to the turnover of the membranal components themselves. The proteins elaborated by the mechanisms described above are dynamic substances which undergo continual synthesis and degradation. The alteration of the turnover rate by physiological and pharmacological substances does influence the steady-state level of these proteins.

The steady-state level of proteins, whether they function as enzymes or in a structural capacity, is greatly influenced by physiological factors, by ontogenic factors, by exogenous factors as in nutrition, by pathological circumstances, and by environmental substances. Thus, proteins undergo marked qualitative and quantitative changes as a result of the normal development from fetus to adult; of the normal circadian rhythm associated with hormone release; of the acquisition of new nutritional life styles; of the chronic exposure to atmospheric components, e.g., soots that act as inducers or suppressors; and the onset of one of a number of pathological events, e.g., liver disease. This chapter considers the influence of several environmental factors upon a specific protein network that functions in detoxication, the cytochrome P-450 monooxygenases (see this volume chapters 5 and 8). Particular emphasis is placed on the influence directed toward several of the regulatory mechanisms described above. Although similar effects on other enzymes are noted throughout treatise, the cytochrome P-450 monooxygenases have received the est attention and serve as a good example of the induction process.

Xenobiotics as Inducers

Most organisms exhibit a profound ability to adjust to changes in their environment. Nowhere is this better illustrated than in the case of the machinery for detoxication, activation, solubilization, and alteration of properties of a wide variety of xenobiotics. In this treatise, these processes are referred to as detoxication. Many of these reactions require the activity of the NADPH-dependent cytochrome P-450 monooxygenases, i.e., the mixed function oxidases, and the increased concentration of such materials as polycyclic aromatic hydrocarbons in the atmosphere results in a rather profound elevation in the activity of these enzymes.

It was the report of Richardson and Cunningham[3] which first revealed this most interesting environmental adjustment. After the intravaginal instillation of 3-methylcholanthrene, a polycyclic aromatic hydrocarbon, rats that were fed the aminoazo dye, 3'-methyl-4-dimethylaminoazobenzene, showed considerably less liver cancer than those fed the aminoazo dye alone. Miller et al.[4] were subsequently able to show a marked reduction in the concentration of the aminoazo dye in the blood and in the liver of rats to which 3-methylcholanthrene had been administered. These reductions in blood and tissue levels were due to the enhancement in the metabolism of the dye as a result of an increased intracellular N-demethylase and azo dye reductase following the administration of the polycyclic hydrocarbon. These two enzymes are intimately involved in the detoxication of the aminoazo dyes. Thereby was born the phenomenon of induction of mixed function oxidases by a polycyclic hydrocarbon.

II. INDUCTION OF MIXED FUNCTION OXIDASES

It soon became apparent that polycyclic hydrocarbons were not the only substances to effect an increase in the activity of the monooxygenases. While investigating barbiturate tolerance in animals, Remmer noted the marked proliferation of the endoplasmic reticulum, a major location of the drug-metabolizing enzyme system in liver.[5,6] One of the most potent of the barbiturates to effect this induction is phenobarbital, which causes an increase in a large number of both microsomal and cytosolic enzymatic reactions.[7] Although overlap in the enzymatic systems that are induced by 3-methylcholanthrene and phenobarbital is sometimes noted, these two agents generally affect different systems. As will be apparent, the difference is largely due to the type of cytochrome P-450 which is affected.

A third type of induction of mixed function oxidases has been reported.

Certain steroids, e.g., pregnenolone 16-α-carbonitrile, can effect an extensive increase in smooth endoplasmic reticulum membranes of liver after chronic exposure as well as give rise to a concomitant increase in the activity of a number of mixed function oxidases.[8,9] Subsequently, a large number of agents have been found to induce cytochrome P-450 including insecticides, polyhalogenated biphenyls, and various chemotherapeutic drugs.

A.   Increase in Cytochrome P-450

Cytochrome P-450 is the terminal oxidase of the microsomal electron transport system which is responsible for the metabolism of xenobiotics. This important hemoprotein is discussed more fully in this volume, chapter 6. Over the years, substantial evidence has accumulated for the definition of multiple forms of this cytochrome. Furthermore, treatment of animals with various inducers results in the appearance of different types of cytochrome P-450. For example, the administration to rats of a polycyclic hydrocarbon such as 3-methylcholanthrene results in the induction of a particular type of hemoprotein referred to as cytochrome P-448 or P_1-450 or P-450c, whereas treatment with phenobarbital is attendant by an increase in the amount of cytochrome P-450b. The evidence for the multiplicity of the cytochrome P-450 is based upon observations of different spectral and catalytic forms; different carbon monoxide spectra, ethyl isocyanide difference spectra, substrate specificity, and sodium dodecyl sulfate-gel electrophoresis patterns have been reported. The final piece of evidence relating to multiple forms is based upon immunochemical evidence with monospecific antibody prepared against several of the types of cytochrome P-450.[10]

Evidence will be presented in a later section of this chapter which bears upon the induction of the particular types of cytochrome P-450 as a result of administration of different types of agents.

B.   Inhibitor Studies

With the use of inhibitors, a significant contribution has been made in the definition of the action of barbiturates and polycyclic hydrocarbons upon the cytochrome P-450 monooxygenase system. Conney et al.[11] first established that a methionine analogue, ethionine, will prevent the rise in the activity of an azo dye N-demethylase in the livers of rats that had received 3-methylcholanthrene. Furthermore, macromolecular binding of the aminoazo dye could be significantly reduced by the concomitant administration of this analogue along with 3-methylcholanthrene.[12] Although the mechanism of action of ethionine has not been elucidated, this analogue does affect protein synthesis in a quantitative manner. These initial studies, therefore, implicated new protein synthesis as a mechanism in the elevation of monooxygenase activity after the administration of polycyclic aromatic hydrocarbons. A more specific inhibitor of protein synthesis, puromycin, was used in later studies[13,14]: the increase in N-demethylase and aryl hydrocarbon hydroxylase activities which would be attendant to the administration of 3-methylcholanthrene was abolished.

The necessity for new protein synthesis in the induction phenomenon was also demonstrated by Cutroneo and Bresnick[15] by using a substance, β-naphthoflavone, which produces a pattern of induction similar to polycyclic hydrocarbons. In their studies, the concomitant administration of a potent inhibitor of peptidyltransferase, cycloheximide, abolished the β-naphthoflavone-mediated induction of aryl hydrocarbon hydroxylase. In a similar series of experiments,[15] phenobarbital-mediated induction of microsomal enzymes was similarly blocked by cycloheximide.

The need for new RNA synthesis in both the polycyclic hydrocarbon- and phenobarbital-mediated induction of monooxygenase has also been established by the use of various inhibitors. Actinomycin D, a potent inhibitor of transcription, prevents polycyclic hydrocarbon-mediated induction.[14,16,17] The induction of aryl hydrocarbon hydroxylase activity in fetal rat liver explants is similarly prevented by mercapto-(pyridethyl)benzimidazole, an inhibitor of RNA synthesis.[15] Jacob et al.[18] have also demonstrated that α-amanitin, a potent blocker of RNA polymerase II, when administered up to 8 h after the injection of phenobarbital, prevents the induction of cytochrome P-450 and ethylmorphine N-demethylase activity; beyond this time, RNA synthesis was not required for induction.

III. CHROMATIN ACTIVATION

The administration of 3-methylcholanthrene to rats is attended by a significant increase in the activity of the "aggregate" nuclear RNA polymerase system in liver.[19] These results were extended to include the effect of phenobarbital administration.[20] Such an increase in the aggregate enzyme system could conceivably result from enhanced enzyme synthesis, decreased catabolism of enzyme, or an increased efficiency of the chromatin template for RNA synthesis. The investigations by Madix and Bresnick[21] and by Piper and Bousquet[22] have suggested that the efficacy of the liver chromatin as a template is significantly improved after administration of the inducers. In subsequent studies, the difference in template efficiency could be obviated by the treatment of the chromatin with agents that solubilize and consequently remove the basic proteins, i.e., 2 M sodium chloride,[20-23] suggesting that the effect of the inducers might be related to altering the protein arrangement on the deoxyribonucleoprotein complex and thereby increasing its availability for transcription.

Since the nearest neighbor frequency of the RNA transcribed *in vitro* with *Escherichia coli* RNA polymerase and the chromatin indicated a substantial difference between treated and control systems, different RNA molecules must have been elaborated as a result of the induction phenomenon. It is of interest in this report[24] that the transcripts elaborated with chromatin prepared from 3-methylcholanthrene- and phenobarbital-treated rats differed markedly in their nearest neighbor frequency analysis, suggesting a difference in the mechanism of action of these inducers.

Effect on Nuclear Proteins

As noted, the composition of the sodium chloride extracts from chromatin was of importance since the elimination of these proteins from chromatin obviated the difference in template activity. It was shown that the profile of the saline-extractable proteins differed in the chromatins prepared from control or 3-methylcholanthrene-pretreated rats.[25] The stimulation of incorporation of amino acid precursors into nuclear proteins after phenobarbital administration have also been reported.[26-28] Furthermore, the saline-extractable nuclear proteins showed an increased stimulation in at least one of their components after such treatment.[26]

Histone components are known to be able to undergo posttranslational modifications, such as phosphorylation or acetylation, and such processes markedly alter the binding between nuclear proteins and the DNA. Procaccini and Bresnick[29] investigated the posttranslational processes and reported that the acetylation of histones in liver was significantly increased at 2 h after administration of either 3-methylcholanthrene or phenobarbital to rats. The nonhistone nuclear proteins also showed rather distinct changes in liver after administration of certain inducers with increased incorporation of amino acids into these substances after pretreatment of rats with phenobarbital.[27] The administration of this inducer was also attended by a marked increase in the posttranslational phosphorylation as early as 2–6 h later. The phosphorylated moiety was associated with a serine molecule in protein. It is of interest that such effects were not observable after administration of 3-methylcholanthrene.

These results all suggested that a striking alteration in chromatin proteins may precede changes in RNA synthesis and that this event may be related to the genomic activation produced after administration of either phenobarbital or 3-methylcholanthrene. In part, these alterations may result from posttranslational modifications.

IV. EFFECTS ON RNA POLYMERASE

The effect of polycyclic hydrocarbon administration upon the qualitative and quantitative patterns of the RNA polymerases has been investigated in liver.[30] In these studies, the spectrum of RNA polymerases was determined after their solubilization and separation by the method of Roeder and Rutter.[31] Under these conditions, 3-methylcholanthrene administration effected a 50–90, 40–50, and 0% increase in the activities of RNA polymerase I, II, and III, respectively. The enhanced activity of the RNA polymerases was manifested as early as 6 h after administration of the polycyclic hydrocarbon and reached a maximum at approximately 12–15 h thereafter. These studies then indicate that the administration of polycyclic hydrocarbons is attended not only by an increased efficiency of chromatin as a template for transcription but also by an increase in the enzyme responsible for this process.

V. EFFECTS ON COMPONENTS ON PROTEIN SYNTHESIS

The following discussion will center about the events relating to protein synthesis that transpire after the administration of an inducer to rats. In this regard, we will consider the effects upon ribosomes, upon initiation factor activity, and, finally, upon messenger RNA synthesis and function.

A. Ribosomal Components

An overwhelming body of evidence indicates that the intracellular number of ribosomes is a rate-limiting factor for protein synthesis in mammalian systems. The steady-state level of these suborganelles is a reflection of their rates of synthesis and degradation. Indeed, both these kinetic parameters have been investigated as a function of administration of an inducer of the monooxygenases.

An increase in the amount of ribosomal RNA has been reported to take place in liver after injection of 3-methylcholanthrene.[32] The increase was transitory, and, after reaching a maximum at 12 h, returned to control values by approximately 48 h after administration of the inducer. The incorporation of the RNA precursor, orotic acid, into 45 S cytoplasmic RNA has been studied after administration of this polycyclic hydrocarbon and a significant stimulation into the 45 S species was observed by 16 h.[33] At that time, a marked elevation into both 18 and 28 S ribosomal RNA was also demonstrated. The turnover of both membrane bound and free polyribosomes has been studied as a function of time after administration of this inducer.[34] In these experiments, the polyribosomes were labeled by prior administration of orotic acid and the half-life of retention of label was followed; 3-methylcholanthrene administration markedly increase the half-life of the liver ribosomes resulting in a greater stability of these organelles.

Phenobarbital administration also is attended by a marked effect upon the turnover of liver ribosomes. By 16 h after administration of the barbiturate, an increased incorporation of a variety of precursors into ribosomal RNA has been observed.[35-37] An even more profound effect of the barbiturate has been noted upon the catabolism of the ribosomes. Enhanced stability of these organelles, unrelated to the increased rate of synthesis, has been reported by several investigators.[35,37]

Since ribosomal RNA synthesis is a complex process which involves a number of posttranscriptional modifications including methylation, control of the synthesis of this macromolecule may occur at a number of steps. It has been suggested by several groups[38-40] in this regard that methylation, particularly at the 2'-hydroxyl of the sugar component of a nucleotide, will confer a degree of protection against nuclease action. The methylation of liver nuclear RNA, a good portion of which represents preribosomal RNA precursors, was found to be significantly elevated in treated rats as a function of phenobarbital administration.[41,42] The nucleolar RNA methyltransferase, the enzyme which catalyzes the methylation of preribosomal RNA, is also significantly elevated soon after phenobarbital administration; it is contended that such methylation renders the ribosomal RNA less susceptible to intranuclear metabolism and therefore effects an elevation in the steady-state level of ribosomal RNA.[43]

B. Activity of Initiation Factors in Protein Synthesis

Key determinants in the rate of protein synthesis are the initiation factors, proteins that are loosely associated with ribosomes and which function in the recognition of messenger RNA, in the attachment of the initiating transfer RNA to the appropriate ribosomal subunit, and in the formation of the ribosomal complex. The effects of inducers such as 3-methylcholanthrene upon the activity of the initiation factors has been studied in some detail in Bresnick's laboratory.[44-46] In these studies, the formation of the initiation complex was shown to be accelerated after administration of the inducer to rats. The increase in the activity of the initiation factors was evident as early as 2 h after administration of the inducer. The elongation and termination processes, on the other hand, were unaltered in liver by pretreatment of the animals with 3-methylcholanthrene. These studies indicate that the initiation in protein synthesis, a potent site of regulation, is strikingly stimulated in liver by pretreatment of rats with 3-methylcholanthrene at a time prior to the manifestation of the induction of the cytochrome P-450 monooxygenases. Consequently, the induction phenomenon is, at least in part, the result of such a stimulation.

C. Messenger RNA

That phenobarbital treatment can increase liver messenger RNA activity was first suggested by the experiments of Kato et al.[47] in which microsomes obtained from the livers of pretreated rats demonstrated greater protein synthetic activity than similarly prepared organelles from control animals. The microsomes from either control or treated rats appeared equally effective in the translation of the synthetic messenger, polyuridylic acid.

The early experiments from Bresnick's laboratory[48] suggested an increase in messenger RNA activity in liver as a result of 3-methylcholanthrene pretreatment of rats. This conclusion was based upon the increase in the incorporation of ^{32}P into poly(A)-cytoplasmic RNA occurring at 6 h after an injection of 3-methylcholanthrene. By 24 h, a 33-fold increase in such incorporation was observed. Since polyadenylation is a required posttranscriptional process for the synthesis of many messenger RNA's, these experiments suggest the increased formation of the latter.

More definitive experiments from this laboratory have shown an increase in messenger RNA activity resulting after either phenobarbital or 3-methylcholanthrene pretreatment. In these studies, the messenger RNA responsible for the elaboration for the particular cytochrome, i.e., cytochrome P-450b or P-450c after phenobarbital or 3-methylcholanthrene treatment, respectively, has been studied in some detail. Cytoplasmic RNA was isolated from the livers of control and 3-methylcholanthrene-treated rats and the messenger RNA activity contained therein was assessed using a rabbit reticulocyte system with [^{35}S]methionine as the amino acid precursor. 3-Methylcholanthrene pretreatment was associated with a significant increase in messenger RNA activity as defined by this reticulocyte system. The translational products were immunoprecipitated with monospecific antibody to cytochrome P-450c, the immunoprecipitate was dissolved and subjected to gel electrophoresis and the dried gel was subjected to radioautography. Only with the RNA obtained from the livers of 3-methylcholanthrene-pretreated rats was a significant cytochrome P-450c region demonstrated on the gels.[30] In some studies, using the reticulocyte system and [^{3}H]leucine as the amino acid precursor, the immunoprecipitate was also subjected to SDS-polyarylamide gel electrophoresis. These studies clearly indicated the presence of a region of radioactivity that was coincident with the standard cytochrome P-450c.[30]

The synthesis of liver cytochrome P-450b has also been examined with cell-free protein synthesizing systems. Liver polysomes isolated from rats treated with phenobarbital, had a fivefold greater rate of incorporation [^{3}H]leucine into total protein than did control polysomes.[49] The translational products were immunoprecipitated, and polysomes from the induced animals were demonstrated to synthesize cytochrome P-450b at a rate almost seven times that of the polysomes from control animals. The increased cytochrome P-450b synthesis was detectable as early as 6 h after phenobarbital administration. By 24 h after administration of the barbiturate, synthesis of cytochrome P-450b was almost at control values.

The studies are certainly in concert with the idea of elevation in messenger RNA as the result of either 3-methylcholanthrene or phenobarbital administration. Whether such an increase is due to an increase in synthesis or decrease in degradation is unknown and will be ascertained only after production of complementary DNA to purified messenger RNA fractions. Such studies should be completed within the near future.

VI. COMMENTS

A number of the regulatory steps in the induction of the cytochrome P-450 monooxygenases have been briefly considered. It is apparent that a selection mechanism must exist for the elevation of specific forms of cytochrome P-450 by specific inducers. It would be interesting if this induction involved a similar mechanism to that of specific immunoglobulin elaboration with variable and constant regions being present as part of the primary structure of the cytochromes. At the present, this thought is only speculation which awaits experimentation.

ACKNOWLEDGMENTS

This author's research, cited herein, has been supported by grants from the National Institutes of Health (CA 20711 and ES 01974).

REFERENCES

1. Schimke, R. T. (1975). Control of enzyme levels in mammalian tissues. *Adv. Enzymol.* **37**, 135–187.
2. Gelehrte, T. D. (1979). Enzyme induction in mammals—an overview. *In* "Induction of Drug Metabolism" (R. W. Esterbrook and E. Lingenlaub, eds.), pp. 7–24. Schottauer, Stuttgart.
3. Richardson, H. L., and Cunningham, L. (1951). The inhibitory action of methylcholanthrene on rats fed the azodye 3'-methyl-4-dimethylaminoazobenzene. *Cancer Res.* **11**, 274.
4. Miller, E. C., Miller, J. A., and Brown, R. R. (1952). On the inhibitory action of certain polycyclic hydrocarbons on azo dye carcinogenesis. *Cancer Res.* **12**, 282–283.
5. Remmer, H. (1957). The acceleration of evipan oxidation and the demethylation of methylaminopyrine by barbiturates. *Naunyn-Schmiedebergs Arch. Exp. Pathol. Pharmakol.* **237**, 296–307.
6. Remmer, H., and Merker, H. J. (1963). Drug-induced changes in the liver endoplasmic reticulum: Association with drug-metabolizing enzymes. *Science* **142**, 1657–1658.
7. Conney, A. H. (1966). Pharmacological implications of microsomal enzyme induction. *Pharmacol. Rev.* **19**, 317–366.
8. Garg, B. D., Kovacs, K., Blascheck, J. A., and Selye, H. (1970). Ultrastructural changes induced by pregnenolone nitrile in the rat liver. *J. Pharm. Pharmacol.* **22**, 872–873.
9. Solymoss, B., Werringloer, J., and Toth, S. (1971). The influence of pregnenolone-16α-carbonitrile on hepatic mixed-function oxygenases. *Steroids* **17**, 427–433.
10. Thomas, P. E., Lu, A. Y. H., Ryan, D., West, S. B., Kawalek, J., and Levin, W. (1976). Multiple forms of rat liver cytochrome *P*-450. *J. Biol. Chem.* **251**, 1385–1391.
11. Conney, A. H., Miller, E. C., and Miller, J. A. (1956). The metabolism of methylated aminoazo dyes. V. Evidence for induction of enzyme synthesis in the rat by 3-methylcholanthrene. *Cancer Res.* **16**, 450–459.
12. Gelboin, H. V., Miller, J. A., and Miller, E. C. (1958). The *in vitro* formation of protein-bound derivatives of aminoazo dyes by rat liver preparations. *Cancer Res.* **19**, 975–985.
13. Conney, A. H., and Gilman, A. G. (1963). Puromycin inhibition of enzyme induction by 3-methylcholanthrene and phenobarbital. *J. Biol. Chem.* **238**, 3682–3685.
14. Gelboin, H. V., and Blackburn, N. R. (1963). The stimulatory effect of 3-methylcholanthrene on microsomal amino acid incorporation and benzpyrene hydroxylase activity and its inhibition by actinomycin D. *Biochim. Biophys. Acta* **72**, 657–660.
15. Cutroneo, K. R., and Bresnick, E. (1973). Induction of benzpyrene hydroxylase in fetal liver explants by flavones and phenobarbital. *Biochem. Pharmacol.* **22**, 675–687.
16. Gelboin, H. V., and Blackburn, N. R. (1964). The stimulatory effect of 3-methylcholanthrene on benzpyrene hydroxylase activity in several rat tissues: Inhibition by actinomycin D and puromycin. *Cancer Res.* **24**, 356–360.
17. Jervell, K. F., Cristoffersen, T., and Morland, J. (1965). Studies on 3-methylcholanthrene induction and carbohydrate repression of rat liver dimethylaminoazobenzene reductase. *Arch. Biochem. Biophys.* **111**, 15–22.
18. Jacob, S. T., Schars, M. B., and Vesell, E. S. (1974). Role of RNA in induction of hepatic microsomal mixed-function oxidase. *Proc. Natl. Acad. Sci. U.S.A.* **71**, 704–707.
19. Bresnick, E. (1966). Ribonucleic acid polymerase activity in liver nuclei from rats pretreated with 3-methylcholanthrene. *Mol. Pharmacol.* **2**, 406–410.

20. Gelboin, H. V., Wortham, J. S., and Wilson, R. G. (1967). 3-Methylcholanthrene and phenobarbital stimulation of rat liver RNA polymerase. *Nature* (*London*) **214**, 281–283.
21. Madix, J. C., and Bresnick, E. (1967). Increased efficacy of liver chromatin as a template for RNA synthesis after administration of 3-methylcholanthrene. *Biochem. Biophys. Res. Commun.* **28**, 445–452.
22. Piper, W. N., and Bousquet, W. F. (1968). Phenobarbital and methylcholanthrene stimulation of rat liver chromatin template activity. *Biochem. Biophys. Res. Commun.* **33**, 602–605.
23. Bresnick, E. (1970). Enhanced initiation of RNA chains in liver after administration of 3-methylcholanthrene. *Biochim. Biophys. Acta* **217**, 204–206.
24. Bresnick, E., and Mossé, H. (1969). Activation of genetic transcription in rat liver chromatin by 3-methylcholanthrene. *Mol. Pharmacol.* **5**, 219–226.
25. Yee, M., and Bresnick, E. (1971). Effect of administration of 3-methylcholanthrene on the salt-extractable chromatin proteins of rat liver. *Mol. Pharmacol.* **7**, 191–198.
26. Ruddon, R. W., and Rainey, C. H. (1970). Stimulation of nuclear protein synthesis in rat liver after phenobarbital administration. *Biochem. Biophys. Res. Commun.* **40**, 152–160.
27. Ruddon, R. W., and Rainey, C. H. (1971). Comparison on effects of phenobarbital and nicotine on nuclear protein synthesis in rat liver. *FEBS Lett.* **14**, 170–174.
28. Blankenship, J., and Kuehl, L. (1973). Effects of phenobarbital and 3-methylcholanthrene on amino acid incorporation into rat liver chromosomal proteins. *Mol. Pharmacol.* **9**, 247–258.
29. Procaccini, R. L., and Bresnick, E. (1975). Histone modification in liver after administration of inducers of mixed function oxidase activity. *Chem.-Biol. Interact.* **11**, 523–533.
30. Bresnick, E., Liberator, P., Brosseau, M., Thomas, P. E., Ryan, D. E., and Levin, W. (1979). Transcription and translation in liver from 3MC-treated rats. *Fed. Proc., Fed. Am. Soc. Exp. Biol.* **38**, 402.
31. Roeder, R. G., and Rutter, R. J. (1969). Multiple forms of DNA-dependent RNA polymerase in eukaryotic organisms. *Nature* (*London*) **224**, 234–237.
32. Argyris, T. S., and Heinemann, R. (1975). Ribosome accumulation in 3-methylcholanthrene-induced liver growth in adult male rats. *Exp. Mol. Pathol.* **22**, 335–341.
33. Bresnick, E., Synerholm, M. E., and Tizard, G. T. (1968). Changes in cytoplasmic RNA after administration of 3-methylcholanthrene. *Mol. Pharmacol.* **4**, 218–223.
34. Hopkinson, J., and Bresnick, E. (1975). Stabilization of total and free ribosomes associated with 3-methylcholanthrene-induced adult rat liver growth. *Biochem. Biophys. Pharmacol.* **24**, 435–437.
35. Cohen, A. M., and Ruddon, R. W. (1971). Stability of polyribosomes isolated from rat liver after phenobarbital administration. *Mol. Pharmacol.* **7**, 484–489.
36. McCauley, R., and Couri, D. (1971). Early effects of phenobarbital on cytoplasmic RNA in rat liver. *Biochim. Biophys. Acta* **238**, 233–244.
37. Smith, S. J., Hill, R. N., Gleeson, R. A., and Vesell, E. S. (1972). Evidence for post-transcriptional stabilization of ribosomal precursor ribonucleic acid by phenobarbital. *Mol. Pharmacol.* **8**, 691–700.
38. Vaughan, M. H., Jr., Soeiro, R., Warner, J. R., and Darnell, J. E. (1967). The effects of methionine deprivation on ribosome synthesis in HeLa cell. *Proc. Natl. Acad. Sci. U.S.A.* **58**, 1527–1534.
39. Darnell, J. E., Jr. (1968). Ribonucleic acids from animal cells. *Bacteriol. Rev.* **32**, 262–290.

40. Maden, B. E., and Salim, M. (1974). The methylated nucleotide sequences in HeLa cell ribosomal RNA and its precursors. *J. Mol. Biol.* **88,** 133–152.

41. Wold, J. S., and Steele, W. J. (1972). Alterations in liver nuclear ribonucleic acid of rats treated with phenobarbital. *Mol. Pharmacol.* **3,** 681–690.

42. Smith, S. J., Jacob, S. T., Liu, D. K., Duceman, B., and Vesell, E. S. (1974). Further studies on post-transcriptional stabilization of ribosomal precursor ribonucleic acid by phenobarbital. *Mol. Pharmacol.* **10,** 248–256.

43. Smith, S. J., Liu, D. K., Leonard, T. B., Duceman, B. W., and Vesell, E. S. (1976). Phenobarbital-induced increases in methylation of ribosomal precursor ribonucleic acid. *Mol. Pharmacol.* **12,** 820–831.

44. Lanclos, K. D., and Bresnick, E. (1972). Enhanced initiation factor activity in rats after administration of 3-methylcholanthrene or β-naphthoflavone. *Res. Commun. Chem. Pathol. Pharmacol.* **4,** 421–432.

45. Lanclos, K. D., and Bresnick, E. (1973). Initiation factor activity in liver after 3-methylcholanthrene administration. *Drug. Metab. Dispos.* **1,** 239–247.

46. Hopkinson, J., Prichard, P. M., and Bresnick, E. (1974). Translational control at the level of initiation of protein synthesis. The effect of 3-methylcholanthrene administration on the activity of rat liver IF-M$_2$A, IF-M$_2$B and IF-M$_3$ *Biochim. Biophys. Acta* **366,** 1–10.

47. Kato, R., Jondorf, W. R., Loeb, L. A., Ben, T., and Gelboin, H. V. (1966). Studies on the mechanism of drug-induced microsomal enzyme activities. V. Phenobarbital stimulation of endogenous messenger RNA and polyuridylic acid-directed L-^{14}C-phenylalanine incorporation. *Mol. Pharmacol.* **2,** 171–186.

48. Lanclos, K. D., and Bresnick, E. (1975). The formation of PolyA-containing RNA in rat liver after administration of 3-methylcholanthrene. *Chem.-Biol. Interact.* **12,** 341–348.

49. Colbert, R. A., Bresnick, E., Levin, W., Ryan, D. E., and Thomas, P. E. (1979). Synthesis of liver cytochrome *P*-450b in a cell-free protein synthesizing system. *Biochem. Biophys. Res. Commun.* **91,** 886–891.

Chapter 5

Comparative Aspects of Detoxication in Mammals

JOHN CALDWELL

I. INTRODUCTION*

When chemicals foreign to the energy-yielding metabolism of living organisms (xenobiotics) enter the organism, they may undergo one or more of three fates; they may (a) remain unchanged, (b) break down spontaneously, or (c) undergo enzymatic metabolism. Of these pos-

* In this chapter the terms drug, xenobiotic and foreign compound are used interchangeably, as are drug metabolism, xenobiotic metabolism, and foreign compound metabolism.

ENZYMATIC BASIS OF DETOXICATION, VOL. I

sibilities, it is enzymatic metabolism which is by far most frequently encountered. The products of metabolism are in general more water soluble and thus more easily eliminated than their precursors. The metabolic products of xenobiotics are frequently less active biologically than their parent compounds, and these processes have been termed "detoxication mechanisms."[1,2] R. T. Williams has pointed out that this term is a misnomer, and many examples can be quoted that show that activation of xenobiotics can occur as a consequence of metabolism.[3,4] It may in fact be more reasonable to view the processes of xenobiotic metabolism as programmed generally to increase the water solubility of the molecule rather than to decrease its toxicity, although this is not always true.[4]

The metabolic reactions of foreign compounds may be classified very broadly into four types occurring in two distinct phases.[5] The first phase of metabolism involves either oxidation, reduction, and hydrolysis, followed by the second phase in which the original compound or a metabolite is linked with an endogenous molecule in a synthetic or conjugation reaction. Not all compounds undergo both phases of metabolism, so that some products of Phase I metabolism are eliminated as such, while other compounds only undergo Phase II metabolism. Additionally, most compounds are eliminated unchanged to some extent. Four types of products can be excreted after the ingestion of a xenobiotic (a) the unchanged compound, (b) Phase I metabolites, (c) Phase II metabolites, and (d) metabolites arising from both Phase I and II metabolism.

An early observation was the very marked species differences encountered in the metabolic fate of foreign compounds, a finding which is not surprising in view of the great diversity of species even within the order Mammalia (see Table I).[6] The current literature contains an enormous number of illustrations of such variation, most of it dealing with *in vivo* studies. Inevitably, such work requires the consideration of many other processes, such as absorption, distribution, and excretion, all going on at the same time and influencing metabolism as well as exhibiting species variation. Experience shows that these other variables are of much less significance than the metabolic one, but an example can be quoted from intermediary metabolism to illustrate the point. Uric acid, a product of purine catabolism, is excreted as such by man, apes and the Dalmatian coach hound, but undergoes extensive further metabolism in other mammals.[7] In the cases of man and the apes, this is due to the absence of the enzyme uricase which cleaves uric acid to allantoin. However, the liver of the Dalmatian dog contains the same amount of uricase as those of other breeds of dog, and this apparent metabolic defect arises from a specific defect in the renal tubular reabsorption of urate.[8]

TABLE I

Zoological Classification of Placental Mammals[a]

Cohort	Order	Representative species
Unguiculta	Insectivora	Hedgehogs, moles
	Chiroptera	Bats
	Dermoptera	Flying lemur
	Edentata	Armadillos, sloths
	Pholidota	Pangolins
	Primates	Monkeys, man
Glires	Rodentia	Rats, mice
	Lagomorpha	Rabbits
Mutica	Cetacea	Whales, dolphins
Ferungulata	Carnivora	Dogs, cats
	Pinnepedia	Seals
	Tubulidentata	Aardvark
	Hyracoidea	Hyraxes
	Proboscidea	Elephants
	Sirenia	Sea-cows
	Perissodactyla	Horses, rhinos
	Artiodactyla	Cattle, sheep, pigs

[a] Based on Young.[6]

This chapter will concern itself with the influence of animal species on the pathways of metabolism of xenobiotics in mammals. Other factors potentially contributing to interspecies variation will only be mentioned where appropriate. It is important to appreciate that animal species is only one of a huge number of variables which may influence the metabolism of a foreign compound (see Table II). The literature drawn upon is concerned, as far as is possible, with genuinely comparative studies, where

TABLE II

Some Variables Which Have Been Shown to Influence Drug Metabolism *in Vivo*

Species	Diet
Strain	Bedding
Age	Lighting
Sex	Pregnancy
Altitude	Enzyme inducers
Time of day	Enzyme inhibitors
Season	Dose size
Gravity	Dose vehicle
Fever	Route of administration

the influence of factors such as strain, dose size, route of administration, sex, and environmental chemicals leading to inhibition or induction of the metabolizing enzymes has been minimized. It is intended to explore four aspects of the comparative biochemistry of foreign compounds: (a) the origins of species differences in metabolism, (b) the possibility that species patterns of metabolism of groups of compounds exist, (c) the significance of the comparative biochemistry of foreign compounds for animal taxonomy, and (d) the enzymatic basis of species differences in metabolism.

II. THE ORIGINS OF SPECIES DIFFERENCES IN FOREIGN COMPOUND METABOLISM

As has been mentioned, the occurrence and extent of interspecies differences in foreign compound metabolism is very well documented. It is not intended to repeat here examples of this, which have been reviewed by Williams[9] and by Parke and Smith.[10]

Given that species differences do occur, it is possible to suggest that they arise from one or more of three origins, (a) the existence of species defective in individual metabolic reactions, (b) the restriction of certain reactions to an individual species or group of species, and (c) variations in the relative extents of two or more competing reactions which a compound may undergo. These possibilities will be discussed in turn.

A. Species Defects in Drug Metabolism Reactions

A considerable number of examples are recorded of species that are unable to carry out metabolic reactions which are otherwise widespread in their occurrence through the order Mammalia. These are listed in Table III,[5,11-15] from which it will be seen that most of the examples involve conjugation reactions, and only two instances of defects of oxidative reactions have thus far been discovered. Both of these involve N-hydroxylations, the guinea pig and Steppe lemming being unable to N-hydroxylate the arylacetamide, 2-acetamidofluorene, and the rat and marmoset being unable to N-hydroxylate the aliphatic amine chlorphentermine. Species differences in these reactions are shown in Table IV.

Probably the best known of the defects listed in Table III is the inability of the domestic cat to form glucuronides of many compounds that are conjugated with glucuronic acid in virtually all other mammals[9]; this defect extends to other cat-like species, such as the lion, civet, and genet.[16] That these species are not congenitally jaundiced suggests that

TABLE III

Species Defects in Foreign Compound Metabolism

Reaction	Defective species	Reference
Aliphatic amine N-hydroxylation	Rat, marmoset	11
Arylacetamide N-hydroxylation	Guinea pig, Steppe lemming	12
Arylamine N-acetylation	Dog	5
Glucuronidation	Cat, lion, lynx	13
Sulfation	Pig, opossum	13
Hippuric acid formation	African fruit bat	14
Mercapturic acid formation	Guinea pig	15

they are able to conjugate bilirubin with glucuronic acid, and this certainly is the case in the cat.[17] Data in Table V, culled from this laboratory and the literature, show that the cat is very markedly defective in its ability to form the glucuronides of small, water-soluble substrates. With increasing molecular size, and presumably lipid solubility, the extent of glucuronidation increases to that found in other animal species. Data for the rat are included for comparison. This pattern is observed for the formation of both ester and ether glucuronides.

At least two examples exist of species defects in N-acetylation reactions. A number of different nitrogen-containing functional groups may be acetylated, including aromatic and aliphatic amines, hydrazines and hydrazides, the sulfonamido group and the amino group of S-substituted cysteines that arise from the breakdown of glutathione conjugates.[5,18] The inability of the dog to N-acetylate aromatic amines, hydrazines and hy-

TABLE IV

Species Defects in N-Hydroxylation Reactions

Species	Dose undergoing N-hydroxylation (%)	
	Chlorphentermine[a]	2-Acetamidofluorene[b]
Rat	0	8
Guinea pig	50	0
Rabbit	42	21
Marmoset	0	—
Rhesus monkey	76	2
Man	44	9

[a] Data from Caldwell et al.[11]

[b] Data from Weisburger et al.[12]

TABLE V

Influence of Substrate on the Glucuronidation Defect in the Cat[a]

Compound	Excretion as glucuronide (% in 24h)	
	Cat	Rat
Ether glucuronides		
Phenol	1	44
1-Naphthol	1	47
2-Naphthol	3	52
Acetaminophen	3	24
Phenolphthalein	60	98
Ester glucuronides		
Benzoic acid	0	<1
1-Naphthylacetic acid	0	51
Hydratropic acid	41	79
Clofibric acid	2	82
Diphenylacetic acid	76	95

[a] Based on Williams[9] and Hirom et al.[13] and references therein.

drazides is well documented, and examples are listed in Table VI.[5,19-23] A single study of the fate of sulfanilamide in the fox shows that this defect in the dog apparently extends to this related species.[24] The second defect of N-acetylation occurs in the guinea pig, which is unable to N-acetylate S-substituted cysteines. Table VII[15,25] shows that the guinea pig is unable

TABLE VI

Species Variations in the N-Acetylation of Various Nitrogen Centers in the Dog and the Rat

Nitrogen center	Dose N-acetylated (%)		Reference
	Dog	Rat	
p-Aminobenzoic acid (aromatic amine)	0	47	19
Sulfanilamide (N^4-amino)	0	36	5
Sulfadimethoxine (N^4-amino)	0	46	20
Isoniazid (hydrazide)	0	50	21
4-Aminobiphenyl (aromatic amine)	0	+++	22
Ethacrynic acid mercapturate (S-arylcysteine)	25	18	23

TABLE VII

Species Variations in the Excretion of Mercapturic Acids after the Administration of Organic Halogen and Nitro Compounds[a]

Compound	Dose excreted as mercapturic acid (%)		
	Guinea pig	Rat	Rabbit
Chlorobenzene	<1	18	20
Benzyl chloride	4	27	49
p-Chlorobenzyl chloride	5	37	23
3,4-Dichloronitrobenzene	3	19	45
2,3,4,6-Tetrachloronitrobenzene	2	32	37
2,3,5,6-Tetrachloronitrobenzene	0	11	14

[a] From Bray et al.[15] and Corner and Young.[25]

to form mercapturic acids of compounds generally metabolized to such conjugates. It is known that the guinea pig is able to form glutathione conjugates, which are the precursors of the mercapturic acids,[26] and the defect recorded in Table VII originates from the failure of this species to carry out the final step in the transformation of a glutathione conjugate to a mercapturic acid.[15]

With respect to sulfate conjugation, it appears that the pig and the opossum are defective in their ability to conjugate phenols in this way. Table VIII[27,28] shows that this defect in the pig is far more specific in terms of substrate than those reported above. The other examples of species defects in conjugation presented in Table III occur with only a very few substrates and will not be discussed further.

TABLE VIII

Species Variations in the Sulfation of Phenols in the Pig and Rat[a]

Compound	Dose excreted as sulfate (%)	
	Pig	Rat
Phenol	6	54
1-Naphthol	32	53
2-Naphthol	6	48
Acetaminophen	11	52

[a] From Capel et al.[27] and Smith and Timbrell.[28]

B. Reactions of Restricted Species Occurrence

For many years, there has been a great deal of interest in the possibility that certain reactions are restricted in their occurrence to a particular species or group of species. Specifically, the close similarity between man and primate species led to an interest in the possibility that some reactions only occur in these species. This originated from the discovery that phenylacetic acid is conjugated with glutamine in man, rather than with glycine, as is the case in subprimate mammals. To date, four reactions restricted in their occurrence to man and nonhuman primates have been found, namely, the glutamine conjugation of arylacetic acids, the aromatization of quinic acid, the N^1-glucuronidation of sulfadimethoxine and the O-methylation of 4-hydroxy-3,5-diiodobenzoic acid.

The nature of the amino acid used in the conjugation of phenylacetic acid is species dependent, and glutamine is encountered only in man, apes, and Old and New World monkeys.[29] Subprimate species do not, however, have an absolute defect in the glutamine conjugation of arylacetic acids since 2-naphthylacetic acid forms a glutamine conjugate in the rat, rabbit, hamster, ferret, and gerbil.[30,31] It is apparent, however, that this reaction is much less versatile in terms of substrate in the subprimate species, since it has only been reported for this one acid.

The aromatization of quinic acid occurs by a series of dehydroxylation reactions performed by the gut flora yielding benzoic acid, which is conjugated with glycine and excreted as hippuric acid.[32] The reaction is apparently only important in man and Old World monkeys, being at a low level in other primates and subprimates.[33]

The major routes of metabolism of the sulfonamides involve N-acetylation of the aromatic amino group (N^4), which occurs with all sulfonamides, and conjugation of the sulfonamido nitrogen (N^1) with glucuronic acid, this being found for only a few congeners.[2,33] This has been studied most intensively with sulfadimethoxine, and N^1-glucuronidation occurs to a significant extent in man and other primate species, with N^4-acetylation predominating in subprimate species.[34]

The O-methylation of 4-hydroxy-3,5-diiodobenzoic acid occurs in man and Old and New World monkeys, but not in the rat or the rabbit.[35]

Table IX[36] shows a zoological classification of the primates, and it is clear that the four examples of reactions restricted to these species differ greatly in their zoological distribution.

Another example of a reaction restricted to a group of species is seen in the case of the taurine conjugation of arylacetic acids. Although their taurine conjugation is a reaction of widespread occurrence, it is particularly well developed in the carnivores. The restricted occurrence of

TABLE IX

Zoological Classification of the Primates[a]

Suborder	Family	Representative species
Prosimii	Tupaiidae	Tree shrews
	Lemuridae ⎫	
	Indriidae ⎬	Malagasy lemurs
	Daubentonidae ⎭	
	Lorisidae	Lorises, galages
	Tarsiidae	Tarsiers
Anthropoidae	Callitrichidae	Marmosets ⎫ New World
	Cebidae	Monkeys ⎭ Monkeys
	Cercopithecidae	Old World monkeys
	Hylobatidae	Lesser apes
	Pongidae	Great apes
	Hominidae	Man

[a] Adapted from Smith and Caldwell.[36]

taurine conjugation of some arylacetic acids to the dog as compared with the rat is seen clearly with phenylacetic acid,[29] 1-naphthylacetic acid,[37] fenclofenac,[38] *p*-cyclopropyl(carbonyl)phenylacetic acid,[39] and isoxepac.[40]

C. Species Differences in the Relative Extents of Competing Reactions for a Given Molecule

Most xenobiotics which undergo metabolism do so by more than one pathway, either involving the same functional group or different regions of the molecule. Barring the examples listed above, it is generally true that all mammalian species have the potential to carry out all of the metabolic options open to a given molecule. However, interspecies variations commonly exist in the relative extents of the various reactions which a compound may undergo.

This may be seen with a molecule as simple as phenol, which is metabolized by conjugation of the hydroxyl group either with glucuronic acid or with sulfate. The relative extents of these two conjugation options varies very greatly between species, with man and Old World monkeys being very different from New World monkeys and subprimate mammals. This is shown in Table X,[41] which also illustrates the species defects in these two reactions in the pig (sulfation) and cat (glucuronidation), which account for the great differences between these species and the others quoted.

TABLE X

Species Variations Due to Competing Reactions—The Conjugation of Phenol[a]

| Species | Excretion (% in 24h) conjugated with | | Ratio S/G |
	Sulfate (S)	Glucuronic acid (G)	
Man and Old World monkeys	80	12	7
New World monkeys	25	50	0.5
Rat and mouse	45	40	1
Cat	93	1	93
Pig	2	95	<0.1

[a] Data taken from Capel et al.[41]

The fate of two amphetamine congeners has been examined in a representative range of species and found to exhibit considerable variation in the relative extents of competing metabolic reactions. Amphetamine itself undergoes either aromatic hydroxylation giving 4'-hydroxyamphetamine, or side chain degradation to benzoic acid.[42] Chlorphentermine, which differs from amphetamine in having a 4'-chloro substituent in the aromatic ring and a second methyl group α to the amino group, is only metabolized at the nitrogen atom. This may be by N-hydroxylation to the corresponding hydroxylamine, further oxidized to the nitro analogue, or by a conjugation. The nature of the latter reaction remains unknown, but the product is easily cleared to chlorphentermine, and is not N-acetyl- or N-succinylchlorphentermine, nor its N-sulfate or N-glucuronide.[11] The relative extents of both of the competing reactions for both compounds are very variable between species, data being listed in Table XI. In the case of amphetamine, the compound is metabolized extensively in most species, but the nature of the metabolites formed is variable. In the rat ring hydroxylation predominates, in the guinea pig chain breakdown is the major route, and in other species both routes are significant, with side chain breakdown generally more important.

With chlorphentermine, the N-conjugation reaction occurs in the rat and marmoset, N-hydroxylation in the guinea pig, rhesus monkey, and man, while both pathways are extant in the rabbit.

Although the data for these two compounds appear unrelated at first sight, it is interesting to note the close correlation between the extent of side chain breakdown of amphetamine in a given species and its ability to N-hydroxylate chlorphentermine (Table XI). There is still controversy over the relative importance of the N- and C-oxidation routes known to be involved in deamination of amphetamines and the subsequent breakdown

TABLE XI

Species Variations Due to Competing Reactions—Fate of Amphetamine and Chlorphentermine[a]

	Excreted material (%) as that metabolite					
	Amphetamine			Chlorphentermine		
	Unchanged	4′-Hydroxy	Benzoic acid	Unchanged	N-Hydroxy	Conjugate
Rat	13	82	5	65	0	35
Guinea pig	23	0	77	50	50	0
Rabbit	6	9	85	58	42	34
Marmoset	79	7	13	48	0	52
Rhesus monkey	42	16	42	24	76	0
Man	40	10	50	56	44	0

[a] Taken from Caldwell,[42] and references therein.

of the side chain.[43] These results support the view[11,44] that the N-oxidative route is by far the more important.

Meperidine is a very interesting compound to consider in the present context, since it is metabolized by five different reactions to a total of eight metabolites.[45] Its metabolism is illustrated in Fig. 1. The two major routes are N-demethylation and ester hydrolysis, giving rise to nor-meperidine and normeperidinic and meperidinic acids, and the two acids are extensively conjugated with glucuronic acid. Two oxidations, one in the aromatic ring yielding 4′-hydroxymeperidine and one on the nitrogen producing meperidine N-oxide, are minor routes. It is important to appreciate that normeperidinic acid is a secondary metabolite, arising from the hydrolysis of the N-demethylation product, normeperidine; meperidinic acid is not N-demethylated *in vivo* (Notarianni & Caldwell, unpublished data).

Meperidine metabolism has been studied in the rat, rabbit, guinea pig, vervet, patas and mona monkeys, mangabey, and man. Table XII gives data presented in terms of the individual metabolites excreted and also expressed as the proportion of the dose being metabolized by the various pathways, making allowance for the secondary metabolite and for the initial distribution of meperidine metabolism between N-demethylation and ester cleavage to meperidinic acid.[45]

In the previous section, the conjugation of various arylacetic acids with glutamine and taurine was discussed, and these compounds are worthy of consideration under the present subheading. Most arylacetic acids are

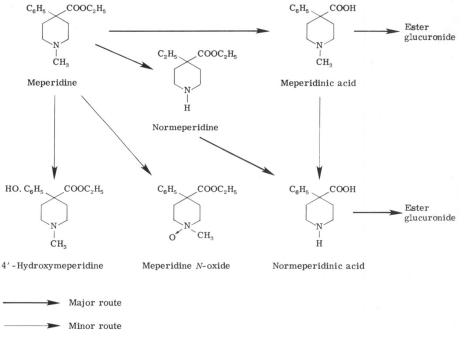

Fig. 1. Metabolic routes of meperidine in mammalian species.

metabolized by conjugation of their carboxyl group either with glucuronic acid or an amino acid.[37,46,47] Available data are listed in Table XIII from which it is clear that two factors emerge as governing the fate of a particular arylacetic acid. The structure of the acid, notably the nature of the aryl moiety, and the presence or absence of substituents on the carbon alpha to the carboxyl group, determines the relative extents of the glucuronic acid and amino acid conjugations which it may undergo. The species of mammal governs the nature of the amino acid used, with glycine used principally by rodents and lagomorphs, taurine by carnivores especially, and glutamine generally restricted to the primates. Additionally, the glucuronidation defect of the cat toward some of these acids further alters the influence of chemical structure on metabolic pattern.

III. DO "SPECIES PATTERNS" OF FOREIGN COMPOUND METABOLISM EXIST?

With our present knowledge of the metabolic pathways of xenobiotics, it may be possible to predict potential metabolites of a compound by

TABLE XII

Species Variations in the Exents of Competing Reactions in the Metabolism of Meperidine[a]

				Excretion (% in 24h) present in that form in urine of				
	Rat	Rabbit	Guinea pig	Vervet	Patas monkey	Mona monkey	Mangabey	Man
Meperidine	3	23	10	38	19	19	7	11
Normeperidine	20	7	19	10	19	30	10	17
Meperidine N-oxide	1	0.5	1	1	3	1	0.3	2
4'-Hydroxymeperidine	13	12	6	1	—	—	—	3
Meperidinic acid free	6	3	5	2	1	1	5	8
glucuronide	50	26	45	16	9	7	22	17
Normeperidinic acid free	1	5	1	3	19	9	10	12
glucuronide	6	23	23	29	30	33	46	30
% Total N-demethylation (normeperidine + normeperidinic acid)	27	35	33	42	68	72	66	59
% Initial ester hydrolysis (meperidinic acid)	56	29	50	18	10	8	27	25
% Glucuronidation of acids (total glucuronides/total acids)	88	86	91	90	76	84	82	70

[a] From Caldwell *et al.*[45]; L. J. Notarianni and J. Caldwell, unpublished data.

TABLE XIII

Variations in the Fate of Arylacetic Acids with Species and Chemical Structure[a]

$$R^1—CH—COOH$$
$$|$$
$$R^2$$

	Excretion (% in 24h) conjugated with			
Acid	Glycine	Taurine	Glutamine	Glucuronic acid
Phenylacetic acid (R^1—C_6H_5; R^2—H)				
Man	0	6	93	0
Rhesus monkey	1	23	32	0
Cat	98	1	0	0
Rabbit	97	1	0	0
Rat	99	1	0	0
1-Naphthylacetic acid (R^1—$C_{10}H_7$; R^2—H)				
Man	0	6	0	94
Rhesus monkey	0	2	2	89
Cat	59	40	0	0
Rabbit	6	0	0	87
Rat	23	1	0	51
Diphenylacetic acid (R^1R^2—C_6H_5)				
Man	0	0	0	70
Rhesus monkey	0	0	0	35
Cat	0	0	0	30
Rabbit	0	0	0	77
Rat	0	0	0	45

[a] Data from James et al.[29] and Dixon et al.[37,46]

inspection. However, it is still very hard to anticipate which routes will predominate *in vivo* in a specific animal. Such anticipation is only possible when the influence of both species and chemical structure on the metabolism of a series of closely related compounds have been examined. This has been done in only a few instances, but it is possible to draw some conclusions for three series of compounds, the amphetamines, the benzodiazepines, and the arylacetic acids, and available information will be discussed for each series.

A. Amphetamines and Related Compounds

Amphetamine (2-amino-1-phenylpropane) is the parent compound for several series of interesting drugs, including anorectics and CNS stimulants. The general structure of these compounds is shown in Fig. 2, and the structures and relative lipid solubilities of those congeners discussed

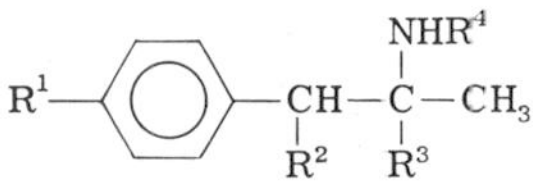

$$R^1-\underset{R^2}{\overset{}{\bigcirc}}-\underset{R^2}{\overset{NHR^4}{CH}}-\underset{R^3}{\overset{|}{C}}-CH_3$$

Fig. 2.   General structure of the amphetamines. R^1 to R^4 refer to Table XIV.

here are given in Table XIV. Their structure would indicate a number of metabolic options, including aromatic ring hydroxylation, aliphatic carbon oxidation of the side chain, N-oxidation, deamination, N-dealkylation (if *N*-alkyl substituents are present), and N-conjugation. Of these possibilities, only oxidation of the terminal side chain carbon does not occur, and it appears that N-oxidation and hydroxylation of the carbon α- to the amino group are both involved in the deamination process. The α-amino alcohols are sufficiently unstable to preclude their isolation as such, while the N-oxidative route is seen only when deamination is prevented by the absence of a proton on the α-carbon, as in the case of the phentermines, in which this proton is replaced by a methyl group. The pathways of the metabolism of the amphetamines have been reviewed extensively elsewhere.[42,44,48]

The fate of seven amphetamine congeners has been examined in the rat; four in the guinea pig; five in the rabbit; and four each in the dog, in the marmoset, and in man. It is clear that species patterns of metabolism do exist for this series of compounds, with aromatic hydroxylation predominating in the rat, side chain breakdown being the major route in the guinea pig, and mixed patterns occurring in the other species. The relative lipid solubilities of the compounds have an influence of their fate, with the

TABLE XIV

Structures and Relative Lipid Solubilities of Amphetamine Congeners[a]

	R^1	R^2	R^3	R^4	Relative lipid solubility[c]
Amphetamine	H	H	H	H	1
Methamphetamine	H	H	H	CH_3	2.3
Mephentermine	H	H	CH_3	CH_3	2.5
Chlorphentermine	Cl	H	CH_3	H	8.3
Norephedrine	H	OH	H	H	0.002
Ephedrine	H	OH	H	CH_3	0.03
Phenmetrazine[b]	H	O⟍ CH_2	H	CH_2⟋	32.5

[a] See structure on Fig. 2 for R^1, R^2, R^3, and R^4.
[b] Phenmetrazine is 3-methyl-2-phenylmorpholine.
[c] Taken from Caldwell.[43]

more water-soluble compounds (notably the ephedrines) being less extensively metabolized, although this is more apparent in man than in animal
species. The species patterns are summarized in Table XV and appear to
provide a useful predictive guide for the fate of new compounds in this
series. However, when structural modification of the basic backbone
renders impossible the preferred metabolic routes of a species, new options may reveal themselves. This is seen in the case of chlorphentermine
in the rat and marmoset. The structure of this molecule prevents aromatic
hydroxylation, and leaves N-oxidation as the only route of metabolism in
most species.[11,44] Both the rat and marmoset are deficient in N-oxidation
of aliphatic amines (see Section II,A) and since aromatic hydroxylation is
obviated, an apparently novel N-conjugation occurs.[11]

B. The Benzodiazepines

Over the last 10 to 15 years there has been intense interest in these CNS
active drugs, and, as a result, there has built up considerable knowledge
of their fate in man and animals.[49] Metabolism of the benzodiazepines is
generally complex, since many products arise from more than one reaction, and, additionally, many congeners are related by metabolism, either
having common metabolites or giving metabolites that are drugs in their
own right. Figure 3 shows the general structure of the 1,4-
benzodiazepines, and indicates the major routes of metabolism, namely,
N^1-demethylation, deamination or oxidation of C-2, hydroxylation at C-3,
N^4-oxidation (or N^4-oxide reduction), hydroxylation of the 5-phenyl substituent, and ring opening of the benzodiazepine ring between N^1 and C-2,

TABLE XV

Summary of Species Variations in the Metabolism of Amphetamines[a]

	Relative extent of pathway in each species			
Species	Excretion of unchanged drug	N-Dealkylation	4'-Hydroxylation	Side-chain breakdown
Rat	++	+++	++++	+
Guinea pig	++	++	−	++++
Rabbit	+	++	+	++++
Dog	+++	+++	++	++
Marmoset	++++	−	+	+
Man	+++	+	++	+++

[a] Drawn from Caldwell.[42]

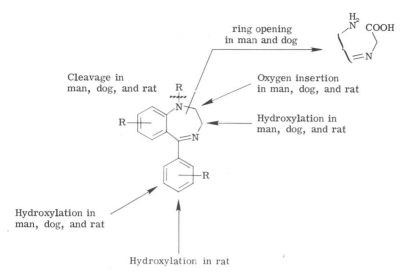

Fig. 3. General structure and main routes of metabolism of the 1,4-benzodiazepines.

which only occurs with compounds with a keto group at C-2. The fate of six benzodiazepines will be considered in man, dog, and rat, and the structures of these compounds are given in Fig. 4.

Diazepam (**I**) is extensively metabolized in man, dog and rat. In man, the major metabolites are 3-hydroxydiazepam, N^1-desmethyldiazepam and 3-hydroxydesmethyldiazepam (oxazepam), while in dog, oxazepam predominates. In the rat, aromatic hydroxylation is the major single metabolic route, with 4'-hydroxy-3- hydroxydiazepam, 4'-hydroxy-N^1-desmethyldiazepam, and 4'-hydroxyoxazepam all being formed, and some 3-hydroxydiazepam is also produced.[50-52]

Chlordiazepoxide (**II**) shows a similar pattern of metabolism to that seen above, with transformations of the benzodiazepine ring occurring in the dog and man, but with extensive metabolism of the aromatic ring occurring in the rat. In man and dog, the major reaction is removal of the N-methyl group at C-2, leading to the lactam and the corresponding ring opened compound (an ω-amino acid). In the rat, the main products are 4'-hydroxychlordiazepoxide, 4'-hydroxy- N-desmethylchlordiazepoxide, the 4'-hydroxylactam, and 4'-hydroxy-N-desmethyldiazepam, the formation of which involves reduction of the N-oxide function.[53,54]

Oxazepam (**III**) has a far simpler fate. In man and dog, the only metabolite is its O-glucuronide at the 3-hydroxy group. In the rat, several metabolites are formed as a result of hydroxylation of the 5-phenyl group.[55]

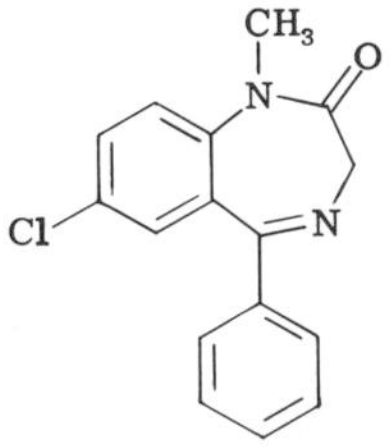

Fig. 4. Structures of the six benzodiazepines discussed in the text. **I**, Diazepam, **II**, chlordiazepoxide; **III**, oxazepam; **IV**, lorazepam; **V**, medazepam; **VI**, bromazepam.

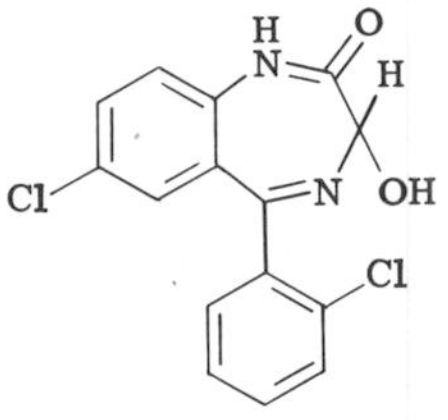

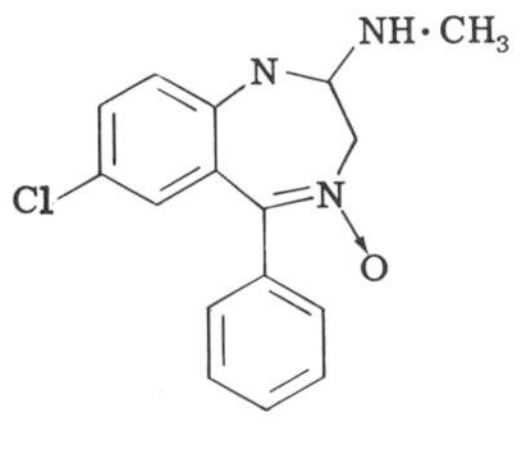

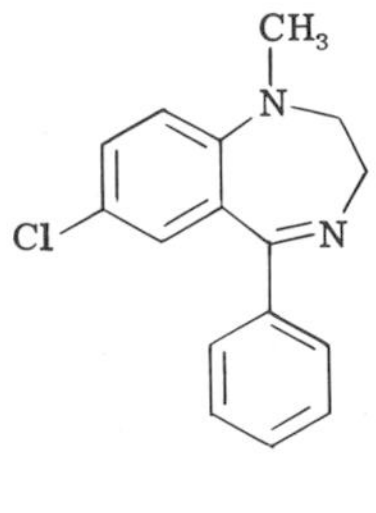

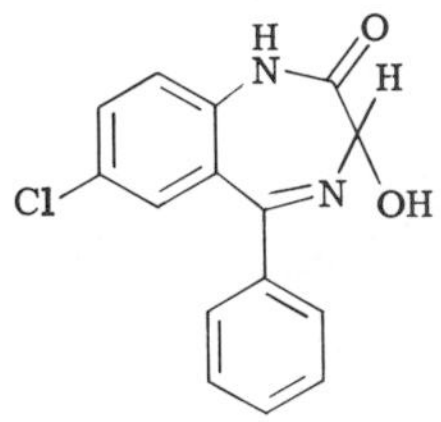

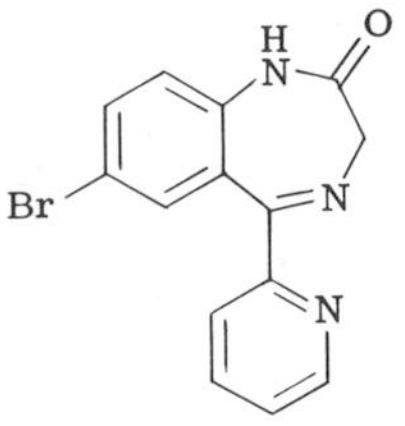

Lorazepam (**IV**) differs from oxazepam only in the incorporation of a 2′-chloro substituent in the 5-phenyl ring, and the fate of the two drugs is very similar. Thus, lorazepam is conjugated with glucuronic acid through its 3-hydroxy group in dog and man, while in the rat the major metabolites are 3′- and 4′hydroxylorazepam.[56]

Medazepam (**V**) undergoes N-demethylation, 3-hydroxylation, lactam formation, and ring opening in man. In dog, all four of these routes are evident with oxazepam, the product of N-demethylation, 3-hydroxylation and lactam formation, as the major metabolite and two ring-opened products as minor metabolites. However, in the rat the main excretion products are the result of aromatic hydroxylation and lactam formation, namely, 4'hydroxydiazepam and 4'-hydroxy-3-hydroxydiazepam.[7]

The fate of bromazepam (VI) in man and dog involves 3-hydroxylation and subsequent opening of the benzodiazepine ring. The rat also forms these metabolites, and additional products in this species include 4'-hydroxybromazepam and a ring-opened compound arising directly from bromazepam.[58,59]

The results for each of these compounds are presented in Table XVI, in terms of the metabolic routes used by each species. From this, it is clear that broad species patterns of metabolism exist for this series of compounds. The dog and man show close similarities, with metabolism in the benzodiazepine ring. In the rat, with every compound examined, there is extensive aromatic hydroxylation of the 5-phenyl group in addition to metabolism in the benzodiazepine ring. However, the routes used for this latter conversion by the rat are frequently different from those employed by the dog and man. The data summarized in Table XVI do provide a general guide to the metabolic routes that might occur in these three species for a novel benzodiazepine.

C. Arylacetic Acids

The arylacetic acids are of contemporary significance since many of them are antiinflammatory agents, by virtue of inhibition of prostaglandin synthesis. The fate of a number of arylacetic acid congeners has been studied in animals and man, and it is possible to deduce the existence of species patterns of metabolism for these compounds. In order to see these clearly, it is first necessary to classify the chemical nature of the substrate, by the nature of the aryl moiety and by the presence or absence of substitution on the carbon alpha to the carboxyl group.[47] The compounds which have been studied thus far fall into five groups as follows: (1) single ring; (2) single ring with substituent(s); (3) twin fused rings; (4) twin fused rings with substituent(s); (5) triple fused rings, and these have been subdivided into subgroup A with protons on the α-carbon and subgroup B with α-carbon substitution.

The metabolic options of the arylacetic acids involve oxidation, reduction, or hydrolysis of the aryl moiety and conjugation of the carboxyl group with glucuronic acid or an amino acid, as has been discussed earlier in Sections II,B and C.

It is inappropriate here to give full details of the fates of the arylacetic acids used to make up this classification, since these have been fully described.[47,60] Table XVII summarizes the fate of some 15 arylacetic

TABLE XVI

Summary of Species Differences in the Routes of Metabolism of the Benzodiazepines[a,b]

Drug	N¹-Dealkylation	C-2 Lactam formation	C-3 Hydroxylation	Ring opening	5-Phenyl hydroxylation
Diazepam (I)					
Man	+	—	+	0	0
Dog	+	—	+	0	0
Rat	+	—	+	0	+
Chlordiazepoxide (II)					
Man	—	+	0	+	0
Dog	—	+	0	+	0
Rat	—	+	+	0	+
Oxazepam (III)					
Man	—	—	—[c]	0	0
Dog	—	—	—[c]	0	0
Rat	—	—	—[c]	0	+
Lorazepam (IV)					
Man	—	—	—[c]	0	0
Dog	—	—	—[c]	0	0
Rat	—	—	—[c]	0	+
Medazepam (V)					
Man	+	+	+	+	0
Dog	+	+	+	+	0
Rat	0	0	+	0	+
Bromazepam (VI)					
Man	—	—	+	+	0
Dog	—	—	+	+	0
Rat	—	—	+	+	+

[a] Data drawn from references quoted in Section III,B.

[b] + indicates occurrence of that route, — that the structure prevents that route occurring, and 0 that the route does not occur.

[c] These compounds excreted as their 3-O-glucuronides.

TABLE XVII

Summary of Species Patterns of Metabolism of the Arylacetic Acids

Group	Species	Metabolic route[a] Subgroup A Phase I	Gly	Tau	Gln	GA	Phase I[b]	Subgroup B Gly	Tau	Gln	GA
1	Rodents	−	++++	+	−	−	−	++	−	−	+++
	Carnivores	−	+++	++	−	−	−	++	++	−	+
	Primates	−	+	+	++++	−	−	+	+	+	+++
2	Rodents	++	+++	−	−	+++	++	+	−	−	+++
	Carnivores	++	++	+++	−	++	++	+	++	−	++
	Primates	++	+	−	++	++	++	−	−	−	+++
3	Rodents	−	+	−	−	++++	−	−	−	−	++++
	Carnivores	−	++	++	−	++	−	−	−	−	++++
	Primates	−	+	−	+	++++	−	−	−	−	++++
4	Rodents	++	−	−	−	++++	−	−	−	−	++
	Carnivores	++	−	−	−	++++	−	−	−	−	++
	Primates	++	−	−	−	++++	−	−	−	−	++
5	Rodents	+++	−	−	−	+++	+++	−	−	−	+++
	Carnivores	++	−	−	−	+++	++	−	−	−	+++
	Primates	+++	−	−	−	+++	+++	−	−	−	+++

[a] The number of pluses indicates the extent of that metabolic route. Drawn from Caldwell,[47,60] and references therein. Phase I represents all possible oxidative, hydrolytic, and reductive reactions; Gly, glycine conjugation; Tau, taurine conjugation; Gln, glutamine conjugation; GA, glucuronic acid conjugation. Groups and subgroups of acids as explained in the text.

acids, classified by their structure as above and dividing their metabolism into the Phase I options open to them, their amino acid conjugation and their glucuronic acid conjugation. It is clear from this that structure and species patterns for their metabolism exist. Thus, Phase I metabolism is only seen with acids in groups 2A, 2B, 4A and 5A, in rodents, carnivores, and primates. Amino acid conjugation is seen with groups 1A, 2A, and 3A, and the major pathways involve glycine in rodents, taurine with some glycine in carnivores, and glutamine in man. This metabolic option is not seen in rodents and primates with any subgroup B substrates, but in carnivores amino acid conjugation occurs with group 1B acids and is more extensive for group 2A and 3A acids than in other species.

Glucuronic acid conjugation is not seen with the group 1A acid nor with some group 2A acids with small substituents (e.g., 4'-chloro, 4'-nitro). However, with groups 2A, 3A, 4A, and 5A glucuronidation becomes progressively more important and occurs with all subgroup B acids it is seen in rodents and primates. Since the carnivores have more versatile amino acid conjugation mechanisms, glucuronidation is quantitatively less important in these species with the smaller acids (groups 1B, 2A, 2B, 3A), and this is especially true in the cat (see Section II,A). However, with larger acids (some of groups 2A and 3A, and all of 4A and 5A) extensive glucuronidation occurs in carnivores as well as the other species.

IV. THE SIGNIFICANCE OF THE COMPARATIVE BIOCHEMISTRY OF FOREIGN COMPOUNDS FOR ANIMAL TAXONOMY

In 1976, Williams[61] coined the term pharmacotaxonomy for the possible contributions which a knowledge of the drug metabolizing characteristics of a species might make to the overall problem of its zoological classification, and it is of some interest to examine whether or not such contributions can be made at present.

It has been obvious for some time that there is generally no evolutionary basis behind the particular drug metabolizing ability of a particular species. Indeed, among rodents and primates, zoologically closely related species exhibit markedly different patterns of metabolism, some of which have been illustrated earlier, e.g., the metabolism of amphetamines in the rat is very different from that in the guinea pig (Section III,C), and the marmoset is very different from the rhesus monkey.[42,62] However, in the study of species defects in drug metabolism some evolutionary influences have been discerned, and it is of interest to see whether these defects can act as biochemical markers of families or superfamilies.

Within the order Carnivora there exist two very profound species defects in the metabolism of foreign compounds, namely, of the glucuronidation in the cat and of N-acetylation in the dog. Table XVIII[63] gives the zoological classification of the carnivores, and it will be noted that the domestic cat (family Felidae, superfamily Feloidea) lies at the opposite end of the order to the domestic dog (family Canidae, superfamily Canoidea). The zoological distribution of these defects have been explored with four substrates, phenol, benzoic acid, and 1-naphthylacetic acid (for glucuronidation) and sulfadimethoxine (for N-acetylation), and the results are presented in Table XIX.

Table XIX shows clearly that the defect of the glucuronidation of phenol, benzoic acid, and 1-naphthylacetic acid in the cat extends to other cat-like species, the lion and lynx (both Felidae) and the civet and genet (both Viverridae). The hyena (Hyaenidae) does not excrete phenylglucuronide but does form 1-naphthylacetylglucuronide, while the ferret (Mustelidae) and dog both form all three glucuronides. A reverse of the above pattern is seen with N-acetylation, with N^4-acetylsulfadimethoxine being formed in the cat, lion, civet, genet, and ferret, but not in the dog or hyena. It is thus apparent that the glucuronidation defect of the cat is shared by some other members of the superfamily Feloidea, but not by the hyena, while the N-acetylation defect of the dog is shared by one member of the Feloidea, the hyena. The hyena, therefore, has two biochemically dog-like characteristics, some capacity for glucuronidation and an absence of N-acetylation.

In Table XVIII, the Hyaenidae have been placed in the superfamily Feloidea, but there is some dispute among taxonomists as to whether they

TABLE XVIII

Zoological Classification of the Carnivores Order: Carnivora[a]

Superfamily	Family	No. of species	Examples
Canoidea	Canidae	36	Dog, fox
	Ursidae	7	Bears
	Procyonidae	17	Racoons, pandas
	Mustelidae	68	Ferrets, badgers
Feloidea	Viverridae	72	Civet, genet
	Hyaenidae	4	Hyena
	Felidae	38	Cat, lion, lynx

[a] Based on Ewer.[63]

TABLE XIX

Conjugation of Benzoic Acid, Phenol, 1-Naphthylacetic Acid, and Sulfadimethoxine in Carnivores[a,b]

	Excreted material (%) as that conjugate								
	Benzoic acid		Phenol		1-Naphthylacetic acid			Sulfadimethoxine	
Species	Gly	GA	S	GA	Gly	Tau	GA	N^4-Ac	N^1-GA
Cat	100	0	97	0	59	39	0	18	0
Lion	100	0	97	0	94	0	0	48	0
Lynx	—	—	87	0	—	—	—	—	—
Civet	86	0	99	0	74	6	0	66	0
Genet	71	0	97	0	70	18	0	50	0
Ferret	22	70	58	40	5	49	20	27	0
Hyena	—	—	90	0	46	11	40	0	4
Dog	18	82	82	18	7	65	26	0	19

[a] Drawn from Caldwell et al.[16]

[b] Gly, glycine conjugate; S, sulfate; GA, glucuronide; Tau, taurine conjugate; N^4-Ac, N^4-acetyl conjugate.

should more correctly be placed among the Canoidea[63]; the results of these studies suggest the latter.

V. THE ENZYMATIC BASIS OF SPECIES DIFFERENCES IN FOREIGN COMPOUND METABOLISM

Allowing for the caveats expressed in Section I, it is to be expected that species variations in drug metabolism are a reflection of differences in the activities of the enzymes responsible for the various transformations. Such variations could arise from differences in (a) the absolute activities of the enzymes, (b) the amounts of any endogenous inhibitors present, or (c) the extents of any reverse reactions which might occur.

It must be stated at the outset that the presumed enzymatic basis of species differences in drug metabolism is very poorly understood. This is principally due to the fact that species differences are generally studied *in vivo,* while enzyme studies must of necessity involve *in vitro* work. Unfortunately, the data obtained *in vitro* need bear little relevance to the *in vivo* situation. Thus, Litterst et al.[64] have studied the hepatic microsomal metabolism of model substrates by preparations from rhesus monkey, squirrel monkey, tree shrew, pig, and rat and showed only slight differences among them. Since it is clear from the data quoted previously that enormous differences exist in the fate of foreign compounds in these species *in vivo*, it is important to consider the difficulties in extrapolation from *in vivo* to *in vitro* studies.

Most *in vitro* studies use a single subcellular fraction, frequently the microsomes, from homogenates of liver, fortified with cofactors appropriate to the reaction being studied. Usually only one reaction is studied. However, in order to be metabolized *in vivo*, compounds must pass several membranes to reach the metabolizing enzymes, and in only a few cases are only one set of enzymes involved. Although the oxidation of foreign compounds occurs principally (but not exclusively) in the microsomes, as does the hydration of epoxides and glucuronic acid conjugation, many reductases are present in the cytosol, as are the sulfate conjugating enzymes. Amino acid conjugation occurs in the mitochondria. Additionally in the whole animal many organs other than the liver can contribute to metabolism, sometimes catalyzing reactions that cannot occur in the liver, e.g., dog kidney can conjugate benzoic acid with glycine while the liver cannot[60] and the gut flora can perform several reactions not carried out by the tissues.[65] The preparation of subcellular fractions by homogenization and centrifugation may also change the properties of cell membranes and their associated enzymes. When all of these factors are considered, it should not be surprising that *in vitro* data is rarely directly applicable to events *in vivo*.

Nevertheless it is sometimes possible to attribute *in vivo* differences between species in the metabolism of xenobiotics to differences in the nature of the metabolizing enzymes present. Clearly, the species defects in metabolism are most amenable to such investigation. The glucuronidation defect of the cat has been illustrated earlier (Section II,A and Table V), and many studies point to this being due to the absence of a form of glucuronyltransferase responsible for the conjugation of low molecular weight compounds from the endoplasmic reticulum of cat liver.[17] In most species, 1-naphthol is conjugated with glucuronic acid and sulfate in approximately equal amounts, but only with sulfate in the cat. This is a reflection of the properties of the two conjugating enzyme systems in the livers of cat and rat, with the K_m for sulfation of 1-naphthol in cat liver being lower than in the rat and the reverse occurring for glucuronidation.

Since the majority of species differences in the metabolism of foreign compounds arise from differences in the relative extents of competing reactions for the compound, this provides a situation which is obviously harder to study *in vitro* than are species defects. Most *in vitro* studies, as mentioned, follow a given reaction in a single organelle, whereas a pattern of competing reactions will involve several options, probably in more than one organelle and possibly in more than one organ. With conjugation reactions, it is still possible to examine the enzymatic basis of this source of variation, as illustrated by Heirwegh's work on bilirubin-IXα conjugation. This degradation product of heme can be conjugated by microsomal preparations with glucose or glucuronic acid, and the ratio of glucoside to glucuronide formed by a species correlate with the ratio of glucosyltransferase to glucuronyltransferase present in the microsomes.[66]

Examples such as those quoted are deceptively simple, however. Among the metabolic reactions of interest to many investigators are the Phase I processes, notably the oxidations. While these reactions are mostly carried out by the cytochrome P-450 system of the microsomes and have the same cofactor requirements, many compounds have more than one oxidation option open to them. It is currently thought that these are carried out by different discrete forms of cytochrome P-450[67] but although several of these have been separated (see chapter 6, this volume), the interpretation of their significance for *in vivo* patterns of metabolism will require much more work.

Smith[68] has opined that species differences in the pattern of metabolism due to competing reactions are relatively unpredictable and are likely to remain so. In the writer's opinion, this is not necessarily the case. If foreign compounds can be characterized chemically as to their structures and their metabolic pathways, and if various animal species can be characterized in terms of the nature and amounts of each of the enzymes of drug metabolism they possess, then it may be possible to correlate this information and predict the metabolism of novel compounds in distinct species.

VI. COMMENTS

The overall pattern of drug metabolism in mammals is similar despite the very considerable species variations within that pattern. The origins of this variation have been classified and illustrated, and data presented to show that species patterns of metabolism emerge when series of related compounds are evaluated. The difficulty in interpreting many studies on drug metabolism in different animals is due to the fact that these frequently involve compounds which cannot readily be related chemically to others investigated in the past. From the results of studies in carnivorous species, it is seen that the examination of the comparative biochemistry of foreign compounds can contribute to animal taxonomy, a field that may well expand. The problems of the enzymatic basis of species variations in foreign compound metabolism at present are very great, but it is anticipated that a fuller understanding of this basis will be a key to the prediction of drug metabolism patterns in animals and man.

REFERENCES

1. Williams, R. T. (1947). "Detoxication Mechanisms," Chapman & Hall, London.
2. Williams, R. T. (1959). "Detoxication Mechanisms," 2nd ed. Chapman & Hall, London.
3. Jollow, D. J., Kocsis, J. J., Snyder, R., and Vainio, H., eds. (1977). "Biological Reactive Intermediates." Plenum, New York.
4. Caldwell, J. (1980). The conjugation reactions and their significance in biochemical pharmacology and toxicology. In "Concepts in Drug Metabolism" (P. Jenner and B. Testa, eds.). Dekker, New York (in press).
5. Williams, R. T. (1967). Comparative patterns of drug metabolism. Fed. Proc., Fed. Am. Soc. Exp. Biol. 26, 1029–1039.
6. Young, J. Z. (1962). "The Life of Vertebrates," 2nd ed. Oxford Univ. Press, London and New York.
7. Friedman, M., and Byers, S. O. (1948). Observations concerning the causes of excess excretion of uric acid in the dalmatian dog. J. Biol. Chem. 175, 727–735.
8. Yu, T. S., Berger, L., Kupfer, S., and Gutman, A. B. (1960). Tubular secretion of urate in the dog. Am. J. Physiol. 199, 1199–1204.
9. Williams, R. T. (1974). Inter-species variations in the metabolism of xenobiotics. The 8th CIBA Medal lecture. Biochem. Soc. Trans. 2, 359–377.
10. Parke, D. V., and Smith, R. L., eds. (1976). "Drug Metabolism from Microbe to Man," Taylor & Francis, London.
11. Caldwell, J., Köster, U., Smith, R. L., and Williams, R. T. (1975). Species variations in the N-oxidation of chlorphentermine. Biochem. Pharmacol. 24, 2225–2232.
12. Weisburger, J. H., Grantham, P. H., Vanhorn, E., Steigbigel, N. H., Rall, D. P., and Weisburger, E. K. (1964). Activation and detoxification of N-2-fluorenylacetamide in man. Cancer Res. 24, 475–487.
13. Hirom, P. C., Idle, J. R., and Millburn, P. (1976). Comparative aspects of the biosynthesis and excretion of xenobiotic conjugates by non-primate mammals. In "Drug Metabolism from Microbe to Man" (D. V. Parke and R. L. Smith, eds.), pp. 299–329. Taylor & Francis, London.
14. Bababunmi, E. A., Smith, R. L., and Williams, R. T. (1973). The absence of hippuric acid synthesis in the Indian fruit bat. Life Sci. 12, 317–326.
15. Bray, H. G., Franklin, T. J., and James, S. P. (1959). The formation of mercapturic acids. 3. N-Acetylation of S-substituted cysteines in the rabbit, rat and guinea pig. Biochem. J. 73, 465–473.
16. Caldwell, J., Williams, R. T., Bassir, O., and French, M. R. (1977). Drug metabolism in "exotic" animals. Eur. J. Drug Metab. Pharmacokinet. 3, 61–66.
17. Dutton, G. J., Wishart, G. J., Leakey, J. E. A., and Goheer, M. A. (1976). Conjugation with glucuronic acid and other sugars. In "Drug Metabolism from Microbe to Man" (D. V. Parke and R. L. Smith, eds.), pp. 71–90. Taylor & Francis, London.
18. Weber, W. W. (1971). Acetylating, deacetylating and amino acid conjugating enzymes In "Handbuch der experimentellen Pharmakologie" (O. Eichler, ed.), Vol. 28, Part 2, pp. 564–583. Springer-Verlag, Berlin and New York.
19. Idle, J. R. (1976). Species differences in the conjugation of some aromatic acids. Ph.D. Thesis, University of London.
20. Kibby, M. R. (1966). The metabolism of some heterocyclic sulphonamide drugs in man and other animals. Ph.D. Thesis, University of London.
21. Peters, J. H. (1960). Metabolism of isoniazid. Am. Rev. Respir. Dis. 81, 485–497.
22. Miller, J. A., Wyatt, C. A., Miller, E. C., and Hartmann, H. A. (1961). The

N-hydroxylation of 4-acetylaminobiphenyl by the rat and dog and the strong carcinogenicity of *N*-hydroxy-4-acetylaminobiphenyl in the rat. *Cancer Res.* **21,** 1465–1473.

23. Klaasen, C. D., and Fitzgerland, T. J. (1974). Metabolism and biliary excretion of ethacrynic acid. *J. Pharmacol. Exp. Ther.* **191,** 548–552.

24. Bridges, J. W., and Williams, R. T. (1963). Species differences in the acetylation of sulphamilamide. *Biochem. J.* **87,** 19P.

25. Corner, E. D. S., and Young, L. (1954). Biochemical studies of toxic agents. 7. The metabolism of naphthalene in animals of different species. *Biochem. J.* **58,** 647–655.

26. Jerina, D., and Bend, J. R. (1977). Glutathione S-transferases. *In* "Biological Reactive Intermediates" (D. J. Jollow, J. J. Kocsis, R. Snyder, and H. Vainio, eds.), pp. 207–232. Plenum, New York.

27. Capel, I. D., Millburn, P., and Williams, R. T. (1974). The conjugation of 1- and 2-naphthols and other phenols in the cat and pig. *Xenobiotica* **4,** 601–610.

28. Smith, R. L., and Timbrell, J. A. (1974). Factors affecting the metabolism of phenacetin. I. Influence of dose, chronic dosage, route of administration and species on the metabolism of [1-¹⁴C-acetyl]phenacetin. *Xenobiotica* **4,** 489–497.

29. James, M. P., Smith, R. L., Williams, R. T., and Reidenberg, M. (1972). The conjugation of phenylacetic acid in man, subhuman primates and some non-primate species. *Proc. Ry. Soc. London, Ser. B* **182,** 25–35.

30. Emudianughe, T. S., Caldwell, J., and Smith, R. L. (1977). Structure-metabolism relationships of arylacetic acids: The metabolic fate of 2-naphthylacetic acid *in vivo*. *Biochem. Soc. Trans.* **5,** 1006–1008.

31. Emudianughe, T. S., Caldwell, J., Dixon, P. A. F., and Smith, R. L. (1978). Studies on the metabolism of arylacetic acids. 5. The metabolic fate of 2-naphthylacetic acid in the rat, rabbit and ferret. *Xenobiotica* **8,** 525–534.

32. Adamson, R. H., Bridges, J. W., Evans, M. F., and Williams, R. T. (1970). Species differences in the aromatization of quinic acid *in vivo* and the role of gut bacteria. *Biochem. J.* **116,** 437–441.

33. Bridges, J. W., Kibby, M. R., and Williams, R. T. (1965). The structure of the glucuronide of sulphadimethoxine formed in man. *Biochem. J.* **96,** 829–836.

34. Adamson, R. H., Bridges, J. W., Kibby, M. R., Walker, S. R., and Williams, R. T. (1970). The fate of sulphadimethoxine in primates compared with other species. *Biochem. J.* **118,** 41–46.

35. Wold, J. S., Smith, R. L., and Williams, R. T. (1973). Species variations in the *O*-methylation of *n*-butyl 4-hydroxy-3,5-diiodobenzoate. *Biochem. Pharmacol.* **22,** 1865–1871.

36. Smith R. L., and Caldwell, J. (1976). Drug metabolism in nonhuman primates. *In* "Drug Metabolism from Microbe to Man" (D. V. Parke and R. L. Smith, eds.), pp. 331–356. Taylor & Francis, London.

37. Dixon, P. A. F., Caldwell, J., and Smith, R. L. (1977). Studies on the metabolism of arylacetic acids. 1. The metabolic fate of 1-naphthylacetic acid and its variation with species and dose. *Xenobiotica* **7,** 695–706.

38. Jordan, B. J., and Rance, M. J. (1974). Taurine conjugation of fenclofenac in the dog. *J. Pharm. Pharmacol.* **26,** 359–360.

39. Lan, S. J., El-Hawey, A. M., Dean, A. V., and Schreiber, E. C. (1975). Metabolism of *p*-(cyclopropylcarbonyl) phenylacetic acid (SQ 20, 650): Species differences. *Drug Metab. Dispos.* **3,** 171–179.

40. Illing, H. P. A., and Fromson, J. M. (1978). Species differences in the disposition and metabolism of 6, 11-dihydro-11-oxodibenz(*be*)oxepin-2-acetic acid (Isoxepac) in rat, rabbit, dog, rhesus monkey and man. *Drug Metab. Dispos.* **6,** 510–517.

41. Capel, I. D., French, M. R., Millburn, P., Smith, R. L., and Williams, R. T. (1972). The fate of [^{14}C]phenol in various species. *Xenobiotica* **2**, 25–31.

42. Caldwell, J. (1976). The metabolism of amphetamines in mammals. *Drug Metab. Rev.* **4**, 219–280.

43. Caldwell, J. (1978). Proceedings of round-table discussion on mechanisms of deamination (Chairman's report). *In* "Biological Oxidation of Nitrogen" (J. W. Gorrod, ed.), pp. 495–500.

44. Caldwell, J. (1978). Comparative biochemistry of *N*-oxidation of aliphatic amines: studies with chlorphentermine. *In* "Biological Oxidation of Nitrogen" (J. W. Gorrod, ed.), pp. 57–64. Elsevier, Amsterdam.

45. Caldwell, J., Notarianni, L. J., Smith, R. L., Fafunso, M. A., French, M. R., Dawson, P., and Bassir, O. (1979). Non-human primate species as metabolic models for the human situation: Comparative studies on meperidine metabolism. *Toxicol. Appl. Pharmacol.* **48**, 273–278.

46. Dixon, P. A. F., Caldwell, J., and Smith, R. L. (1977). Studies on the metabolism of arylacetic acids. 3. The metabolic fate of diphenylacetic acid and its variations with species and dose. *Xenobiotica* **7**, 717–726.

47. Caldwell, J. (1978). Structure-metabolism relationships in the amino acid conjugations. *In* "Conjugation Reactions in Drug Biotransformation" (A. Aitio, ed.), pp. 111–120. Elsevier, Amsterdam.

48. Cho, A. K., Sum, C. Y., Jonsson, J., and Lindeke, B. (1978). The role of N-hydroxylation in the metabolism of phentermine and amphetamine by liver preparations. *In* "Biological Oxidation of Nitrogen" (J. W. Gorrod, ed.), pp. 15–23. Elsevier, Amsterdam.

49. Dring, L. G. (1976). Species variations in pre-conjugation reactions of non-primate mammals. *In* "Drug Metabolism from Microbe to Man" (D. V. Parke and R. L. Smith, eds.), pp. 281–298. Taylor & Francis, London.

50. Schwartz, M. A., Koechlin, B. A., Postma, E., Palmer, S., and Krol, G. (1965). Metabolism of diazepam in rat, dog and man. *J. Pharmacol. Exp. Ther.* **149**, 423–429.

51. Schwartz, M., Bommer, P., and Vane, F. M. (1967). Diazepam metabolites in rat: Characterization by high resolution mass spectrometry and nuclear magnetic resonance. *Arch. Biochem. Biophys.* **121**, 508–511.

52. Ruelius, H., Lee, J. M., and Alburn, H. E. (1965). Metabolism of diazepam by dogs: Transformation to oxazepam. *Arch. Biochem. Biophys.* **111**, 376–381.

53. Koechlin, B. A., Schwartz, M. A., Krol, G., and Oberhausli, W. (1965). The metabolic fate of ^{14}C-chlordiazepoxide in man, in the dog and in the rat. *J. Pharmacol. Exp. Ther.* **148**, 399–405.

54. Schwartz, M. A., Vane, F. M., and Postma, E. (1968). Chlordiazepoxide metabolites in the rat. Characterization by high resolution mass spectrometry. *Biochem. Pharmacol.* **17**, 695–704.

55. Walkenstein, S. S., Wiser, R., Gudmunsen, C. H., Kimmel, H. B., and Corradino, R. A. (1964). Absorption, metabolism and excretion of oxazepam and its succinate half ester. *J. Pharm. Sci.* **53**, 1181–1184.

56. Schillings, R. T., Schrader, S. R., and Ruelius, H. W. (1971). Urinary metabolites of 7-chloro-5-(*o*-chlorophenyl)-1,3-dihydro-3-hydroxy-2*H*-1,4-benzodiazepin-2-one (lorazepam) in humans and four animal species. *Arzneim.-Forsch.* **21**, 1059–1963.

57. Schwartz, M. A., and Carbone, J. J. (1970). Metabolism of ^{14}C-medazepam hydrochloride in dog, rat and man. *Biochem. Pharmacol.* **19**, 343–353.

58. Schwartz, M. A. (1973). Pathways of metabolism of the benzodiazepines. *In* "The

Benzodiazepines'' (S. Garattini, E. Mussini, and L. O. Randall, eds.), pp. 53–74. Raven, New York.

59. Sawada, S., and Hara, A. (1975). Studies on the metabolism of bromazepam. Identification of new urinary metabolites and their excretion pattern in various animal species. *Yakugaku Zasshi* **95**, 430–437.

60. Caldwell, J., Idle, J. R., and Smith, R. L. (1980). The amino acid conjugations. *In* ''The Extrahepatic Metabolism of Drugs and Other Xenobiotics'' (T. E. Gram, ed.). Spectrum Publications, New York (in press).

61. Williams, R. T. (1976). Future developments. *In* ''Drug Metabolism from Microbe to Man'' (D. V. Parke and R. L. Smith, eds.), pp. 433–435. Taylor & Francis, London.

62. Caldwell, J., Dring, L. G., Franklin, R. B., Köster, U., Smith, R. L., and Williams, R. T. (1977). Comparative metabolism of the amphetamine drugs of dependence in man and monkeys. *J. Med. Primatol.* **6**, 367–375.

63. Ewer, R. F. (1973). ''The Carnivores.'' Weidenfeld & Nicholson, London.

64. Litterst, C. L., Gram, T. E., Mimnaugh, E. G., Leber, P., Emmerling, B., and Freudenthal, R. L. (1976). A comprehensive study of *in vitro* drug metabolism in several laboratory species. *Drug Metab. Dispos.* **4**, 203–207.

65. Caldwell, J., and Smith, R. L. (1977). Metabolism of drugs and the route of administration. *In* ''Formulation and Preparation of Dosage Forms'' (J. Polderman, ed.), pp. 169–180. Elsevier, Amsterdam.

66. Heirwegh, K. P. M. (1978). Formation, metabolism and significance of bilirubin. IX. Glycosides. *In* ''Conjugation Reactions in Drug Biotransformation'' (A. Aitio, ed.), pp. 67–76. Elsevier, Amsterdam.

67. Levin, W. (1977). Purification of liver microsomal cytochrome P-450: Hopes and promises. *In* ''Microsomes and Drug Oxidation'' (V. Ullrich, ed.), pp. 735–747. Academic Press, New York.

68. Smith, R. L. (1978). Extrapolation of animal results to man. *In* ''Drug Metabolism in Man'' (J. W. Gorrod and A. H. Beckett, eds.), pp. 97–106. Taylor & Francis, London.

Part II

Mixed Function Oxygenase Systems

Chapter 6

Microsomal Cytochrome *P*-450: A Central Catalyst in Detoxication Reactions

MINOR J. COON and ANDERS V. PERSSON

I. Introduction

The cytochrome *P*-450-containing enzyme system of liver microsomes is remarkably versatile in the types of chemical reactions it catalyzes and in its choice of substrates. Similar enzyme systems prepared from other tissues and organelles are apparently more specific but still capable of acting on a number of lipophilic substrates and, in particular, of binding many such substances as potential inhibitors. The types of chemical transformations attributed to cytochrome *P*-450 in hepatic microsomes include oxidative reactions in which an atom of molecular oxygen is

117

ENZYMATIC BASIS OF DETOXICATION, VOL. I
Copyright © 1980 by Academic Press, Inc.
All rights of reproduction in any form reserved.
ISBN 0-12-380001-3

inserted into an organic molecule, such as aliphatic and aromatic hydroxylations; N-oxidation; sulfoxidation; epoxidation; N-, S-, and O-dealkylations; and desulfurations and deaminations; reductive reactions involving direct electron transfer, such as reduction of azo, nitro, N-oxide, and epoxide groups; and, by a not yet well understood mechanism, dehalogenations.[1,2]

P-450$_{LM}$* brings about chemical changes both in physiologically important substrates, such as fatty acids, steroids, and prostaglandins, and also in a host of foreign substances, such as petroleum products, drugs, pesticides, anesthetics, and chemical carcinogens, as well as miscellaneous organic substances commonly found on the laboratory shelf.[1-7] In addition, new compounds of medicinal or other commercial importance are manufactured each year, many of which are eventually found to undergo chemical alteration by this microsomal enzyme system. With such a myriad of substrates and such a variety of reactions catalyzed, it has not been possible to select a more suitable name for cytochrome P-450 based on its function.

In the past decade the enzyme system under discussion has been solubilized, resolved into its components (NADPH-cytochrome P-450 reductase, cytochrome P-450, and diacyl-GPC), and reconstituted to give a functional system when supplemented with NADPH under aerobic conditions, as reviewed elsewhere.[8] In addition, this and other laboratories have been concerned with the purification, characterization, and interactions of the components as well as the basis for the broad substrate specificity and the mechanism by which molecular oxygen is "activated," that is, converted to a more reactive species capable of attacking a variety of lipophilic, rather inert substrates.

Although the inducibility and certain other properties of the hydroxylation system have been studied profitably with intact microsomal suspensions, it is clear that many difficult questions cannot be answered without isolation of the individual enzymes. A case in point is the question of whether the numerous activities attributed to P-450$_{LM}$ reside in one or more forms of this pigment. The variable enzyme activities observed in microsomes of animals treated with different inducing agents suggested that numerous forms of the cytochrome might be involved.[1,6,9,10] However, kinetic data obtained with microsomal suspensions[11-13] as well as with the reconstituted system[14] showed that a number of substrates behave as mutually competitive inhibitors, thereby apparently indicating

---

* The following abbreviations are used: P-450$_{LM}$, liver microsomal cytochrome P-450; P-450$_{LM_2}$ and P-450$_{LM_4}$, phenobarbital- and 5,6-benzoflavone-inducible forms, respectively, of rabbit P-450$_{LM}$, so designated according to their electrophoretic properties; and diacyl-GPC, diacylglyceryl-3-phosphorylcholine (phosphatidylcholine).

that they are acted on by a single site in a single enzyme. Spectral evidence was reported suggesting the occurrence of two forms of cytochrome P-450 in liver microsomes, induced either by phenobarbital or polycyclic aromatic hydrocarbons,[15-19] but the results could also be explained in terms of multiple environments for the same protein. The genetic regulation also supported the involvement of a distinct enzyme for aryl hydrocarbon hydroxylation.[20] Fractionation procedures were applied to liver microsomes following the administration of different inducing agents to animals,[21] and it was found that the substrate specificity resides in the different cytochrome fractions as tested in reconstituted systems.[22-24] Finally, the separation and purification of multiple forms of cytochrome P-450 from rabbit liver microsomes permitted the enzymes to be characterized definitively as distinct proteins.[25]

In the present brief review, no attempt will be made to give a detailed historical account of the contributions made by many laboratories to our knowledge of mammalian cytochrome P-450. We shall emphasize, instead, those developments in enzyme purification and characterization related to our own research interests. Rabbit liver microsomes apparently contain as many as six forms of P-450$_{LM}$, two of which, phenobarbital-inducible P-450$_{LM_2}$ and 5,6-benzoflavone-inducible P-450$_{LM_4}$, have been obtained in an electrophoretically homogeneous state and characterized in a number of ways.[25-29] The nomenclature of these forms is based on numerical designation of the cytochrome according to the general recommendation of the Nomenclature Committee of the International Union of Biochemistry.[30] The various forms are designated by their behavior upon sodium dodecyl sulfate-polyacrylamide gel electrophoresis, and the individual protein bands are numbered according to increasing molecular weight and decreasing mobility as P-450$_{LM_1}$, P-450$_{LM_2}$, etc.[8,25] The isolation of the cytochrome from 3-methylcholanthrene-treated rats and rabbits and from phenobarbital-treated rats has been reported by Ryan et al.[31] and Kawalek et al.,[32] and from phenobarbital- and 3-methylcholanthrene-treated rabbits by Imai and Sato[33] and Hashimoto and Imai.[34] Cytochrome P-450 has also been purified from liver microsomes of neonatal rabbits induced by 2,3,7,8-tetrachlorodibenzo-p-dioxin[35] and from rabbit lung microsomes,[36,37] as well as from a number of other sources as recently reviewed by Lu and West.[38]

II. PURIFICATION PROCEDURES

When this laboratory undertook the solubilization and resolution of the fatty acid ω-hydroxylation system of liver microsomes, we found the activity was quite labile, and examined the possible usefulness of various agents such as thiols and polyols known to stabilize enzymes. Glycerol proved to be particularly effective, and was therefore added routinely to the deoxycholate-solubilizing solution.[39,40] Only after the enzyme system had been resolved by column chromatography on DEAE-cellulose did we realize that cytochrome P-450 was an essential component of the system. Subsequent experience has shown that glycerol both improves the solubility of purified P-450$_{LM}$ and prevents its conversion to cytochrome P-420.

The detailed procedures adopted by this laboratory for the isolation of P-450$_{LM_2}$ from liver microsomes of phenobarbital-induced rabbits and P-450$_{LM_4}$ from liver microsomes of 5,6-benzoflavone-induced, control, and phenobarbital-induced rabbits are presented elsewhere[28,29] and need not be given in detail here. The essential steps are pyrophosphate extraction of microsomes to remove some contaminating proteins, precipitation of P-450$_{LM}$ by polyethylene glycol 6000, column chromatography on DEAE-cellulose, and column chromatography on hydroxyapatite-silica gel; excess detergent is then removed by treatment with calcium phosphate gel. The preparations are solubilized initially with cholate, and with Renex 690 (polyoxyethylene[10]nonylphenyl ether) during chromatography on DEAE–cellulose and subsequent steps. P-450$_{LM_2}$ is obtained in an overall yield of 9 to 17% (prior to detergent removal by calcium phosphate gel), based on total P-450$_{LM}$ in pyrophosphate-treated microsomes of phenobarbital-induced animals, and P-450$_{LM_4}$ is obtained in an overall yield of about 10% (prior to detergent removal) from pyrophosphate-treated microsomes of 5,6-benzoflavone-induced animals. Obviously, the yields would be considerably higher if calculated on the basis of the starting content of P-450$_{LM_2}$ and P-450$_{LM_4}$ in the microsomal membrane, but those levels are not known with accuracy. The final preparations appear to be homogeneous as judged by SDS-polyacrylamide gel electrophoresis.

A particular advantage of the procedure described is that NADPH-cytochrome P-450 reductase is recovered in the supernatant fraction from polyethylene glycol fractionation, and the reductase and the cytochromes can therefore be isolated from the same preparation of microsomes. Our earlier studies showed that the detergent-solubilized reductase is active toward P-450$_{LM}$, whereas protease-solubilized reductase preparations are not.[7,41] This flavoprotein was extensively purified from rat liver microsomes several years ago, and it was shown that the ratio of activities toward P-450$_{LM}$ and cytochrome c remained constant throughout the purification procedure.[42] A very useful improvement in the procedures for isolation of the detergent-solubilized reductase was the introduction of affinity chromatography with the use of Sepharose derivatized with ADP[43] or NADP.[44] We have therefore incorporated a step involving column chromatography on ADP-agarose into our procedures for isolation of the enzymes from rat[45] and rabbit liver microsomes.[46] The reductase is isolated in over 60% yield from pyrophosphate-treated microsomes obtained from phenobarbital-induced rabbits; following solubilization by Renex 690, DEAE-cellulose and ADP-agarose column chromatography steps are carried out in the presence of this detergent and glycerol. Alternatively, the polyethylene glycol step may be carried out after cholate solubilization to permit isolation of the various forms of P-450$_{LM}$, but the overall yield of the reductase is then reduced to 30%. Exposure of the reductase preparations to light is avoided where possible, especially in the presence of a high salt concentration, which causes loss of FMN.[47] We have found that the reductase from rabbit liver microsomes is somewhat subject to proteolysis, but that with care one can avoid the formation of degradation products during the purification procedure. The recommended precautions include rapid, careful handling of the microsomes and crude extracts, avoidance of bacterial growth by including 0.01% sodium azide during the early fractionation steps, and sterilization of the purified enzyme by passage through a Millipore filter (pore size, 0.45 μm).

III. ENZYME ASSAYS

A. Cytochrome P-450

In view of the large number of substrates for P-450$_{LM}$, a variety of reactions may be used to assay this enzyme. The general reaction is as follows, where RH represents a substrate and ROH the corresponding product

$$RH + O_2 + NADPH + H^+ \rightarrow ROH + H_2O + NADP^+ \qquad (1)$$

Although in theory one could simply measure oxygen uptake with an oxygen electrode or NADPH disappearance at 340 nm with a spectrophotometer in place of product formation, such determinations are complicated by the occurrence of a cytochrome P-450-catalyzed NADPH oxidase reaction in the presence or absence of substrate

$$O_2 + NADPH + H^+ \rightarrow H_2O_2 + NADP^+ \qquad (2)$$

The sum of Eqs. (1) and (2) shows the expected stoichiometry, with the sum of ROH and H_2O_2 produced being equimolar with respect to O_2 consumed or NADPH oxidized.[48] A further complication is that the presence of different substrates alters the rate of the endogenous oxidase activity to a varying extent. Accordingly, substrate-dependent NADPH oxidation or O_2 disappearance gives only an approximate measure of the extent to which a particular substrate is undergoing modification, and for more reliable results one should measure substrate disappearance or product formation directly.

Benzphetamine has proved to be a useful substrate because of the relatively high V_{max} and because the formaldehyde formed is conveniently measured by colorimetric[49,50] or radiometric procedures.[51]

Benzphetamine (N-benzyl-N,α-dimethylphenylethylamine) + O_2 + NADPH + H^+
$\rightarrow$ benzylamphetamine (desmethylbenzphetamine) + CH_2O + H_2O + $NADP^+$ (3)

In assays involving the reconstituted enzyme system, purified P-450$_{LM}$, purified reductase, and dilauroyl-GPC are premixed in a concentrated solution, to which buffer and benzphetamine are added, and then NADPH is introduced as the final component to initiate the reaction. The reaction mixture is incubated at 30°, and at some time interval (usually 10 min) the reaction is stopped and the formaldehyde formed is determined. The detailed procedure is presented elsewhere.[26,52] Under the conditions employed, formaldehyde formation, NADPH oxidation, and O_2 consumption are linear with time up to about 20 min. In all such assays P-450$_{LM}$ is the rate-limiting component. Accordingly, the rates are linear with respect to the cytochrome concentration and one is justified in expressing the results as turnover numbers, i.e., moles of formaldehyde formed per mole of P-450$_{LM}$ per minute. It should be added that the proposed intermediate in the reaction, the N-hydroxymethyl compound, is not isolated but decomposes spontaneously and quantitatively to give desmethylbenzphetamine and formaldehyde.

Alternatively, purified P-450$_{LM}$ may be assayed in reaction mixtures in which a peroxy compound is substituted for NADPH, NADPH-cytochrome P-450 reductase, and O_2.[53] The reaction is as follows, where XOOH represents a hydroperoxide or peracid and XOH the corresponding alcohol or carboxylic acid

$$RH + XOOH \rightarrow ROH + XOH \qquad (4)$$

The rate of the peroxide-dependent, cytochrome P-450-catalyzed reaction is close to that measured in the usual reconstituted system, but the cytochrome is destroyed at varying rates by the oxidants used, and the reaction may therefore be linear with time over only a short period when H_2O_2, cumene hydroperoxide, or related compounds are employed.

B. NADPH-Cytochrome P-450 Reductase

The reductase is usually assayed by its ability to reduce cytochrome c with NADPH as the electron donor; 0.3 M phosphate buffer is used because the activity is maximal at this concentration. As already indicated, however, activity toward cytochrome c does not necessarily indicate that a reductase preparation will be functional toward P-450$_{LM}$. In the assay for NADPH-cytochrome P-450 reductase activity, saturating levels of P-450$_{LM_2}$, diacyl-GPC, and benzphetamine are added, and the rate of NADPH oxidation is measured at 340 nm.[45,46] The rate of the reaction is linear with respect to the reductase concentration over a limited range. With the purified rat liver reductase the turnover numbers toward P-450$_{LM_2}$ and cytochrome c at 30° are 92 and 4010, respectively, and with the rabbit liver reductase the corresponding values are 93 and 4030, respectively.

IV. CHARACTERIZATION OF PURIFIED ENZYMES

A. Multiple Forms of Cytochrome P-450

Some of the properties of the two forms of rabbit liver microsomal cytochrome P-450 which are homogeneous by the criteria of electrophoretic behavior,[28] immunological properties,[54] and end group analysis[28,55] are presented in Table I. Other minor forms, which have been partially

TABLE I

Properties of Purified Cytochrome P-450

| | Form of cytochrome | |
Property determined	P-450$_{LM_2}$	P-450$_{LM_4}$
Inducing agent	Phenobarbital	5,6-Benzoflavone
Heme content (per polypeptide chain)	1	1
Carbohydrate content (per polypeptide chain)	1 glucosamine, 2 mannose	1 glucosamine, 2 mannose
C-Terminal residue	Arginine	Lysine
N-Terminal residue	Methionine	(Blocked)
Absorption maxima		
Oxidized	568, 535, 418 nm	645, 394 nm
Reduced	544, 413 nm	542, 411 nm
CO complex	552, 451 nm	550, 448 nm
Molecular weight		
Apparent	300,000	500,000
Minimal	49,000	55,000

purified and exhibit different electrophoretic behavior, have not yet been characterized in detail. P-450$_{LM_2}$ is present in significant amounts only when the animals have been induced by phenobarbital. In contrast, P-450$_{LM_4}$ is present in both uninduced and phenobarbital-induced animals at the same level and upon the administration of 5,6-benzoflavone is approximately doubled in amount. In immunochemical studies,[54] no cross-reactions observable by precipitin band formation were detected between anti-LM$_2$ serum and P-450$_{LM_4}$, or between anti-LM$_4$ serum and P-450$_{LM_2}$. Competitive binding studies with radiolabeled cytochromes confirmed that rabbit anti-LM$_2$ does not cross-react with P-450$_{LM_4}$, but slight cross-reactions were detected by this technique between goat anti-LM$_2$ and P-450$_{LM_4}$, and between goat anti-LM$_4$ and P-450$_{LM_2}$. In general, our immunological studies have led to the conclusion that these two forms of rabbit P-450$_{LM}$ have significant structural differences. The molecular weights of the polypeptide chains are 49,000 and 54,000 for P-450$_{LM_2}$ and P-450$_{LM_4}$, respectively, as judged by calibrated sodium dodecyl sulfate-polyacrylamide gel electrophoresis. In contrast, without added detergent the apparent molecular weights determined by gel exclusion chromatography are considerably higher, as is typical of a number of membrane proteins having a tendency to aggregate. In contrast, the camphor-metabolizing cytochrome of *Pseudomonas putida,* called P-450$_{cam}$, is a typical soluble protein with little tendency to aggregate.[56]

The best preparations of the liver microsomal cytochromes contain one molecule of heme per polypeptide chain, but in some instances the heme is partially lost by dissociation during purification. The two enzymes have a low carbohydrate content (apparently corresponding to the same three sugar residues), but they differ in their terminal amino acyl residues. Whereas P-450$_{LM_4}$ appears to have a blocking group at the N-terminus, P-450$_{LM_2}$ has an N-terminal methionine, and 17 of the first 20 residues are hydrophobic, including two clusters of five leucines each.[55] The composition and sequence, as shown in Table II, are similar to those of the short-lived hydrophobic amino-terminal precursor segments of certain other proteins, especially myeloma immunoglobulin light chains[57] and pancreatic zymogens.[58] Blobel and Doberstein[59] have proposed that such precursor segments may be responsible for binding of the nascent polypeptide to the endoplasmic reticulum or other membranous sites during transport. Our results provided the first example of such a hydrophobic amino-terminal segment in a "mature" protein. The biological usefulness of this segment in a mature protein is as yet unknown, but it may function in the binding of the cytochrome to the endoplasmic reticulum or in its orientation in functional complexes with other components of the hydroxylation system such as NADPH-cytochrome P-450

TABLE II

Amino-Terminal Sequence of Rabbit and Rat Liver $P\text{-}450_{LM}$

Cytochrome	1				5					10					15					20
										Sequence										
$P\text{-}450_{LM_2}$	Met	Glu	Phe	Ser	Leu	Leu	Leu	Leu	Ala	Phe	Leu	Ala	Gly	Leu	Leu	Leu	Leu	Leu	Leu	Phe
$P\text{-}450_a$	Met	Leu	Asp	Thr	Gly	Leu	Leu	Leu	Val	Ile	Leu	Ala	Thr	Leu	Thr	Val	Met	Leu	Leu	
$P\text{-}450_b$	Glu	Pro	Thr	Ile	Leu	Leu	Leu	Ala	Leu	Leu	Val	Gly	Phe	Leu	Leu	Leu	Leu	Val	Arg	
$P\text{-}450_c$	Ile	Thr	Val	Tyr	Gly	Phe	Pro	Ala	Phe	Thr	Ala	Ser	Glu	Leu	Leu	Leu	Leu	Val	Val	

reductase and phospholipid. We are continuing with the sequencing of $P\text{-}450_{LM_2}$ to determine whether it has variable as well as fixed regions, which would be indicative of the presence of isozymes. As yet, however, all of the data available from sequencing and other analytical procedures show no indication of more than a single protein in purified $P\text{-}450_{LM_2}$ preparations.

The N-terminal sequences of three forms of rat $P\text{-}450_{LM}$ as recently reported by Botelho, Ryan, and Levin[60] are also presented in Table II. Their data indicate that cytochromes $P\text{-}450_a$, $P\text{-}450_b$, and $P\text{-}450_c$ from rat liver microsomes are separate gene products and not posttranslational modifications of one primary gene product. Of particular interest, greater sequence homology in the amino-terminal region is evident between the cytochromes isolated from the two species treated with phenobarbital ($P\text{-}450_{LM_2}$ and $P\text{-}450_b$) than among the three rat liver enzymes.

Also shown in Table I are the absorption maxima of the two major forms of $P\text{-}450_{LM}$ purified from rabbit liver microsomes.[28] $P\text{-}450_{LM_2}$ has a Soret band at 418 nm characteristic of the low-spin state, whereas in $P\text{-}450_{LM_4}$ the corresponding band is shifted to the blue at 394 nm and is characteristic of the high-spin state. Such differences are observed clearly only when detergent has been removed from the latter cytochrome. $P\text{-}450_{LM_4}$ is converted fairly completely from the 394 nm form to the 418 nm form by addition of Renex 690 at a concentration of 0.5% or at higher protein concentrations. Hashimoto and Imai[34] have reported the isolation of $P\text{-}448$ (that is, $P\text{-}450_{LM_4}$) from 3-methylcholanthrene-induced rabbits with this polycyclic hydrocarbon still bound. However, since $P\text{-}450_{LM_4}$ isolated in our laboratory from 5,6-benzoflavone-induced, phenobarbital-induced, and uninduced rabbits has the same spectral properties, it is not clear how bound substrate could account for the spectrum observed.[28] Furthermore, exposure to Renex during the purification procedure might be expected to remove bound substrate, whether administered to the animal or endogenous in nature. In the substrate-free state, purified $P\text{-}450_{LM}$ exhibits a spectrum typical of low-spin ferric hemeproteins, with a Soret band at 418 nm and definite α- and β-absorbances. This hexacoordinate form of $P\text{-}450_{LM_2}$ is reversibly converted to the pentacoordinate, high-spin state upon binding of substrate with Soret transition to higher energy (394 nm). When either of the two purified cytochromes is reduced by chemical or enzymatic means in the absence of oxygen, the Soret band loses intensity and moves to about 412 nm, and the α- and β-bands collapse to a single broad band. The most striking difference between $P\text{-}450$ and other heme proteins is seen when carbon monoxide is added to the ferrous form; the Soret band of ferrous carbonyl $P\text{-}450_{LM_2}$ is at 451 nm, and that of $P\text{-}450_{LM_4}$ at 448 nm.

B. NADPH-Cytochrome P-450 Reductase

As already described, the use of detergents rather than proteases as solubilizing agents has permitted the isolation of the reductase in a native form capable of reducing P-450$_{LM}$. The properties of this enzyme, which is now available in highly purified state from both rat[42-45] and rabbit liver microsomes,[46] are summarized in Table III. The reductase is usually isolated from phenobarbital-induced animals, in which the level is increased about twofold. This flavoprotein is unusual in containing both FMN and FAD, in accord with the earlier finding that both flavins are present in trypsin-solubilized preparations.[61] The absorption spectra are virtually identical for the rat and rabbit enzymes, but the latter is slightly lower in minimal molecular weight as shown by calibrated sodium dodecyl sulfate-polyacrylamide gel electrophoresis and by amino acid analysis.[45,46] As is the case with purified P-450$_{LM}$, the purified reductase readily aggregates except in the presence of detergents.

Iyanagi et al.[62] found that the air-stable semiquinone form of the purified trypsin-solubilized reductase was a one-electron reduced form, and it is now generally agreed that the same is true of the purified detergent-solubilized reductase.[45,46,63] We have recently turned our attention to the identification of the high and low potential flavins of this enzyme.[64] FMN was selectively removed from the purified rat liver reductase and the properties were compared with those of the native enzyme. Whereas the FMN-depleted enzyme lost the ability to transfer electrons to P-450$_{LM}$ in the reconstituted system as well as to certain artificial electron acceptors such as cytochrome c, these activities were restored when the depleted enzyme was incubated with FMN. A series of spectral experiments were carried out to determine whether the properties of the FMN-depleted enzyme correspond to those of the high or low potential flavin of the

TABLE III

Properties of Purified NADPH-Cytochrome P-450 Reductase

Property examined	Reductase from rat liver	Reductase from rabbit liver
Inducing agent	Phenobarbital	Phenobarbital
Flavin content (per polypeptide chain)	1 FMN, 1 FAD	1 FMN, 1 FAD
Absorption maxima (oxidized)	456, 384 nm	456, 382 nm
Identity of air-stable semiquinone	1-electron-reduced form	1-electron-reduced form
Molecular weight		
Apparent	500,000	500,000
Minimal	76,000	74,000

native enzyme; the redox potentials (E_0' values) for these flavins are
-0.190 and -0.328 V, respectively, as shown by Iyanagi *et al.*[62] Addition
of NADP to our fully reduced, FMN-depleted reductase resulted in significant oxidation of flavin, indicating a midpoint potential for FAD near
that of the pyridine nucleotide couple. The semiquinone form of the
FMN-depleted reductase, which was produced during air oxidation of
NADPH-reduced enzyme or during stepwise photochemical reduction of
oxidized enzyme under anaerobic conditions, had spectral characteristics
similar to those of the semiquinone of the low potential flavin of the native
enzyme and was readily oxidized under aerobic conditions. Addition of
FMN to one-electron-reduced FMN-depleted reductase under anaerobic
conditions produced an enzyme species with properties similar to those of
the one-electron-reduced form of the native enzyme, thus showing that
electron transfer from FAD to FMN is thermodynamically favorable.
From such experiments we conclude that the low and high potential
flavins of the reductase are FAD and FMN, respectively. The pattern of
electron flow in the complete system may be as follows, NADPH $\rightarrow$ FAD
$\rightarrow$ FMN $\rightarrow$ *P*-450, but the precise redox state of each flavin during the
transfer of the first and second electrons to the cytochrome during substrate hydroxylation remains to be established.

V. SUBSTRATE SPECIFICITY

The very broad specificity of *P*-450$_{LM}$ is apparently not due solely to the
occurrence of multiple forms. Instead, the picture which has emerged is
that the individual forms exhibit only partial selectivity among a variety of
substrates and may all be described as partially specific but overlapping in
their activities.[28] For example, *P*-450$_{LM_2}$ appears to be the most active
form in benzphetamine demethylation and in the hydroxylation of
biphenyl in the 4-position and testosterone in the 16α-position, but also
exhibits activity toward ethylmorphine, aniline, and other substrates. A
mixture of *P*-450$_{LM_1}$ and LM$_7$ is particularly effective in the hydroxylation
of testosterone in the 6β position but also attacks a variety of other
substrates. More recently, we have demonstrated partial regio- and
stereoselectivity by various forms of *P*-450$_{LM}$ in the metabolism of
radioactive benzo[*a*]pyrene and (-)*trans*-7,8-dihydroxy-7,8-dihydro-
benzo[*a*]pyrene in collaboration with Gelboin and his associates.[65] In
this work, a correlation was observed between product formation and
binding to DNA. In other studies carried out with Fasco and Kaminsky,[66]
the metabolism of the *R* and *S* enantiomers of warfarin by these cyto-

chromes was examined. The results indicated that P-450$_{LM_4}$ acts on warfarin only when it is oriented to approach the enzyme catalytic site at the coumarin side of the molecule, and preferentially when warfarin is in the R configuration. P-450$_{LM_2}$ also acts on the coumarin side of warfarin, but preferentially when it is in the S configuration, and in addition attacks the phenyl ring, but preferentially when warfarin is in the R configuration. The metabolite patterns seen with microsomes and with the reconstituted system indicated that the regio- and stereoselectivities of the cytochromes were maintained when removed from the microsomal membrane and purified.

Although the possibility may be considered that an almost unlimited number of isozymes accounts for the versatility of liver microsomal cytochrome P-450, the evidence now available points much more strongly to the occurrence of only five or six distinct forms. The activities of these forms may be modulated to varying extents by the phospholipid environment in the membrane, by the availability of reducing equivalents from the reductase as the limiting enzyme in the system, patterns of induction, and nutritional effects.[2]

VI.   SUMMARY AND COMMENTS ON FACTORS CONTROLLING ACTIVITIES OF CYTOCHROME P-450

Recent studies in this and other laboratories have led to the isolation of liver microsomal cytochrome P-450 (P-450$_{LM}$) and NADPH-cytochrome P-450 reductase in apparently homogeneous form. The various activities attributed to this cytochrome in intact microsomes are observed in a reconstituted system containing the isolated enzymes, phosphatidylcholine, and NADPH under aerobic conditions. P-450$_{LM}$ occurs in several discrete forms with somewhat different but overlapping activities. Phenobarbital-inducible P-450$_{LM_2}$ and 5,6-benzoflavone-inducible P-450$_{LM_4}$ obtained from rabbit liver microsomes have been characterized as distinct proteins by a number of criteria. NADPH-cytochrome P-450 reductase has also been characterized and shown to contain equimolar amounts of FAD and FMN as the low- and high-potential flavins, respectively.

The contributions of the different forms of cytochrome P-450 to the activities observed in intact animals as well as in isolated microsomes may be modified by alterations in the phospholipid environment or in availability of reducing equivalents from NADPH via the reductase, as well as by different patterns of induction and various nutritional effects. In addition, naturally occurring substrates (prostaglandins, steroids, and fatty acids) and foreign substances (drugs, anesthetics, pesticides, car-cinogens, etc.) may act as mutually competitive inhibitors and thereby alter the catalytic activities of the individual cytochromes.

ACKNOWLEDGMENTS

Recent research in this laboratory was supported by Grant PCM76-14947 from the National Science Foundation and Grant AM-10339 from the National Institutes of Health. Anders V. Persson was a Postdoctoral Fellow of the Damon Runyon-Walter Winchell Cancer Research Fund.

REFERENCES

1. Gillette, J. R. (1966). Biochemistry of drug oxidation and reduction by enzymes in hepatic endoplasmic reticulum. *Adv. Pharmacol.* **4**, 219–261.
2. Coon, M. J. (1978). Oxygen activation in the metabolism of lipids, drugs, and carcinogens. *Nutr. Rev.* **36**, 319–328.
3. Talalay, P. (1965). Enzymatic mechanisms in steroid biochemistry. *Annu. Rev. Biochem.* **34**, 347–380.
4. Conney, A. H., Levin, W., Ikeda, M., Kuntzman, R., Cooper, D. Y., and Rosenthal, O. (1968). Inhibitory effect of carbon monoxide on the hydroxylation of testosterone by rat liver microsomes. *J. Biol. Chem.* **243**, 3912–3915.
5. Voigt, W., Thomas, P. J., and Hsia, S. L. (1968). Enzymic studies of bile acid metabolism. I. 6β-Hydroxylation of chenodeoxycholic and taurochenodeoxycholic acids by microsomal preparations of rat liver. *J. Biol. Chem.* **243**, 3493–3499.
6. Conney, A. H. (1967). Pharmacological implications of microsomal enzyme induction. *Pharmacol. Rev.* **19**, 317–366.
7. Lu, A. Y. H., Junk, K. W., and Coon, M. J. (1969). Resolution of the cytochrome *P*-450-containing ω-hydroxylation system of liver microsomes into three components. *J. Biol. Chem.* **244**, 3714–3721.
8. Coon, M. J., Vermilion, J. L., Vatsis, K. P., French, J. S., Dean, W. L., and Haugen, D. A. (1977). Biochemical studies on drug metabolism: Isolation of multiple forms of liver microsomal cytochrome *P*-450. *In* "Drug Metabolism Concepts" (D. M. Jerina, ed.), Symp. Ser. No. 44, pp. 46–71. Am. Chem. Soc., Washington, D.C.
9. Gelboin, H. V. (1967). Carcinogens, enzyme induction and gene action. *Adv. Cancer Res.* **10**, 1–81.
10. Gillette, J. R. (1969). Mechanism of oxidation by enzymes in the endoplasmic reticulum. *FEBS Symp.* **16**, 109–124.
11. Rubin, A., Tephly, T. R., and Mannering, G. J. (1964). Kinetics of drug metabolism by hepatic microsomes. *Biochem. Pharmacol.* **13**, 1007–1016.
12. Tephly, T. R., and Mannering, G. J. (1968). Inhibition of drug metabolism. V. Inhibition of drug metabolism by steroids. *Mol. Pharmacol.* **4**, 10–14.
13. Wada, F., Shimakawa, H., Takasugi, M., Kotake, T., and Sakamoto, Y. (1968). Effects of steroid hormones on drug-metabolizing enzyme systems in liver microsomes. *J. Biochem. (Tokyo)* **64**, 109–113.
14. Lu, A. Y. H., Strobel, H. W., and Coon, M. J. (1970). Properties of a solubilized form of

the cytochrome P-450-containing mixed function oxidase of liver microsomes. *Mol. Pharmacol.* **6**, 213–220.

15. Sladek, N. E., and Mannering, G. J. (1966). Evidence for a new P-450 hemoprotein in hepatic microsomes from methylcholanthrene-treated rats. *Biochem. Biophys. Res. Commun.* **24**, 668–674.

16. Alveres, A. P., Schilling, G., Levin, W., and Kuntzman, R. (1967). Studies on the induction of CO-binding pigments in liver microsomes by phenobarbital and 3-methylcholanthrene. *Biochem. Biophys. Res. Commun.* **29**, 521–526.

17. Hildebrandt, A., Remmer, H., and Estabrook, R. W. (1968). Cytochrome P-450 of liver microsomes—One pigment or many. *Biochem. Biophys. Res. Commun.* **30**, 607–612.

18. Nebert, D. W., and Gelboin, H. V. (1968). Substrate inducible microsomal aryl hydroxylase in mammalian cell culture. II. Cellular responses during enzyme induction. *J. Biol. Chem.* **243**, 6250–6261.

19. Jefcoate, C. R. E., and Gaylor, J. L. (1969). Ligand interactions with hemoprotein P-450. II. Influence of phenobarbital and methylcholanthrene induction processes on P-450 spectra. *Biochemistry* **8**, 3464–3472.

20. Gielen, J. E., Goujon, F. M., and Nebert, D. W. (1972). Genetic regulation of aryl hydrocarbon hydroxylase induction. II. Simple Mendelian expression in mouse tissues in vivo. *J. Biol. Chem.* **247**, 1125–1137.

21. Comai, K., and Gaylor, J. L. (1973). Existence and separation of three forms of cytochrome P-450 from rat liver microsomes. *J. Biol. Chem.* **248**, 4947–4955.

22. Lu, A. Y. H., Kuntzman, R., West, S., Jacobson, M., and Conney, A. H. (1972). Reconstituted liver microsomal enzyme system that hydroxylates drugs, other foreign compounds and endogenous substrates. II. Role of the cytochrome P-450 and P-448 fractions in drug and steroid hydroxylations. *J. Biol. Chem.* **247**, 1727–1734.

23. Lu, A. Y. H., and Levin, W. (1972). Partial purification of cytochromes P-450 and P-448 from rat liver microsomes. *Biochem. Biophys. Res. Commun.* **46**, 1334–1339.

24. Nebert, D. W., Heidema, J. K., Strobel, H. W., and Coon, M. J. (1973). Genetic expression of aryl hydrocarbon hydroxylase induction. Genetic specificity resides in the fraction containing cytochromes P-448 and P-450. *J. Biol. Chem.* **248**, 7631–7636.

25. Haugen, D. A., van der Hoeven, T. A., and Coon, M. J. (1975). Purified liver microsomal cytochrome P-450. Separation and characterization of multiple forms. *J. Biol. Chem.* **250**, 3567–3570.

26. van der Hoeven, T. A., and Coon, M. J. (1974). Preparation and properties of partially purified cytochrome P-450 and reduced nicotinamide adenine dinucleotide phosphate-cytochrome P-450 reductase from rabbit liver microsomes. *J. Biol. Chem.* **249**, 6302–6310.

27. van der hoeven, T. A., Haugen, D. A., and Coon, M. J. (1974). Cytochrome P-450 purified to apparent homogeneity from phenobarbital-induced rabbit liver microsomes: Catalytic activity and other properties. *Biochem. Biophys. Res. Commun.* **60**, 569–575.

28. Haugen, D. A., and Coon, M. J. (1976). Properties of electrophoretically homogeneous phenobarbital-inducible and β-naphthoflavone-inducible forms of liver microsomal cytochrome P-450. *J. Biol. Chem.* **251**, 7929–7939.

29. Coon, M. J., van der Hoeven, T. A., Dahl, S. B., and Haugen, D. A. (1978). Two forms of liver microsomal cytochrome P-450: $P-450_{LM_2}$ and $P-450_{LM_4}$ (rabbit liver). *In* "Methods in Enzymology" (S. Fleisher and L. Packer, eds.), Vol. 52, pp. 109–117. Academic Press, New York.

30. Commission on Biochemical Nomenclature for Enzymes (1972). "Enzyme Nomenclature," p. 24. Am. Elsevier, New York.

31. Ryan, D., Lu, A. Y. H., Kawalek, J., West, S. B., and Levin, W. (1975). Highly purified cytochrome P-448 and P-450 from rat liver microsomes. Biochem. Biophys. Res. Commun. 64, 1134–1141.

32. Kawalek, J. C., Levin, W., Ryan, D., Thomas, P. E., and Lu, A. Y. H. (1975). Purification of liver microsomal cytochrome P-448 from 3-methylcholanthrene-treated rabbits. Mol. Pharmacol. 11, 874–878.

33. Imai, Y., and Sato, R. (1974). A gel-electrophoretically homogeneous preparation of cytochrome P-450 from liver microsomes of phenobarbital-treated rabbits. Biochem. Biophys. Res. Commun. 60, 8–14.

34. Hashimoto, C., and Imai, Y. (1976). Purification of a substrate complex of cytochrome P-450 from liver microsomes of 3-methylcholanthrene-treated rabbits. Biochem. Biophys. Res. Commun. 68, 821–827.

35. Norman, R. L., Johnson, E. F., and Muller-Eberhard, U. (1978). Identification of the major cytochrome P-450 form transplacentally induced in neonatal rabbits by 2,3,7,8-tetrachlorodibenzo-p-dioxin. J. Biol. Chem. 253, 8640–8647.

36. Arinç, E., and Philpot, R. M. (1976). Preparation and properties of partially purified pulmonary cytochrome P-450 from rabbits. J. Biol. Chem. 251, 3213–3220.

37. Guengerich, F. P. (1977). Preparation and properties of highly purified cytochrome P-450 and NADPH-cytochrome P-450 reductase from pulmonary microsomes of untreated rabbits. Mol. Pharmacol. 13, 911–923.

38. Lu, A. Y. H., and West, S. B. (1978). Reconstituted mammalian mixed-function oxidases: Requirements, specificities and other properties. Pharmacol. Ther., Part A 2, 337.

39. Lu, A. Y. H., and Coon, M. J. (1968). Role of hemoprotein P-450 in fatty acid ω-hydroxylation in a soluble enzyme system from liver microsomes. J. Biol. Chem. 243, 1331–1332.

40. Coon, M. J., and Lu, A. Y. H. (1969). Fatty acid ω-oxidation in a soluble microsomal enzyme system containing P-450. In "Microsomes and Drug Oxidations" (J. R. Gillette et al., eds.), pp. 151–166. Academic Press, New York.

41. Coon, M. J., Strobel, H. W., and Boyer, R. F. (1973). On the mechanism of hydroxylation reactions catalyzed by cytochrome P-450. Drug. Metab. Dispos. 1, 92–97.

42. Vermilion, J. L., and Coon, M. J. (1974). Highly purified detergent-solubilized NADPH-cytochrome P-450 reductase from phenobarbital-induced rat liver microsomes. Biochem. Biophys. Res. Commun. 60, 1315–1322.

43. Yasukochi, Y., and Masters, B. S. S. (1976). Some properties of a detergent-solubilized NADPH-cytochrome c (cytochrome P-450) reductase purified by biospecific affinity chromatography. J. Biol. Chem. 251, 5337–5344.

44. Dignam, J. D., and Strobel, H. W. (1977). NADPH-cytochrome P-450 reductase from rat liver: Purification by affinity chromatography and characterization. Biochemistry 16, 1116–1123.

45. Vermilion, J. L., and Coon, M. J. (1978). Purified liver microsomal NADPH-cytochrome P-450 reductase. Spectral characterization of oxidation–reduction states. J. Biol. Chem. 253, 2694–2704.

46. French, J. S., and Coon, M. J. (1979). Properties of NADPH-cytochrome P-450 reductase purified from rabbit liver microsomes. Arch. Biochem. Biophys. 195, 565–577.

47. Vermilion, J. L., and Coon, M. J. (1976). Properties of highly purified detergent-solubilized NADPH-cytochrome P-450 reductase from liver microsomes. In "Flavins and Flavoproteins" (T. P. Singer, ed.), pp. 674–678. Elsevier, Amsterdam.

48. Nordblom, G. D., and Coon, M. J. (1977). Hydrogen peroxide formation and

stoichiometry of hydroxylation reactions catalyzed by highly purified liver microsomal cytochrome P-450. *Arch. Biochem. Biophys.* **180**, 343–347.

49. Nash, T. (1953). The colorimetric estimation of formaldehyde by means of the Hantzsch reaction. *Biochem. J.* **55**, 416–421.

50. Cochin, J., and Axelrod, J. (1959). Biochemical and pharmacological changes in the rat following chronic administration of morphine, nalorphine and normorphine. *J. Pharmacol. Exp. Ther.* **125**, 105–110.

51. Poland, A. P., and Nebert, D. W. (1973). A sensitive radiometric assay of aminopyrine N-demethylation. *J. Pharmacol. Exp. Ther.* **184**, 269–277.

52. Coon, M. J. (1978). Reconstitution of the cytochrome P-450-containing mixed function oxidase system of liver microsomes. *In* "Methods in Enzymology" (S. Fleisher and L. Packer, eds.), Vol. 52, pp. 200–206. Academic Press, New York.

53. Nordblom, G. D., White, R. E., and Coon, M. J. (1976). Studies on hydroperoxide-dependent substrate hydroxylation by purified liver microsomal cytochrome P-450. *Arch. Biochem. Biophys.* **175**, 524–533.

54. Dean, W. L., and Coon, M. J. (1977). Immunochemical studies on two electrophoretically homogeneous forms of rabbit liver microsomal cytochrome P-450: P-450$_{LM_2}$ and P-450$_{LM_4}$. *J. Biol. Chem.* **252**, 3255–3261.

55. Haugen, D. A., Armes, L. G., Yasunobu, K. T., and Coon, M. J. (1977). Amino-terminal sequence of phenobarbital-inducible cytochrome P-450 from rabbit liver microsomes: Similarity to hydrophobic amino-terminal segments of preproteins. *Biochem. Biophys. Res. Commun.* **77**, 967–973.

56. Katagiri, M., Ganguli, B. N., and Gunsalus, I. C. (1968). A soluble cytochrome P-450 functional in methylene hydroxylation. *J. Biol. Chem.* **243**, 3543–3547.

57. Burstein, Y., and Schechter, I. (1977). Amino acid sequence of the NH$_2$-terminal extra piece segments of the precursors of mouse immunoglobulin λ-type and κ-type light chains. *Proc. Natl. Acad. Sci. U.S.A.* **74**, 716–720.

58. Devillers-Thiery, A., Kindt, T., Scheele, G., and Blobel, G. (1975). Homology in amino-terminal sequence of precursors to pancreatic secretory proteins. *Proc. Natl. Acad. Sci. U.S.A.* **72**, 5016–5020.

59. Blobel, G., and Doberstein, B. (1975). Transfer of proteins across membranes. I. Presence of proteolytically processed and unprocessed nascent immunoglobulin light chains on membrane bound ribosomes of murine myeloma. *J. Cell Biol.* **67**, 835–851.

60. Botelho, L. H., Ryan, D. E., and Levin, W. (1979). Amino acid compositions and partial amino acid sequences of three highly purified forms of liver microsomal cytochrome P-450 from rats treated with polychlorinated biphenyls, phenobarbital or 3-methylcholanthrene. *J. Biol. Chem.* **254**, 5635–5640.

61. Iyanagi, T., and Mason, H. S. (1973). Some properties of hepatic reduced nicotinamide adenine dinucleotide phosphate-cytochrome *c* reductase. *Biochemistry* **12**, 2297–2308.

62. Iyanagi, T., Makino, N., and Mason, H. S. (1974). Redox properties of the reduced nicotinamide adenine dinucleotide phosphate-cytochrome P-450 and reduced nicotinamide adenine dinucleotide-cytochrome b_5 reductases. *Biochemistry* **13**, 1701–1710.

63. Yasukochi, Y., Peterson, J. A., and Masters, B. S. S. (1979). NADPH-cytochrome *c* (P-450) reductase: Spectrophotometric and stopped-flow kinetic studies on the formation of reduced flavoprotein intermediates. *J. Biol. Chem.* **254**, 7097–7104.

64. Vermilion, J. L., and Coon, M. J. (1978). Identification of the high and low potential flavins of liver microsomal NADPH-cytochrome P-450 reductase. *J. Biol. Chem.* **253**, 8812–8819.

65. Deutsch, J., Leutz, J. C., Yang, S. K., Gelboin, H. V., Chiang, Y. L., Vatsis, K. P., and Coon, M. J. (1978). Regio- and stereoselectivity of various forms of purified cytochrome *P*-450 in the metabolism of benzo[*a*]pyrene and (-)*trans*-7,8-dihydroxy-7,8-dihydrobenzo[*a*]pyrene as shown by product formation and binding to DNA. *Proc. Natl. Acad. Sci. U.S.A.* **75,** 3123–3127.
66. Fasco, M. J., Vatsis, K. P., Kaminsky, L. S., and Coon, M. J. (1978). Regioselective and stereoselective hydroxylation of R and S warfarin by different forms of purified cytochrome *P*-450 from rabbit liver. *J. Biol. Chem.* **253,** 7813–7820.

Chapter 7

Reactions Catalyzed by the Cytochrome *P*-450 System

PETER G. WISLOCKI, GERALD T. MIWA,
and ANTHONY Y. H. LU

ENZYMATIC BASIS OF DETOXICATION, VOL. I

I. INTRODUCTION

The *in vitro* metabolism of foreign compounds by mammalian tissues via reactions such as dealkylation and deamination was first described by Mueller and Miller,[1,2] Axelrod,[3,4] and Brodie *et al.*[5] These and subsequent studies established that the responsible enzyme system is mainly localized in the endoplasmic reticulum and that the reaction requires both NADPH and molecular oxygen. In 1965, Cooper and co-workers[6] demonstrated that the hydroxylation of acetanilide, the N-demethylation of mono-methylaminopyrine and the O-demethylation of codeine by liver microsomes were inhibited by CO and that this inhibition could be relieved by exposure to light with the maximum reversal of inhibition occurring at 450 nm. Since the resultant photochemical action spectrum was identical to the reduced CO difference spectrum of a microsomal cytochrome, it was concluded that this protein, cytochrome *P*-450, was the liver microsomal oxidase responsible for the metabolism of drugs and various foreign compounds.

Since then, numerous chemicals have been shown to be metabolized by the microsomal cytochrome *P*-450 system of liver and other tissues through a variety of different reactions. In this chapter, we describe the reactions known to be catalyzed by the cytochrome *P*-450 system. Because of the broad scope of the subject, no attempt will be made to cite and review all the relevant references or to cover all aspects of the subject. Instead, the discussion is limited to examples of the different types of reactions catalyzed by cytochrome *P*-450, the reaction intermediates and mechanism, evidence for the involvement of cytochrome *P*-450, and the toxicological implication of the metabolism. The classification of a reaction as one catalyzed by cytochrome *P*-450 does not imply that other enzymes are incapable of also catalyzing the reaction. The reader is directed to several review articles[7-14] and the excellent book by Testa and Jenner[15] for further discussion of this subject.

II. CRITERIA USED TO ESTABLISH THE INVOLVEMENT OF CYTOCHROME *P*-450 IN VARIOUS REACTIONS

Various criteria have been used to implicate microsomal cytochrome *P*-450 in the metabolism of foreign as well as endogenous compounds. While the results from some of the studies can be interpreted with certainty, others remain questionable. The usefulness and limitation of different approaches are briefly discussed in this section. For a specific reaction to be classified as a cytochrome *P*-450 type in this chapter, we have required that at least two of the four criteria be fulfilled. This requirement has eliminated many other examples which may be found in the future to be catalyzed by cytochrome P-450.

A. In vivo Studies with Cytochrome P-450 Inducers and Inhibitors

The induction of a cytochrome P-450 type reaction was first observed by the Millers[16] who found the in vitro N-demethylation of 4-dimethylaminoazobenzene to be increased by pretreating rats with 3-methylcholanthrene. The level of microsomal cytochrome P-450 in the liver can be increased by in vivo administration of inducers, e.g., phenobarbital, 3-methylcholanthrene, benzo[a]pyrene, 5,6-benzoflavone, pregnenolone-16α-carbonitrile, and 2,3,7,8-tetrachlorodibenzo-p-dioxin, and decreased by treatment of animals with inhibitors, e.g., cobalt chloride and carbon tetrachloride. Thus, an increase in metabolism of a compound after animals are treated with appropriate inducers and a decrease in metabolism following in vivo treatment with inhibitors would generally suggest a role of cytochrome P-450 in the metabolism. However, due to the presence of multiple forms of cytochrome P-450 and the differential effect of various inducers and inhibitors on each individual form of cytochrome P-450 (Chapter 6, this volume),[17-21] the enhancement or inhibition of liver microsomal metabolism would depend on the animals or tissues used and the substrates being studied. Therefore, the absence of induction or inhibition does not rule out the involvement of cytochrome P-450 in a reaction.

B. In vitro Inhibition Studies

A number of chemicals can affect the cytochrome P-450 system either by specific interaction with the heme or protein moiety or by serving as competing substrates. Thus, classic inhibitors such as SKF 525-A, metyrapone, 7,8-benzoflavone, 2,4-dichloro-6-phenylphenoxyethylamine (DPEA), and CO have been widely used to evaluate the involvement of cytochrome P-450 in various reactions. While inhibition of metabolism would suggest cytochrome P-450 involvement, a lack of effect must be interpreted with caution due to the presence of different forms of cytochrome P-450. For example, the oxidative metabolism of aminopyrine, hexobarbital, and morphine is inhibited by metyrapone in both untreated and phenobarbital-treated rats, but acetanilide hydroxylation by liver microsomes from both control and phenobarbital-induced rats is enhanced in the presence of metyrapone.[22,23] Similarly, 7,8-benzoflavone stimulates benz[a]pyrene hydroxylation in control rats, has little or no effect on its hydroxylation in phenobarbital-treated rats, and strongly inhibits its metabolism in liver microsomes of methylcholanthrene-treated rats.[24] Differential inhibition has also been demonstrated in reconstituted systems containing different purified cytochrome P-450 species.[25,26]

Among the inhibitors studied, CO is perhaps the most widely used for demonstration of cytochrome P-450 involvement due to its ability to coordinate with the reduced cytochrome P-450 iron. However, a note of caution is warranted since the experimental design of some of the reports is less than desirable. One should not simply replace oxygen by CO in the reaction mixture since inhibition of metabolism under this condition could reflect the absence of sufficient oxygen rather than inhibition by CO. Ideally, the experimental approach of Cooper et al.[6] should be followed, i.e., using a series of gas mixtures of CO and O_2 with different N_2 to O_2 mixtures as controls. Reversal of CO inhibition with monochromatic light at different wavelengths is also a valuable test for cytochrome P-450 involvement.

Antibodies produced against purified cytochrome P-450 provide another useful approach in determining the involvement of cytochrome P-450. Several studies[27-29] have shown that microsomal metabolism can be inhibited by such antibodies indicating that at least part of the cytochrome P-450 is exposed to the hydrophilic environment on the exterior of microsomal membranes. Again, due to the presence of various molecular forms of cytochrome P-450, the extent of antibody inhibition is dependent not only on the source of the antibody and microsomes but also on the specific substrate and the reactions being catalyzed.

C. Spectral Studies

Many compounds can interact with microsomal cytochrome P-450 to perturb the protein and result in characteristic spectral changes that have been labeled as "type I spectra," "type II spectra," and "reversed type I spectra."[30-32] Such spectral interaction has been used as a criterion for the involvement of cytochrome P-450 in metabolism. Since spectral changes usually measure only the interaction between a compound and cytochrome P-450, a conclusion on the catalytic role of cytochrome P-450 must be supported by other studies.

The metabolites of some substrates can form a complex with reduced cytochrome P-450 having an absorbance maximum at 455 nm.[33] This type of cytochrome P-450 metabolite complex can be observed spectrally only in the presence of NADPH indicating that product formation is necessary for the formation of this complex. Observation of this type of complex provides good evidence for the involvement of cytochrome *P*-450 in a specific reaction.

D. Studies Using the Reconstituted System Containing Purified Cytochrome *P*-450

The resolution of the microsomal monooxygenase system[34] into three components (cytochrome *P*-450, NADPH-cytochrome *c* reductase or cytochrome *P*-450 reductase (Chapter 8, this volume), phospholipid) provides an additional approach. Since the two protein components have been purified, an absolute dependency on cytochrome *P*-450 for the metabolism of many substrates can be demonstrated. Because of the existence of multiple forms of cytochrome *P*-450, a negative answer must be interpreted with care since the form(s) of cytochrome *P*-450 used for the metabolism study may not catalyze the reaction under study.

III. OXIDATIVE REACTIONS

A. Aliphatic Hydroxylation

The aliphatic hydroxylation of laurate was one of the first reactions observed to be catalyzed by the reconstituted cytochrome *P*-450 system resolved from rabbit liver.[35] Fatty acids undergo hydroxylation by microsomes at the ω-carbon (CH$_3$ group) and the ω-1-carbon (the next to the last carbon). Evidence that this reaction was mediated by cytochrome *P*-450 came from the observations that CO inhibited the hydroxylation of laurate by 80%, while phenobarbital treatment of rats induced aliphatic hydroxylation activity by liver microsomes.[36] Lu and Coon[35] solubilized and resolved the cytochrome *P*-450 system from rabbit liver into three components: cytochrome *P*-450, NADPH cytochrome *c* reductase, and phosphatidylcholine. When the three fractions were combined, they were capable of mediating the hydroxylation of lauric acid. Other aliphatic hydroxylations which could be performed by this reconstituted mixed function oxidase system included the hydroxylation of hexane, cyclohexane, and the series of fatty acids from hexanoate to palmitoleate.[37] The reconstituted system from rat liver microsomes was capable of hydroxylating cyclohexane, hexobarbital and the series of *n*-alkanes from hexane to decane in addition to dodecane, tetradecane, and hexadecane.[37] As is evident, the cytochrome *P*-450 system is capable of hydroxylating a wide variety of aliphatic substances.

Aliphatic hydroxylation is site selective, i.e., certain regions of the molecules are metabolized selectively. The hydroxylation of n-heptane by rat liver microsomes, for example, occurred at all four possible positions, although inducers had a differential effect upon the hydroxylation of n-heptane at the various sites.[38] Thus, phenobarbital treatment of the rats led to a 4- to 5-fold increase in hydroxylation at the 2-, 3-, and 4- positions whereas hydroxylation at the 1-position was increased by only 60%. Treatment of the rats with benzo[a]pyrene, an inducer similar to, heptane methylcholanthrene, led to a decrease in the hydroxylation of heptane at the 1- and 2-positions but increased hydroxylation at the 3- and 4-positions by 4- and 12-fold, respectively. Inhibition of hydroxylation by CO was also different for the reaction at carbons 1 and 2. Cytochrome P-450 fractions from the liver microsomes of control rats and from rats treated with the two inducers showed varying degrees of hydroxylation of testosterone at the 6-, 7-, and 16-positions.[39] Further studies of testosterone hydroxylation, using purified forms of cytochrome P-450, have confirmed the site selectivity of aliphatic hydroxylation by different purified forms of cytochrome P-450.[40-42]

The mechanism by which cytochrome P-450 catalyzes the hydroxylation of aliphatic compounds is unknown. Groves $et\ al.$[43] have proposed that aliphatic hydroxylation is a stepwise process [Eq. (1)] involving the formation of iron peroxide, hydrogen abstraction to yield a free radical, followed possibly by formation of a carbonium ion and the attack on the iron bound carbonium ion or free radical by the substrate.

$$Fe^{3+} - OOH \xrightarrow{\ ^{-}OH\ } Fe^{5+} = O \xrightarrow{\ RH\ } Fe^{4+} - OH + R\cdot] \longrightarrow Fe^{3+} + ROH \quad (1)$$

Data to support a radical abstraction mechanism come from the large isotope effect observed in the hydroxylation of tetradeuterated norbornane by purified cytochrome P-450$_{LM_2}$ isolated from phenobarbital-induced rabbits.[43] An isotope effect of 11.5 was observed in the formation of norborneol,[43] while an isotope effect of 11 was reported in the hydroxylation of 1,3-diphenylpropane-1,1-d_2 by rat liver microsomes.[44] Both laboratories concluded that the large isotope effect was consistent with a radical abstraction mechanism similar to that observed in the hydroxylation of aliphatic compounds catalyzed by metal-containing peroxidases.[45]

Since hydrophobic aliphatic compounds must be hydroxylated in order to be able to be eliminated as polar metabolites by the kidney, their hydroxylation is a detoxication type reaction. Without hydroxylation these compounds would be expected to either accumulate in the body to toxic levels or, as in the case of hexobarbital, exert a pharmacological effect for a much longer time.

B. Aromatic Oxidation

There are two means by which a hydroxyl group can be added to an aromatic system. One method is by direct insertion of the oxygen into the C—H bond; the other is by oxygen addition to the carbon–carbon double bond of the aromatic system and subsequent rearrangement to form a hydroxyl group. The latter method of aromatic hydroxylation is the predominant one and leads to the migration of groups (H, ^{2}H, ^{3}H, CH$_3$, etc.) that are on the carbon atoms being hydroxylated. This migration has been termed the "NIH shift."

In attempting to determine why assays for the enzymatic hydroxylation of aromatic compounds using tritiated substrates were not working as expected, the NIH shift was discovered.[46] Compounds which had been specifically tritiated in the position at which hydroxylation occurred failed to lose the amount of tritium expected, based on the amount of hydroxylated product that was formed. Using deuterated compounds, which could be shown to have all the deuterium in the correct position, deuterium was found to remain in the product but had migrated to an adjacent position. The proposed mechanism is shown in Eq. (2).

$$(2)$$

This shift during hydroxylation is not restricted to isotopes of hydrogen; bromine and chlorine atoms and alkyl groups can also migrate.[46] The amount of NIH shift (deuterium or tritium retention) depends on the substrate.[47] For a substrate that has an ionizable hydrogen on its ring substituent, such as salicylate or aniline, the amount of retention of deuterium is low: 0 and 6%, respectively. However, compounds that do not have an ionizable hydrogen on the side chain retain a large percentage of the deuterium, e.g., anisole (60%) and toluene (54%). The mechanism of the NIH shift predicts a cyclohexadieneoid intermediate [Eq. (3)]

$$\text{(structure with X, HO, D)} + H^+ \longrightarrow \text{(structure with XH, OH)} + D^+ \tag{3}$$

which contributes to the loss of deuterium. Such an intermediate can be formed only when an ionizable hydrogen is present on the side chain whose removal permits conjugation of the electrons of the substituent with that of the aromatic ring. The more acidic the ionizable hydrogen, the less the retention of deuterium observed. Similarly, the higher the pH of the incubation mixture, the lower the retention. The pH of the reaction has no effect on the amount of the NIH shift observed with substrates such as toluene and anisole.

These studies led to the discovery that arene oxides were important intermediates in the hydroxylation of aromatic compounds by drug-metabolizing enzymes and that the arene oxides were chemically reactive intermediates which might be responsible for the toxic effects of aromatic hydroxylation, including carcinogenicity and mutagenicity.[48,49] Cytochrome P-450 enzymes are capable of hydroxylating aromatic compounds both through an arene oxide intermediate, causing a possible NIH shift,[50] as well as by direct insertion of an oxygen atom into the molecule.[51-53] Rahimtula et al.[50] demonstrated an NIH shift in the hydroxylation of acetanilide by purified cytochromes P-450 and P-448. The different forms of cytochrome P-450 also differed with respect to the amount of shift which occurred. These shifts were comparable to the shifts observed in the hydroxylation of acetanilide catalyzed by the microsomes from which the purified enzymes were isolated. It was also shown that the purified enzymes were capable of arene oxide formation using phenanthrene as the substrate.

The metabolism of chlorobenzene (I) was found to involve both arene oxide and direct insertion pathways. The 3- (II) and 4-chlorobenzene oxides (IV) rearrange to o-chlorophenol (III) and p-chlorophenol (V) respectively. The presence of o-, m- and p-chlorophenols indicates that m-chlorophenol (VI) was formed by direct insertion of oxygen. The formation of all three phenols was inhibited in intact microsomes by CO, metyrapone, SKF 525-A, and 7,8-benzoflavone. The reconstituted cytochromes P-450 and P-448 were also capable of forming all three chlorophenols, indicating that the cytochrome P-450 system is capable of direct insertion of oxygen besides just arene oxide formation. Tomaszewski *et al.*[53] observed isotope effects ($k_H/k_D = 1.3-1.75$) for the formation of meta-hydroxylated products of nitrobenzene, methyl phenyl sulfide, and methyl phenyl sulfone *in vivo*. Animals were treated with a 1:1 mixture of the normal substrate and perdeuterated substrate, and urinary metabolites were isolated and analyzed. Under these conditions only the meta-hydroxylated products showed a significant isotope effect. This would be expected since a direct insertion mechanism would have an effect on the C—H bond while the ortho- and para-hydroxylated products, formed by the arene oxide pathway and subsequent nonenzymatic rearrangement, should not have an isotope effect associated with it. Billings and McMahon[51] also found that biphenyl (**VII**) hydroxylation occurred at ortho, meta, and para positions. Hydroxylation at all three

sites was inhibited by CO, SKF 525-A, and DPEA and required NADPH and O_2. Only hydroxylation at the meta position showed a significant isotope effect (1.27–1.43), indicating again a nonarene oxide pathway for meta-hydroxylation. Thus, there is evidence that hydroxylation of several aromatic compounds can occur by a mechanism other than one involving an arene oxide intermediate, i.e., by a direct insertion mechanism which results in a small but significant isotope effect.

However, the major pathway of aromatic hydroxylation does occur by the formation of an arene oxide. The toxicological significance of the arene oxide pathway lies in the fact that these are electrophilic compounds known to react with cellular nucleophiles to cause necrosis, mutations, and carcinogenesis.[48,49] The metabolism of bromobenzene (VIII) illustrates the toxicity of an arene oxide and the role that the different forms of cytochrome *P*-450 play in the metabolism of this compound.[54-57] Bromobenzene is metabolized to its 2,3-oxide (IX) and its 3,4-oxide (X). These epoxides can rearrange to the 2- and

Br   Br   Br

(VIII)   (IX)

O

(X)

4-bromophenols, respectively, be hydrated by epoxide hydrase (Volume II, Chapter 15) to form their respective dihydrodiols or be acted upon by glutathione transferase (Volume II, Chapter 7) to give glutathione conjugates. The 3,4-oxide of bromobenzene (X) is apparently responsible for the liver necrosis caused by bromobenzene.[54] Studies on the effects of inducers of cytochrome *P*-450 systems on the level of liver necrosis revealed that phenobarbital treatment of rats increased their susceptibility to bromobenzene-induced liver necrosis[55] while methylcholanthrene treatment of the rats decreased the level of liver necrosis.[56] Analysis of the urinary metabolites indicated that, after a toxic dose of bromobenzene (VIII), treatment with the inducer caused a 50% decrease in the formation of 4-bromophenol, whereas the amount of 2-bromophenol and bromophenyldihydrodiol increased 5- and 4-fold, respectively, compared to the levels of these compounds in untreated rats.[57] Thus, the reduction in the toxicity of bromobenzene by this treatment may be due to the shift in metabolism away from bromobenzene 3,4-oxide by enhancing 2,3-oxide formation and by increasing the amount of 3,4-dihydrodiol formed. Phenobarbital treatment did not radically change the relative amounts of bromobenzene metabolites, but it did decrease the half-life of bromobenzene by 40%; the increased toxicity in phenobarbital-treated rats could be due to the more rapid formation of the bromobenzene 3,4-oxide which cannot be detoxified sufficiently rapidly by epoxide hydrase or glutathione transferase. The toxicity of bromobenzene appears, therefore, to be the result of an arene oxide whose formation is controlled by levels of the different forms of cytochrome P-450 forming it.

Recent studies on the metabolism of benzo[a]pyrene (BP) (**XI**) are of interest in that they show both the regioselectivity, i.e., the area of the molecule metabolized, and the stereoselectivity, i.e., the formation of optically active metabolites, of cytochrome P-450 catalyzed reactions, with respect to both the aromatic oxidation of BP itself and the alkene oxidation of BP-7,8-dihydrodiol. BP is converted to many compounds

(XI)

including arene oxides at the 4,5-, 7,8-, and 9,10-positions which can be hydrated to the corresponding dihydrodiols by epoxide hydrase or, in the case of the 7,8- and 9,10-oxides, can rearrange spontaneously to 7-hydroxy- and 9-hydroxy-BP, respectively. The metabolism of BP by rat liver microsomes is inhibited by CO; reversal of inhibition follows the action spectra of cytochrome P-450.[58] Purified cytochromes P-448 and P-450 from rat liver microsomes are both capable of metabolizing BP, but cytochrome P-448 (isolated from methylcholanthrene-treated rats) is much more active than is cytochrome P-450 (isolated from phenobarbital-treated rats).

The regioselectivity of BP metabolism is well established. Holder *et al.*[59] used control and induced rat microsomes, and a reconstituted purified cytochrome P-448 system supplemented with epoxide hydrase to metabolize BP to diols, quinones, and phenols. Analysis of the results demonstrated the regioselectivity in the metabolism of BP. For example, while the overall *in vitro* metabolism of BP was increased to 1.5 and 3 times the control value by phenobarbital and 3-methylcholanthrene treatment of the rats, the formation of BP-7,8-dihydrodiol was decreased by 60% in phenobarbital-treated microsomes but increased 6-fold in methylcholanthrene-treated microsomes. Differences were also seen in most of the other metabolite fractions. Yang *et al.*[60] have also shown that the induction of cytochrome P-450 usually does not result in an equal increase among the metabolites formed; the ratios of metabolites formed with methylcholanthrene-induced microsomes to that formed by control microsomes varied by a factor of ten.[60] Accordingly, the amount of the metabolite formed at the different ring locations is determined by the form of cytochrome P-450 used for the metabolism. With different purified forms of cytochrome P-450 isolated from rabbit liver, strong site preferences were observed indicating that the regioselectivity of metabolism resides in the cytochrome P-450.[61]

The stereoselectivity of the metabolism of BP by the cytochrome P-450 system is observed in the formation of the dihydrodiols of BP, which are produced through arene oxide formation followed by hydration by epoxide hydrase. Optical activity of the resultant dihydrodiols could be due both to stereoselectivity of arene oxide formation or to stereoselectivity in the addition of water to the arene oxide. The stereoselectivity of hydration can be determined separately by using arene oxides of known optical activity. Thakker et al.[62] and Yang et al.[63-65] have done such studies using the racemic 4,5-, 7,8-, and 9,10-arene oxides of BP. Only low levels of enantiomer were observed in the formation of BP 7,8-dihydrodiol [8% optical purity, 54% (−)-enantiomer, 46% (+)-enantiomer] and BP 9,10-dihydrodiol (22% optical purity), whereas BP 4,5-dihydrodiol was 78% optically pure.[62] The (−)-enantiomer predominated in all three cases. Since the optical purity of the three dihydrodiols formed from BP by liver microsomes of methylcholanthrene-treated rats was determined to be 92%,[66] the data indicate that the arene oxide precursors of the dihydrodiols are of high optical activity. Yang et al.[64,65] likewise found that the formation of the arene oxide was stereoselective, suggesting the predominantly one-sided attack of oxygen on one plane of the BP molecule. This attack must be determined by the positioning of the compound at the active site of the cytochrome.

Another example of the stereoselective nature of metabolism catalyzed by the different forms of cytochrome P-450 is the metabolism of the BP 7,8-dihydrodiol to diol epoxides. This is an epoxidation of a benzylic double bond and not an aromatic hydroxylation. Yang et al.[63,64,67] presented evidence for high stereoselectivity in the formation of diol epoxides from (−)-BP-7,8-dihydrodiol (XII). With purified cytochrome P-450 isozymes and both enantiomers of BP-7,8-dihydrodiol, (−)-BP-7,8-dihydrodiol (XII) was metabolized to its diol epoxides with high stereoselectivity by cytochrome P-448, forming approximately 76% (+)-diol epoxide 2 (XIII) and 17% (−)-diol epoxide 1 (XIV); (+)-BP-7,8-dihydrodiol (XV) was metabolized to approximately 77% (+)-diol epoxide 1 (XVI), and 2% (−)-diol epoxide 2 (XVII).[66] The liver microsomes from control and induced rats also showed stereoselectivity in the metabolism of the two dihydrodiols which varied in the extent of stereoselectivity depending on the source of microsomes. Deutsch et al.,[68] using purified forms of cytochrome P-450 from rabbit liver, also found large differences between the purified forms with regard to formation of the diol epoxides 1 and 2. The amounts of the diol epoxides formed depend on the enantiomeric composition of the substrate and on the type of cytochrome *P*-450 catalyzing the reaction. The stereoselectivity and regioselectivity of catalysis can have a profound effect on the biological activity of a compound. Site selective metabolism at the 7,8-position of BP and strong stereoselectivity toward the formation of (−)-BP-7,8-dihydrodiol (**XII**) and its stereoselective conversion to (+)-diol epoxide 2 (**XIII**) are probably responsible for the strong carcinogenicity of BP. In the newborn mouse, for example, (+)-diol epoxide 2 (**XIII**) is 34 times more carcinogenic than any of the other diol epoxides.[69]

C.　Alkene Epoxidation

The cytochrome P-450 system is capable of oxidizing carbon–carbon double bonds that are not part of an aromatic system leading to the formation of epoxides as products. Although the diol which is formed from the epoxide upon addition of water is much more polar than the parent compound and therefore more easily excreted, the epoxide is a reactive intermediate. Therefore, the epoxidation of an alkene group usually cannot be considered a detoxication reaction. Among the compounds that are activated to toxic metabolites by alkene epoxidation are some of the most potent carcinogens, aflatoxin B_1 (AFB$_1$) (**XVIII**) and the 7,8-dihydrodiol of BP (**XIX**). AFB$_1$ is a natural product formed by the mold *Aspergillus flavus* and is the most potent carcinogen known causing the formation of liver tumors in many species of animals.[70] Several studies

(XVIII)　　　　　　　　　　　　　　　(XIX)

indicate that the metabolic product which is responsible for the carcinogenicity of AFB$_1$, i.e., the ultimate carcinogen of AFB$_1$, is an epoxide. These include identification of the major product of AFB$_1$ metabolism which binds to nucleic acids and studies of bacterial toxicity and mutagenicity of AFB$_1$ and its structural analogues. The bacterial toxicity studies of Garner *et al.*[71] were the first to suggest that the 2,3-double bond of AFB$_1$ was required for a high level of biological activity. This was interpreted as meaning that a 2,3-oxide was probably the reactive intermediate responsible for the toxicity and most likely the carcinogenicity of AFB$_1$. This finding was later confirmed with mutagenicity tests.[72] The 2,3-dihydrodiol of AFB$_1$ could be isolated from the acid hydrolysate of the AFB$_1$-RNA adduct formed by the liver microsomes from rats and hamsters,[73,74] and an AFB$_1$-guanine adduct was isolated after incubation of AFB$_1$ with DNA in the presence of rat liver microsomes. The latter was identified as 2,3-dihydro-2-(N^7-guanyl)-3-hydroxyaflatoxin B$_1$, again indicating the 2,3-epoxide of AFB_1 as the reactive intermediate.[75] The products bound to nucleic acids are the same whether the reactive intermediate is generated by microsomal metabolism of AFB_1 or by treatment of AFB_1 with *m*-chloroperbenzoic acid, the latter expected to give the 2,3-oxide of AFB_1.[76]

Evidence for the involvement of a cytochrome *P*-450 in the formation of the AFB_1-2,3-oxide came from data showing an increase in the *in vitro* binding of AFB_1 to RNA upon treatment of hamsters and rats with phenobarbital[71,73,77] and to DNA upon phenobarbital or 3-methylcholanthrene treatment.[77,78] This binding required NADPH and molecular oxygen. SKF 525-A inhibited the formation of a glutathione-AFB_1 adduct *in vitro* (the formation of which is presumed to involve the 2,3-oxide of AFB_1[79]) and the binding of AFB_1 to DNA and RNA.[77]

The epoxidation of the 7,8-dihydrodiol of BP (**XIX**) has been discussed under the section on aromatic oxidation (Section III,B). The epoxidation of the benzylic 9,10-double bond of the BP-7,8-dihydrodiol (**XIX**) leads to the formation of very reactive diol epoxides. As reviewed by Levin *et al.*[80] and Yang *et al.*[81], the 7,8-diol 9,10-epoxides of BP (**XX**) are responsible

(XX)

for the biological activity of BP. Further studies on the metabolic activation of polycyclic aromatic hydrocarbons to ultimate carcinogens and mutagens have led to a general acceptance of the bay region theory of polycyclic aromatic hydrocarbon carcinogenesis proposed by Jerina.[82] Jerina and co-workers, on the basis of a quantum mechanical model, proposed that the bay region diol epoxides of polycyclic aromatic hydrocarbons, if formed metabolically, are most likely responsible for the carcinogenicity of these compounds. On this basis, the carcinogenicity of many polycyclic aromatic hydrocarbons probably depends on the oxidation of the bay region nonaromatic double bond by a cytochrome *P*-450 system to form the ultimate carcinogenic metabolite.

There are several other examples of oxidation of alkenes to epoxides of toxicological concern. Vinyl chloride is carcinogenic to humans. Its mutagenicity is increased by phenobarbital pretreatment of the rats whereas vinyl chloride epoxide is mutagenic without further metabolic activation.[83] The insecticides heptachlor and aldrin also undergo epoxidation to form the stable epoxides, heptachlor epoxide and dieldrin,[84] a reaction inhibited by SKF 525-A. Finally, the destruction of cytochrome *P*-450 by allyl containing compounds such as secobarbital, allobarbital, allylisopropylacetylcarbamide, and allylisopropylacetamide is the probable result of the epoxidation of the allyl double bond by cytochrome *P*-450 enzymes.[85,86]

Although alkene epoxidation is the mechanism by which a reactive intermediate is formed in the noted examples, epoxidation of a carbon–carbon double bond does not always lead to a more toxic product. For example, dieldrin is somewhat less toxic than aldrin and the *trans*-α,β-epoxide of 4-acetylaminostilbene is nonmutagenic in contrast to its parent, 4-acetylaminostilbene.[87] Nevertheless, in general, the epoxidation of an alkene double bond must be looked upon as a mechanism by which reactive intermediates are formed.

D. Nitrogen Dealkylation and Oxidative Deamination

The oxidative N-dealkylation of drugs has been known for 40 years.[88] Since the initial report of the N-dealkylation of dimethylbarbital *in vivo*, the reaction has become recognized as a major pathway in drug metabolism for a variety of primary, secondary, and tertiary amines.[89] With the exception of *N-tert*-butylnorchlorcyclizine [Eq. (4)],[90] the mechanism of cytochrome *P*-450-catalyzed N-dealkylation has been investigated in terms of either the initial oxidation of the amine nitrogen or the oxidation of the α-carbon atom. Since reviews are available on N-oxidation of

$$\cdots \longrightarrow \cdots + \quad C=O \;+\; CO_2 \quad (4)$$

primary and secondary amines[12] and on the role of tertiary amine *N*-oxides in drug metabolism,[91] (Volume I, Chapter 9 and Volume II, Chapter 2) these topics will be treated here only to the extent that they relate to representative N-dealkylation reactions. Both N-dealkylation of tertiary amines and the deamination of amphetamine have been extensively investigated and represent excellent studies on the mechanism of the N-dealkylation. Finally, the potential significance of the N-dealkylation reaction in toxicity and carcinogenicity is demonstrated in the metabolic activation of dimethylnitrosamine.

Tertiary amines undergo oxidative N-dealkylation by microsomal and purified cytochrome *P*-450 systems in the presence of O_2 and NADPH to form aldehyde and secondary amine products [Eq. (5)]. These products could result from either α-carbon oxidation to form a carbinolamine intermediate or through nitrogen oxidation to form a tertiary amine *N*-oxide intermediate.

$$R_2NCH_2R \xrightarrow[\substack{P\text{-}450,\ O_2,\\ NADPH}]{} R_2NH + OHCR \tag{5}$$

N-Oxides have been produced after incubation with microsomes and there is precedence for tertiary amine *N*-oxide dealkylations in the chemical literature. For example, the Polonovski reaction[92] has been proposed as a chemical model for the enzymatic N-dealkylation of tertiary amine *N*-oxides.[93] N-Dealkylation by the Polonovski reaction requires acylation of the *N*-oxide by acetic anhydride. Since acetic anhydride is unknown in biological systems, other proposals for a biological analogy of the Polonovski reaction have been advanced[93-95] including some iron-containing model systems of cytochrome *P*-450. Ferris *et al.*[95] have demonstrated the Fe(II)-catalyzed N-demethylation of trimethylamine *N*-oxide by the following mechanism.

$$(CH_3)_3NO + H^+ \rightarrow (CH_3)_3\overset{+}{N}OH \tag{6}$$

$$(CH_3)_3\overset{+}{N}OH + Fe(II) + H^+ \rightarrow (CH_3)_3\overset{+}{N}\cdot + H_2O + Fe(III) \tag{7}$$

$$(CH_3)_3\overset{+}{N}\cdot + Fe(III) \rightarrow (CH_3)_2\overset{+}{N}{=}CH_2 + Fe(II) + H^+$$

$$\xrightarrow{H_2O} CH_2O + (CH_3)_2\overset{+}{N}H_2 \tag{8}$$

$$(CH_3)_3\overset{+}{N}\cdot + Fe(II) + H^+ \rightarrow (CH_3)_3\overset{+}{N}H + Fe(III) \tag{9}$$

The tertiary amine *N*-oxide undergoes iron-catalyzed N-demethylation through Eq. (6)–(8). Furthermore, this mechanism provides a pathway for the Fe(II)-catalyzed reduction of tertiary amine *N*-oxides to tertiary amines [Eq. (6), (7), and (9)], products that have been observed under anaerobic conditions with Fe(II) cytochrome *P*-450.[96]

Alternatively, N-dealkylation of tertiary amines could proceed through initial α-carbon oxidation to form an unstable carbinolamine intermediate which would dissociate to the dealkylated secondary amine and aldehyde products [Eq. (10)]. Carbinolamine intermediates have been inferred, in certain cases, from product structures such as in the cyclization products

$$R_2NCH_2R \xrightarrow[\substack{P\text{-}450,\ O_2,\\ NADPH}]{} R_2NCH(OH)R \longrightarrow R_2NH + OHCR \tag{10}$$

resulting from Lidocaine [Eq. (11)][97] and Bisolvon [Eq. (12)][98] metabolism. Stable carbinolamines are rare and have usually been identified from

compounds giving cyclic carbinolamines, such as nicotine [Eq. (13)][99] and medazepam [Eq. (14)].[100] However there have been reports of stable acyclic carbinolamines,[101,102] which were invariably stabilized through nitrogen conjugation with an adjacent carbonyl or aromatic system as has been shown for N-methylcarbazole [Eq. (15)].[102]

In addition, small deuterium isotope effects ($k_H/k_D < 1.5$) have been observed in the N-dealkylation of N-trideuteriomethylethylmorphine (**XXI**)[103] and lidocaine-d_4 (**XXII**)[104] and interpreted as primary kinetic isotope effects indicative of a rate-limiting α-carbon oxidation in the formation of a carbinolamine intermediate.

(XXI)   (XXII)

It has been argued, however, that the small isotope effects observed for many cytochrome *P*-450-catalyzed N-dealkylation reactions could be the result of a secondary deuterium isotope effect consistent with an *N*-oxide intermediate.[95] This is based on the observation of a similar isotope effect in the oxidation of trimethylamine by chlorine dioxide, a reaction known to proceed by electron abstraction from nitrogen rather than α-carbon oxidation.[105] It should be noted, however, that a small isotope effect is insufficient evidence for the absence of a primary isotope effect in a multistep enzyme-catalyzed reaction[106] or even in a single step chemical reaction,[107] and, therefore, the isotope effect data alone cannot distinguish between these two mechanisms.

The iron-catalyzed *N*-oxide dealkylation pathway [Eq. (6)–(8)] leads to an iminium ion intermediate which undergoes hydration followed by dissociation to amine and aldehyde products [Eq. (8)]. In this case, the solvent water is the source of oxygen in the aldehyde product. In contrast, direct α-carbon oxidation produces a carbinolamine intermediate and results in the incorporation of oxygen from molecular oxygen in the resulting aldehyde product. Unfortunately, rapid oxygen exchange between the aldehyde and solvent water has usually prevented direct assignment of the aldehyde oxygen source through $^{18}O_2$ studies. McMahon *et al.*[108] have ingeniously overcome much of this difficulty by trapping the aldehyde oxygen from exchanging with water by the rapid reduction of the carbonyl to the nonexchangeable alcohol. The *N*-benzyl-4-phenyl-4-carbethoxypiperidine substrate (**XXIII**) was incubated with liver microsomes, NADPH, and $^{18}O_2$ and the benzaldehyde oxygen (**XXV**) was prevented from exchanging with water by reduction to the alcohol (**XXVI**) catalyzed by alcohol dehydrogenase.

(XXIII)   (XXIV)   (XXV)

$$\text{(XXV)} \quad \text{—CH}^{18}\text{O} \xrightarrow[\text{NADH}]{\text{ADH}} \text{—CH}_2{}^{18}\text{OH} \quad \text{(XXVI)}$$

The deamination of amphetamine (**XXVII**) originally described by Axelrod[110] has been extensively investigated and as a result has contributed greatly to our understanding of the N-dealkylation reaction (for review, see Cho and Wright[111]). The deamination of amphetamine to phenylacetone (**XXIX**) is an example of a primary amine N-dealkylation reaction. As with tertiary amine N-dealkylation, our understanding of the mechanism of amphetamine deamination has prospered because of a controversy over whether initial α-carbon oxidation (**XXVIII**) or nitrogen oxidation (**XXX**) results in the formation of the N-dealkylated product, phenylacetone (**XXIX**).

(XXVII)   (XXVIII)   (XXIX)

(XXX)   (XXXI)

Brodie *et al.*[112] originally proposed the carbinolamine (**XXVIII**) as an intermediate in amphetamine deamination, although this compound has never been identified. On the other hand, N-hydroxyamphetamine (**XXX**) and the oxime (**XXXI**) are known metabolites of amphetamine.[113-116] Furthermore, Hucker *et al.*[116] have suggested that the oxime is enzymatically converted to phenylacetone; this proposal has been verified.[117,118] On the other hand, there is evidence that the conversion of the oxime to phenylacetone is a minor pathway in amphetamine deamination since isotope dilution experiments indicate that phenylacetone must be formed substantially through a pathway not involving the hydroxylamine intermediate.[119] Furthermore, a substantial deuterium-isotope effect ($k_H/k_D \sim 4$) has been observed for amphetamine-2-*d* in the formation of both phenylacetone and the oxime.[109] This isotope effect is consistent with the rate-limiting carbon–hydrogen bond cleavage involved in the formation of the carbinolamine (**XXVIII**).[109] In addition, this α-deuterium substitution increases hydroxylamine formation consistent with the metabolic switching away from the carbinolamine to the hydroxylamine (**XXX**). Parli and McMahon have also demonstrated that only about 30% ^{18}O enrichment occurs in the phenylacetone formed from the deamination of amphetamine under an $^{18}O_2$ atmosphere suggesting that most of the phenylacetone is derived through hydrolysis of the imine (**XXXII**) rather

$$\text{(structure: phenyl)}-CH_2-\underset{\underset{CH_3}{|}}{C}=NH$$

(**XXXII**)

than by aminolysis of the carbinolamine (**XXVIII**).[109] Despite evidence for the formation of the oxime during amphetamine metabolism, available data indicates that most of the phenylacetone is derived from the carbinolamine rather than from the _N_-hydroxylamine intermediate.

The N-dealkylation reaction has proved to be of great significance in the hepatotoxicity and carcinogenicity of dialkylnitrosamines (for review, see Montesano and Bartsch[120]). For example, dimethylnitrosamine (**XXXIII**) is completely inactive as a mutagen when tested directly in the Ames' _Salmonella_ and other cellular assays but is activated by metabolism to mutagenic and toxic products in the presence of cytochrome _P_-450-containing enzyme systems.[120–122] Hepatic cytochrome _P_-450 has been shown to be involved in the activation of dimethylnitrosamine [Eq. (16)]

$$\underset{(\text{XXXIII})}{\underset{H_3C}{\overset{H_3C}{>}}N-N\overset{}{\underset{O}{\diagdown}}} \longrightarrow \underset{(\text{XXXIV})}{\underset{H_2C}{\overset{H_3C}{>}}N-N\overset{}{\underset{O-H}{<}}O} \longrightarrow \underset{(\text{XXXV})}{\underset{H_3C}{>}N=N-OH} + CH_2O \qquad (16)$$

$$\longrightarrow H_3C^+ + N_2 + HO^-$$

in laboratory animals[121,123] and in human liver microsomes.[122] The carbon monoxide inhibition of microsomal dimethylnitrosamine activation is maximally reversed by 450 nm light providing strong evidence for cytochrome _P_-450 involvement.[122] Furthermore, dimethylnitrosamine demethylase activity has been reconstituted in a cytochrome _P_-450-containing enzyme system isolated from liver microsomes.[124]

Lijinsky _et al._[125] have demonstrated that perdeuterated dimethylnitrosamine is metabolized to a product capable of methylating the guanine of mammalian DNA and RNA. Furthermore, the 7-methylguanine product retains all three deuteriums of the _N_-methyl group of dimethylnitrosamine, consistent with the mechanism illustrated for carbonium ion generation but inconsistent with the diazomethane-carbene mechanism proposed earlier for the methylation of proteins[126] and nucleic acids.[127]

Deuterium isotope effect studies support the α-carbon oxidation intermediate (**XXXIV**) which could then rearrange to form the intermediate (**XXXV**) and the observed N-dealkylated product, formaldehyde. Loss of nitrogen results in the formation of the alkylating carbonium ion species. The *N*-nitrosocarbinolamine intermediate (**XXXIV**) is analogous to the carbinolamine intermediates which have been proposed for N-dealkylation reactions. In addition, kinetic isotope effects have been observed in the rate of formaldehyde formation *in vitro*,[128] mutagenicity *in vitro*,[129] and carcinogenicity *in vivo*,[130] all consistent with the rate-limiting formation of the *N*-nitrosocarbinolamine (**XXXIV**) during the metabolic activation of dimethylnitrosamine. Thus, current evidence indicates that the carbinolamine mechanism, originally proposed by Brodie *et al.*[112] as a general mechanism for N-dealkylation reactions, is a viable mechanism for primary and tertiary amine N-dealkylations and for the activation of *N*-alkylnitrosamines to carcinogenic metabolites.

E. Oxygen Dealkylation

The oxidative O-dealkylation reaction is a common metabolic pathway for drug substrates containing the ether functional group.[89] For example, both indomethacin (**XXXVI**) and phenacetin (**XXXVII**) are extensively metabolized in man by O-dealkylation resulting in the corresponding phenol products.[131,132]

(XXXVI)   (XXXVII)

O-Dealkylation is analogous to the N-dealkylation discussed previously and is also believed to proceed through initial α-carbon oxidation to form the hemiacetal intermediate since this reaction is associated with a primary kinetic isotope effect [Eq. (17)].[133-136] Furthermore, the alcohol and aldehyde products correspond to the dealkylated amine and aldehyde products observed in the N-dealkylation reaction.

$$\langle\!\rangle\!\!-\!\!OCH_2R \longrightarrow \langle\!\rangle\!\!-\!\!\overset{\overset{\displaystyle OH}{|}}{OCHR} \longrightarrow \langle\!\rangle\!\!-\!\!OH \;+\; RCHO \quad (17)$$

Cytochrome *P*-450 has been implicated in O-demethylation of *p*-nitroanisole in perfused liver preparations[137] by CO inhibition and the requirement for NADPH. Microsomal[138] and purified enzyme systems[139] containing cytochrome *P*-450 have more definitely demonstrated the involvement of this enzyme system in O-dealkylation reactions.

Different isozymes of cytochrome *P*-450 demonstrate considerably different capacities to catalyze the O-dealkylation reaction. For example, 7-ethoxycoumarin (**XXXVIII**) is preferentially metabolized by liver microsomes from 3-methylcholanthrene-treated rats containing cytochrome *P*-448.[138] The specific activity of liver microsomes following phenobarbital induction is about 3-fold less than after 3-methylcholanthrene induction, while very low activity is observed with microsomes obtained from untreated rats. Ethoxyresorufin (**XXXIX**) is also preferentially

(XXXVIII)   (XXXIX)

O-dealkylated by the cytochrome *P*-448 isozyme while untreated and phenobarbital treated rat liver microsomes exhibit negligible activity for this substrate.[139,140] These observations have been confirmed with the purified isozymes.[139]

The hair-dye component, 2,4-diaminoanisole, is metabolized by the cytochrome *P*-450 system to reactive metabolites that are mutagenic and are capable of irreversible binding to microsomal proteins.[141] The O-demethylation of this compound appears to be a detoxication pathway since deuterium substitution of the hydrogens in the methyl group increases both the binding and mutagenicity that is observed.

F. Nitrogen Oxidation

There is now evidence that at least two enzyme systems of liver microsomes are capable of catalyzing the oxidation of nitrogen: cytochrome *P*-450 and a flavoprotein oxidase, amine oxidase, that is discussed in this volume, Chapter 9. Since there is more than one type of enzyme with this capability, it is not surprising that the early evidence for the involvement of cytochrome *P*-450 was not strong; attempts at selective inhibition, the usual means of implicating cytochrome *P*-450 in a reaction, yielded ambiguous results. For example, the carcinogenic aromatic amine

2-acetylaminofluorene (AAF) (**XL**) is metabolized by cytochrome P-450 to its hydroxamic acid, N-hydroxy-2-acetylaminofluorene (N-OHAAF) (**XLI**). This conclusion comes from studies of CO inhibition in which

(XL)   (XLI)

N-hydroxylation of AAF was inhibited approximately 50% at a high concentration of CO (9:1, CO:O_2).[142-144] N-Hydroxylation of AAF was induced by methylcholanthrene,[145] was inhibited by antibodies raised against cytochrome P-450 reductase, and inhibited by treatment of mice with cobalt chloride.[144] Reconstitution studies with partially purified cytochrome P-450 fractions from 3-methylcholanthrene-treated rats[146] and hamsters[147] indicated that cytochrome P-450 was responsible.

Since N-OHAAF is the proximate carcinogenic metabolite of AAF,[148] this reaction results in an increase in toxicity. Other N-hydroxyl derivatives of aromatic amines are known to cause methemoglobinemia.[9] In general, N-hydroxylation of aromatic amines by cytochrome P-450 leads to increased toxicity.

The role of N-hydroxylation in the toxicity of phenacetin (**XLII**) and acetaminophen (**XLIII**) is not clear. It has been proposed that

(XLII)   (XLIII)

acetaminophen is metabolized by a cytochrome P-450 system to N-hydroxyacetaminophen which is dehydrated to form the reactive acetimidoquinone.[149] However, a recent study by the same authors found that N-hydroxyacetaminophen could not be isolated after incubation of acetaminophen with hamster liver microsomes, but could be isolated from a similar mixture with N-hydroxyphenacetin.[150] Thus, it appears that N-hydroxyacetaminophen is not a product of the metabolism of acetaminophen in this system. There is, however, good evidence for the N-hydroxylation of phenacetin.[151] This reaction is of the cytochrome P-450 type; NADPH and O_2 are required, and CO inhibits the reaction.

The role of *N*-hydroxyphenacetin in the toxicity of phenacetin is not clear; phenacetin may also be metabolized by epoxidation.[149]

Aliphatic amines are also capable of undergoing N-oxidation. The N-hydroxylation of phentermine (**XLIV**) in rat and rabbit microsomes was inhibited by CO and SKF 525-A to 60 and 50%, respectively,[152] while that of amphetamine (**XLV**) was inhibited by CO and DPEA and induced

(XLIV)   (XLV)

by phenobarbital.[153] Since these substrates would be expected to be metabolized by both cytochrome *P*-450 and amine oxidase,[154] it is not surprising that there are reports based on inhibition studies that phentermine (**XLIV**) is not metabolized by cytochrome *P*-450.[155] Coutts *et al.*[156] reported that *N*-(*N*-propyl)amphetamine was N-hydroxylated by a system that could be inhibited by 25% by SKF 525-A, indicating at least a partial involvement of cytochrome *P*-450. Dibenzylamine (**XLVI**) is N-hydroxylated by the cytochrome *P*-450 system based on the inhibition of the reaction by CO, SKF 525-A and induction by phenobarbital.[157]

(XLVI)

Further evidence for cytochrome *P*-450 in the N-oxidation of amphetamine type compounds comes from the observation that, when these compounds or their *N*-hydroxyl derivatives are metabolized by liver microsomal systems, a metabolite–cytochrome complex with a 455 nm absorption maximum is formed[158,159]; the complex is also formed with reconstituted, purified cytochrome *P*-450.[160] The metabolite responsible for formation of the complex may be the nitroso derivative of the amphetamine[161] or, possibly, a nitroxide or nitrone.[162-164] Thus, it appears that cytochrome *P*-450 is capable of oxidizing primary and secondary aliphatic amines by N-hydroxylation and of oxidizing the N-hydroxyl metabolite to nitroso derivatives (**XLVII**) that are tautomeric with their oximes (**XLVIII**). The oxime of amphetamine has been observed as a metabolite of amphetamine.[116] The oxidation of primary or secondary amines are summarized by [Eq. (18)]. The oxime is a possible

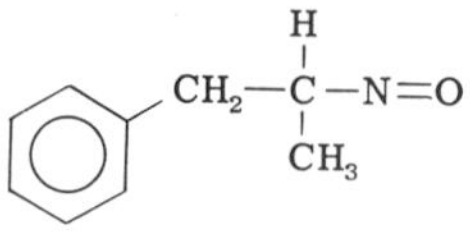

$$
\underset{(XLVII)}{\text{C}_6\text{H}_5-\text{CH}_2-\overset{\text{H}}{\underset{\text{CH}_3}{\text{C}}}-\text{N}=\text{O}}
\qquad
\underset{(XLVIII)}{\text{C}_6\text{H}_5-\underset{\text{CH}_3}{\text{CH}_2\text{C}}=\text{NOH}}
$$

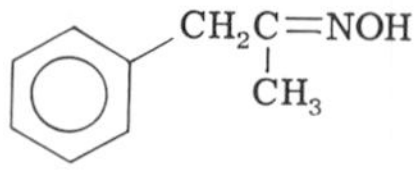

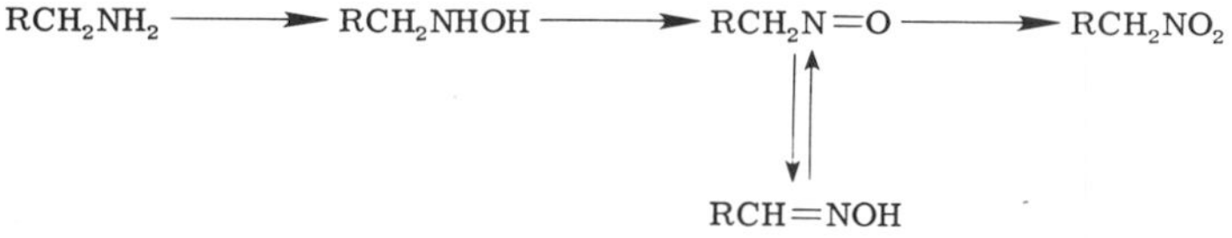

$$
RCH_2NH_2 \longrightarrow RCH_2NHOH \longrightarrow RCH_2N{=}O \longrightarrow RCH_2NO_2 \tag{18}
$$

$$
RCH{=}NOH
$$

metabolite only in the case of primary amines possessing a hydrogen on the α-carbon.

G. Oxidative Desulfuration

Desulfuration of compounds containing C=S or P=S functional groups has been studied for the most part by Neal and his colleagues.[165] This reaction involves the "substitution" of oxygen for sulfur and has been shown to be of the mixed function oxidase type mediated by cytochrome P-450. There is also evidence[166] for amine oxidase catalyzed desulfuration reactions (this volume, Chapter 9). The metabolism of parathion (**XLIX**) to paraoxon (**L**) serves as an example; it requires NADPH, is inhibited by

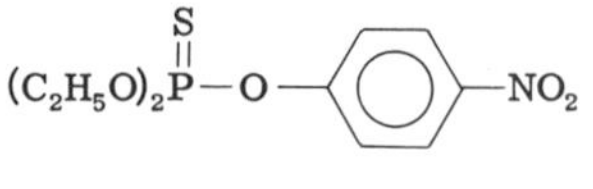

$$
\underset{(XLIX)}{(C_2H_5O)_2\overset{S}{\overset{\|}{P}}-O-C_6H_4-NO_2}
\qquad
\underset{(L)}{(C_2H_5O)_2\overset{O}{\overset{\|}{P}}-O-C_6H_4-NO_2}
$$

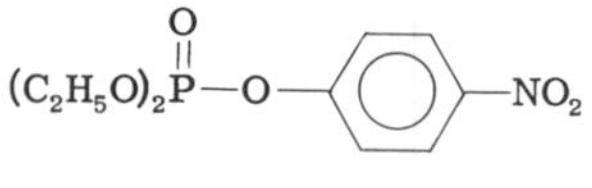

CO,[167] and involves the incorporation of ^{18}O from $^{18}O_2$ into the paraoxon.[168] A reconstituted cytochrome P-450 system can catalyze the reaction which is also inhibited by antibodies to cytochrome P-450.[169] As shown with carbon disulfide [Eq. (19)]

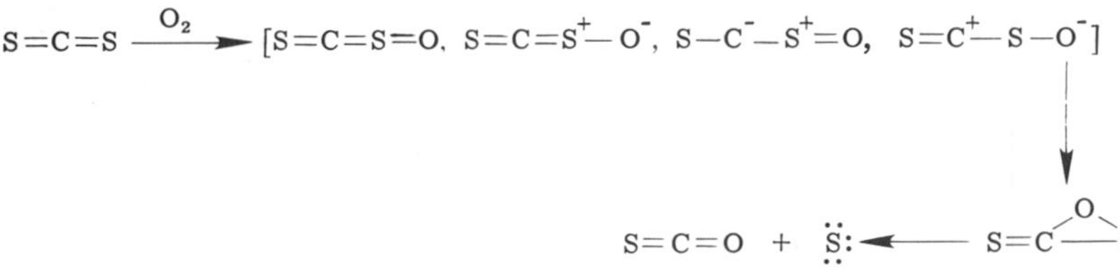

$$
S{=}C{=}S \xrightarrow{O_2} [S{=}C{=}S{=}O,\ S{=}C{=}S^+{-}O^-,\ S{-}C^-{-}S^+{=}O,\ S{=}C^+{-}S{-}O^-] \tag{19}
$$

$$
S{=}C{=}O + \ddot{\text{S}}\text{:} \longleftarrow S{=}C\overset{O}{\underset{}{\diagdown}}S
$$

a cyclic intermediate is formed during this reaction either directly through the attack of oxygen upon the double bond or through the attack of oxygen upon the sulfur atom, to yield carbonyl sulfide and atomic sulfur.[165,170] In support of this proposed mechanism are the findings that parathion (**XLIX**) treated with peroxytrifluoracetic acid (a reagent creating epoxides) also yields paraoxon (**L**).[171] Other compounds which can undergo this type of desulfuration reaction includes methimazole (**LI**),[172] α-naphthylthiourea (**LII**),[173] and carbonyl sulfide.[174]

(L I)

(L II)

Desulfuration does not necessarily lead to detoxication but is rather a mechanism leading to activation and resulting in products capable of covalent binding to tissue macromolecules. It is a rather unique reaction in that the portion of the molecule which binds to the greatest extent is that which has been removed from the main part of the molecule, the sulfur atom. This is not unexpected since the sulfur atom should have a reactivity similar to carbene.[165] Using ^{35}S- and ^{14}C-labeled compounds, 6, 3, and 2.5 times as much ^{35}S was found to be bound to liver microsomes compared with ^{14}C when α-naphthylthiourea (**LII**),[173] parathion (**XLIX**),[167] and methimazole (**LII**),[172] respectively, were incubated with microsomes. The binding of the ^{14}C could be a consequence of the desulfuration reaction. The covalent binding of sulfur also occurs to cytochrome *P*-450 leading to its inactivation.[169] This has been observed *in vivo* as well as in microsomes.[173,175] A large percentage of the ^{35}S bound to microsomes is freed upon treatment of the microsomes with cyanide; treatment with CN⁻ led to the release of 36% of the ^{35}S from methimazole as thiocyanide;[172] with both CS$_2$[176] and parathion[177] 50% of the ^{35}S bound to microsomes was released as SCN⁻. The proposed bound product is a hydrodisulfide, RSSH, which would be expected from the insertion of a sulfur atom.[176] It is clear that the desulfuration of these compounds leads to the formation of a reactive intermediate which is probably related to their toxicity.

Sulfur is also oxidized directly. The catalytic involvement of cytochrome *P*-450 in the catalysis of the oxidation of sulfur is based upon the inhibition of the sulfoxidation of chlorpromazine (**LIII**) by CO[178] and SKF

(L III)

525-A.[179] However, sulfoxidation may also be carried out by a noncyto-chrome P-450 system since the sulfoxidation of diaminodiphenyl sulfide was not inhibited by CO.[178]

H.   Oxidative Dehalogenation

Man's ability to remove halogen metabolically from a compound is taking on increasing importance. The resistance to biodegradation of halogenated organic compounds, once considered an asset for pesticides and herbicides, causes problems for animals at the higher end of the food chain due to bioaccumulation. Halogenated anesthetics, cleaning agents, propellants, and chemical intermediates all challenge the detoxication systems of man. The cytochrome P-450 system is one of several capable of dehalogenating organic compounds.[180]

1,1,2-Trichloroethane was found to be dechlorinated by a pheno-barbital-inducible microsomal enzyme which required NADPH and O_2 but was only slightly inhibited by CO (20%).[181] However, work with the reconstituted system containing cytochrome P-450 indicated that 1,1,2-trichloroethane was dechlorinated in a reaction inhibited by CO.[180,182] The mechanism by which chloroethanes and chloropropanes are dehalogenated is unknown. In determining the ease of dechlorination by rat liver microsomes of a series of chlorinated ethanes, it was observed that chlorine was preferentially removed from that carbon atom bearing two chlorine atoms and a hydrogen atom.[181] From a quantum mechanical viewpoint it was concluded that enzymatic dechlorination was correlated with the electron deficiency of the carbon valance atomic orbital and not with the strength of the C—H or C—Cl bond nor to the charge on the C or Cl atoms.[183] This was interpreted as implying that the activated oxygen attacked the electron deficient carbon orbital displacing chlorine rather than directly cleaving the Cl or H bond.

The metabolism of dihalomethanes has also been studied in some detail. Kubic and Anders[184] found that the debromination of dibromomethane (as determined by CO production) required NADPH and O_2 and was inhib-ited by SKF 525-A at low substrate concentration. The level of debro-mination correlated with the levels of cytochrome P-450 present after phenobarbital treatment and during treatment of rats with cobalt chloride. The relative rates of dehalogenation of the dihalomethanes was $CH_2I_2 > CH_2Br_2 > CH_2BrCl > CH_2Cl_2$. A possible mechanism for dehalogenation of dihalomethanes [Eq. (20)] accounts for the lack of formation of formic acid, formaldehyde, and carbon dixodide and suggests the formation of a formyl halide intermediate.[185]

$$CH_2Br_2 \xrightarrow{[O]} [HOCHBr_2] \longrightarrow H\overset{\overset{\displaystyle O}{\|}}{C}Br + H^+ + Br^-$$

$$(20)$$

$$H\overset{\overset{\displaystyle O}{\|}}{C}Br \longrightarrow C{=}O + H^+ + Br^-$$

By using $^{18}O_2$ the formed carbon monoxide was shown to be $C^{18}O$ indicating that a carbene was not a likely intermediate. When CD_2Cl_2 was used instead of CH_2Cl_2, the V_{max} was 7.7-fold lower indicating the importance of the C—H bond. These observations led to the hypothesis that an alcohol might be an intermediate in this reaction and that a formyl halide would result from the spontaneous decomposition of the alcohol. Formyl halides are themselves unstable and decompose to CO and H$^+$ and X$^-$. An attempt to trap this intermediate using 3,4-dimethylaniline failed with microsomes but the expected product, 3,4-formoxylidide was formed when stannous phosphate was used as a model oxidant. The production of a reactive formyl halide intermediate may account for the protein-bound product observed when dichloromethane is incubated *in vitro* with microsomes.[186] Pohl and Krishna[187] have also found that chloramphenicol undergoes metabolism by the cytochrome *P*-450 system to a metabolite which binds to protein. This binding is inhibited by CO and SKF 525-A

$$O_2N{-}\langle\bigcirc\rangle{-}\overset{\overset{\displaystyle OH}{|}}{C}H{-}\overset{\overset{\displaystyle CH_2OH}{|}}{C}H{-}NH{-}\overset{\overset{\displaystyle O}{\|}}{C}{-}CHCl_2 \longrightarrow R{-}NH{-}\overset{\overset{\displaystyle O}{\|}}{C}{-}\overset{\overset{\displaystyle OH}{|}}{C}HCl_2$$

$$(21)$$

$$R{-}NH{-}\overset{\overset{\displaystyle O}{\|}}{C}{-}\overset{\overset{\displaystyle O}{\|}}{C}OH + HCl \xleftarrow{H_2O} R{-}NH{-}\overset{\overset{\displaystyle O}{\|}}{C}{-}\overset{\overset{\displaystyle O}{\|}}{C}{-}Cl + HCl$$

and occurs only in liver microsomes from rats which have been treated with phenobarbital. The proposed mechanism of activation [Eq. (21)] involves the hydroxylation of the carbon bearing the 2 chlorine atoms and the subsequent dechlorination of the intermediate to form a reactive oxamyl halide. Evidence for the formation of the intermediate comes from the isolation of chloramphenicol oxamic acid after incubations of chloramphenicol with phenobarbital-treated rat liver microsomes[188] and from the base hydrolysis of protein-containing covalently bound chloramphenicol residues. These data, together with structure activity studies, indicate that the oxamyl chloride metabolite of chloramphenicol is most likely responsible for some of the observed binding and may also account for the toxicity in humans.

$$CHCl_3 \rightarrow HOCCl_3 \rightarrow Cl—\overset{\overset{\displaystyle O}{\|}}{C}—Cl + HCl \qquad (22)$$

Pohl et al.[189-191] have also studied the metabolism of chloroform. They proposed that $CHCl_3$ is oxidized by the cytochrome P-450 system to trichloromethanol which is spontaneously converted to phosgene by dehydrochlorination [Eq. (22)]. The evidence for phosgene formation was the isolation of 2-oxothioazolidene-4-carboxylic acid (LIV) from the microsomal incubation of chloroform in the presence of cysteine. This compound is the expected product of the reaction of phosgene and cysteine. A deuterium isotope effect of 2.1 has been observed in the formation of this product both in vivo[191] and in vitro.[190]

$$\underset{\underset{\displaystyle O}{\|}}{\underset{\displaystyle S\diagdown \underset{\displaystyle C}{}\diagup NH}{H_2C\text{——}CHCOOH}}$$

(LIV)

Although cytochrome P-450 is capable of metabolizing halogenated organic compounds, the products formed e.g., carbon monoxide, formyl halides, and phosgene, are obviously toxic. The metabolism of halogenated compounds to reactive intermediates may explain some of the toxicological impact observed with them.

I. Oxidative Denitrification

Ullrich et al.[192] recently published a study on the removal of the nitro group from 2-nitropropane to yield nitrite and acetone, suggesting Eq. (23)

$$CH_3—\overset{\overset{\displaystyle NO_2}{|}}{CH}—CH_3 \xrightarrow{O_2} \left[H_3C—\overset{\overset{\displaystyle NO_2}{|}}{\underset{\underset{\displaystyle OH}{|}}{C}}—CH_3 \right] \longrightarrow H_3C—\overset{\overset{\displaystyle O}{\|}}{C}—CH_3 + HNO_2 \quad (23)$$

as the mechanism. The formation of acetone and an equal amount of nitrite indicates that this is an oxidative rather than a reductive type of denitrification. A similar reaction has been observed with rabbit liver microsomes and 2-nitrophenylpropane, in which phenylacetone was formed in the presence of O_2 and NADPH[193]; the reaction was induced by phenobarbital and inhibited by CO. A reconstituted purified rat liver cytochrome *P*-450 system was also able to carry out this reaction (Jonsson and Miwa, unpublished observations). However, the oxygen incorporated into the phenylacetone was reported to have come from water rather than from molecular oxygen.[193] This was apparently due to exchange of the carbonyl oxygen with the oxygen of water. Conclusive evidence that the denitrification of 2-nitropropane is mediated by cytochrome *P*-450 was presented by Ullrich *et al.*[192] on the basis of inhibition and induction data.

IV. REDUCTIVE REACTIONS

Although cytochrome *P*-450 is classified as a mixed function oxidase, it is also capable of catalyzing reductive reactions under anaerobic conditions. To what extent these reactions occur *in vivo* is unknown. However, the oxygen tension in the liver is not expected to be uniform, and under certain conditions, e.g., anesthesia, the oxygen tension may be even lower than normal.

A. Nitro Reduction

The anaerobic reduction of an aromatic nitro group by liver microsomes was first observed by Fouts and Brodie[194] for chloramphenicol and *p*-nitrobenzoic acid. Oxygen inhibited reduction, although utilization of NADPH still occurred. The extent of reduction of *p*-nitrobenzoic acid to *p*-aminobenzoic acid carried out by liver microsomes was found to be proportional to the levels of cytochrome *P*-450 in the microsomes from phenobarbital or carbon tetrachloride treated animals.[195] Nitroreduction was not proportional to the amount of cytochrome *c* reductase (NADPH-cytochrome *P*-450 reductase) which has also been shown to be able to reduce nitro groups.[196,197] When an atmosphere of N_2 was replaced with one of CO, the reduction of *p*-nitrobenzoic acid to *p*-aminobenzoic acid was inhibited by greater than 80%[195] suggesting the participation of cytochrome *P*-450. However, since reduction is a multistep process [Eq. (24)],

$$RNO_2 \rightarrow RN{=}O \rightarrow RNHOH \rightarrow RNH_2 \qquad (24)$$

cytochrome *P*-450 might catalyze only one reductive step and not the others. Evidence that cytochrome *P*-450 was involved in more than one step came from studies on the metabolism of nitrobenzene. Carbon monoxide inhibited the formation of aniline from nitrobenzene, nitrosobenzene, and phenylhydroxylamine and also blocked the utilization of NADPH in the presence of *p*-nitrobenzoic acid.[195] Inhibition of the reduction implies that cytochrome *P*-450 must be involved in the first reductive step, while the inhibition of aniline formation indicates that it is involved in at least the third step of the reduction of nitrobenzene to aniline. Cytochrome *P*-450 probably functions in all three steps in the reduction of aromatic nitro compounds to aromatic amines.

The inhibitory effect of oxygen upon the reaction was believed to be due to its ability to oxidize the hydroxylamine back to the nitro compound or to oxidize the enzyme responsible for the reduction. Kato *et al.*[198] showed that inhibition by oxygen did not occur in the steps involved in reduction of *p*-hydroxylaminobenzoic acid to *p*-aminobenzoic acid. Mason and Holtzman[199] proposed a third possible explanation for the inhibition by O_2. They presented evidence that the nitroaromatic anion radical ($R\dot{N}O_2^-$) is the first intermediate in the nitroreduction reaction based upon electron spin resonance spectra and the correlation of rates of radical formation with product formation by microsomes. The inhibition of nitroreduction by oxygen may be due to the reaction of the nitroaromatic anion radical with oxygen followed by subsequent regeneration of the nitro group [Eq. (25)]; evidence for the formation of the superoxide anion

$$R\dot{N}O_2^- + O_2 \rightarrow \dot{O}_2^- + RNO_2 \qquad (25)$$

has been presented.[200] However, the evidence presented for the nitroaromatic anion radical was obtained in the presence of flavin-containing enzymes rather than with cytochrome *P*-450; the role of the nitroaromatic anion radical as an intermediate in cytochrome *P*-450 type reactions remains unanswered. It is difficult to gauge the importance of cytochrome *P*-450 in an animal's overall ability to reduce nitro groups. However, the intermediates formed in the reduction of nitro groups to amines, nitroso compounds, hydroxylamines, and radical anions are reactive and therefore are of toxicological concern.

An *N*-hydroxylamine has also been observed to undergo anaerobic reduction. *N*-hydroxyphentermine can be anaerobically reduced to phentermine (**XLIV**) with rat liver microsomes for which cytochrome *P*-450 involvement was indicated by inducibility with phenobarbital and inhibition by carbon monoxide.[201] It is not known whether other N-hydroxylated compounds would serve as substrates for this reaction.

B. Azo Reduction

The azo group, RN=NR, is capable of being reduced both by NADPH cytochrome *P*-450 reductase and by cytochrome *P*-450.[202,203] It was found that the cytochrome *P*-450-dependent reaction was inhibited by CO and O_2 and required NADPH for maximum activity. The amount of the reduction of the azo dye neoprontosil (**LV**) or neotetrazolium which would be inhibited by CO was proportional to the levels of cytochrome _P_-450 induced by phenobarbital or methylcholanthrene. DPEA was observed to inhibit only the CO-sensitive pathway of azo reduction of neoprontosil in rat liver microsomes, whereas SKF 525-A increased the amount of azoreduction.[204] Furthermore, CO inhibited the reduction of amaranth (**LVI**) by 88–94% in

(LV)   (LVI)

control and induced mice.[205] The level of azo reduction was correlated also with the levels of cytochrome _P_-450 in these mice and in the immature Long Evans rat. When the azoreductase activity was plotted against cytochrome _P_-450 concentration, the straight line went through the origin, suggesting that cytochrome _P_-450 was the main microsomal component responsible for azo reduction. The proportion to which cytochrome _P_-450 contributes to azo reduction appears to be related to the species, the strain, and the type of substrate.

C. Tertiary Amine _N_-Oxide Reduction

Reduced cytochrome _P_-450 is itself capable of reducing tertiary amine _N_-oxides to tertiary amines in the absence of O_2.[206,207] Substrates for this reaction include imipramine _N_-oxide (**LVII**), tiaramide _N_-oxide, and _N,N_-dimethylaniline _N_-oxide (**LVIII**). In microsomes this activity appears

(LVII)   (LVIII)

to be completely dependent on cytochrome _P_-450, since it was inhibited 95% by CO.[208] NADPH acts as a better electron donor than NADH for this reaction with a 1:1 stoichiometry between NADPH oxidized and tertiary amine _N_-oxide reduced. The levels of cytochrome _P_-450 in rats of different ages and sex also correlate with the amount of the tertiary amine

N-oxide reduction. FMN and methyl viologen were capable of stimulating the reduction of tertiary amine N-oxide both in microsomes and in the purified reconstituted system.[207,209] This stimulation of 10- and 100-fold, respectively, in the purified system was inhibited by CO. The rate of reduction appeared to be more dependent on the ratio of reduced to oxidized FMN than on the level of reduced FMN. Different forms of cytochrome P-450, including that isolated from camphor grown *Pseudomonas putida*, all reduce the tertiary amine N-oxide with only a 4-fold difference in rate.[206] The lack of high specificity with the different forms of cytochrome P-450, suggest that the substrate interacts with the cytochrome P-450 at the heme moiety. Further support of this hypothesis stems from the finding that EDTA and Fe^{3+} can reduce imipramine N-oxide and that octylamine and DPEA, which bind to heme, inhibit N-oxide reduction.[208] Model studies, discussed in the section on dealkylation, also support this idea.

D. Arene Oxide Reduction

Booth *et al.*[210] found that the incubation of benz[*a*]anthracene 5,6-oxide (**LIX**) with microsomes, O_2, and NADPH led to the formation of a small

(LIX)

amount of benz[*a*]anthracene. The exclusion of oxygen and the addition of cyclohexane oxide to prevent hydration of the oxide by epoxide hydrase, led to an increase in the amount of benz[*a*]anthracene formed. Several arene oxides of polycyclic aromatic hydrocarbons are substrates for this reaction, but none of the epoxides (alkene oxides) derived from polycyclic aromatic hydrocarbons are substrates. Oxide reduction did not occur through the formation of a dihydromonal intermediate, since their inclusion in the assay, replacing the arene oxide, did not lead to the formation of the parent hydrocarbon.

The evidence that cytochrome P-450 is involved in this reduction is due to Kato *et al.*[211] who used BP 4,5-oxide (**LX**) as a substrate and found that CO readily inhibited the reduction of this arene oxide and that NADPH was required. Microsomes from the livers of rats which had been treated with inducers had increased ability to reduce BP 4,5-oxide to BP.

The reduced form of cytochrome P-450 would be the most likely species

(LX)

involved in this reduction. The lack of inhibition by SKF 525-A, which inhibits cytochrome P-450 in its oxidized form, supports this hypothesis. Although the mechanism by which this reaction occurs has not been elucidated, direct abstraction of oxygen is possible. It would be of interest to see whether an ^{18}O atom in an arene oxide could be returned to O_2 or be transferred to another compound. Whether this reaction should be classified as detoxication or one leading to a more toxic intermediate, would depend on the substrate. BP 7,8-oxide which is converted to the potent proximate carcinogen, BP-7,8-dihydrodiol, would be detoxified by arene oxide reductase activity. Other epoxides, those converted to less toxic metabolites, may be reduced to the parent compound which might then be metabolized to a more active intermediate by a different pathway.

E. Reductive Dehalogenation

Under anaerobic conditions carbon tetrachloride (CCl_4) is metabolized by rabbit liver microsomes to chloroform ($CHCl_3$) and to a species which binds to microsomal protein.[212] Both the formation of $CHCl_3$ and the binding of the metabolite require the absence of O_2 and the presence of NADPH. Both phenomenon are inhibited by CO (80%) and metyrapone (70%). SKF 525-A increased the formation of $CHCl_3$ by 25% but inhibited binding by 22%. These data strongly suggest the participation of cytochrome P-450 in the anaerobic dehalogenation of carbon tetrachloride.

Halothane (2-bromo-2-chloro-1,1,1-trifluoroethane) also undergoes reductive defluorination which appears to be cytochrome P-450 dependent.[213] NADPH was required, and the reaction was inhibited by CO. The enzyme was inducible by phenobarbital and polychlorinated biphenyl but not by 3-methylcholanthrene. Since the binding of a metabolite of halothane to microsomes is also observed to increase under anaerobic conditions, anaerobic defluorination may play a part in the binding as well. Its occasional toxicity as an anesthetic might be explained by these observations.

From spectral studies of the binding of halothane and artificially generated trifluoromethylmethylene carbene ($CF_3\ddot{C}H$) to cytochrome P-450, Mansuy et al.[214] proposed that the carbene was the reactive intermediate responsible for the binding to cytochrome P-450 and possibly responsible for the toxicity of halothane. This differs from the previous one since it does not involve defluorination.

V. COMMENTS

As can be seen from the types of reactions discussed, the cytochrome P-450 isozymes are characterized by their ability to metabolize a wide range of substrates. Although the liver microsomal system is the primary site for the oxidation of xenobiotics, the cytochrome P-450 system is also found in other tissues that are exposed to environmental compounds (skin, lung, gastrointestinal tract, and kidney) as well as in the placenta, corpus luteum, lymphocytes, monocytes, pulmonary aveolar macrophages, adrenal, aorta, testis, and brain. The cytochrome P-450 system is not only localized in the endoplasmic reticulum but also is found in the mitochondria of the corpus luteum, kidney, adrenal, and liver and in the nuclear membrane. Thus, the metabolism of a given compound by cytochrome P-450 could occur at several different sites.

Age, sex, diet, and stress, all effect utilization by cytochrome P-450 systems. However, since there are multiple forms of cytochrome P-450, it is difficult to be specific in identifying the effects of these perturbing factors on individual enzymes. The isozymes are affected differently by dietary components and are unevenly distributed among the sexes.[215] Newborn animals usually have only low levels of these enzymes which increase at different rates.

Although there are many isozymes of cytochrome P-450 and many types of reactions, the different isozymes cannot be classified by the reactions that they catalyze. Rather, this group of enzymes is characterized by an overlapping specificity for substrates with certain of them having a greater capacity for a specific reaction.

ACKNOWLEDGMENT

We would like to thank Mrs. Gail Mancinelli for her excellent assistance in the preparation of this manuscript.

REFERENCES

1. Mueller, G. C., and Miller, J. A. (1948). The metabolism of 4-dimethylaminoazobenzene by rat liver homogenates. *J. Biol. Chem.* **176,** 535–544.

2. Mueller, G. C., and Miller, J. A. (1953). The metabolism of methylated aminoazo dyes. II. Oxidative demethylation by rat liver homogenates. *J. Biol. Chem.* **202,** 579–587.
3. Axelrod, J. (1955). The enzymatic deamination of amphetamine. *J. Biol. Chem.* **214,** 753–763.
4. Axelrod, J. (1956). The enzymatic N-demethylation of narcotic drugs. *J. Pharmacol. Exp. Ther.* **117,** 322–330.
5. Brodie, B. B., Axelrod, J., Cooper, J. R., Gaudette, L., La Du, B. N., Mitoma, C., and Udenfriend, S. (1955). Detoxification of drugs and other foreign compounds by liver microsomes. *Science* **121,** 603–604.
6. Cooper, D. Y., Levine, S., Narasimhulu, S., Rosenthal, O., and Estabrook, R. W. (1965). Photochemical action spectrum of the terminal oxidase of mixed function oxidase systems. *Science* **147,** 400–402.
7. Gillette, J. R. (1966). Biochemistry of drug oxidation and reduction by enzymes in hepatic endoplasmic reticulum. *Adv. Pharmacol.* **4,** 219–261.
8. Uehleke, H. (1973). The role of cytochrome *P*-450 in the N-oxidation of individual amines. *Drug Metab. Dispos.* **1,** 299–313.
9. Weisburger, J. H., and Weisburger, E. K. (1973). Biochemical formation and pharmacological, toxicological and pathological properties of hydroxylamines and hydroxamic acids. *Pharmacol. Rev.* **25,** 1–66.
10. Hucker, H. B. (1973). Intermediates in drug metabolism reactions. *Drug Metab. Rev.* **2,** 33–56.
11. Schreiber, E. C. (1974). Metabolically oxygenated compounds: Formation, conjugation and possible biological implications. *J. Pharm. Sci.* **63,** 1177–1190.
12. Coutts, R. T., and Beckett, A. H. (1977). Metabolic N-oxidation of primary and secondary aliphatic medicinal amines. *Drug Metab. Rev.* **6,** 51–104.
13. Jenner, P., and Testa, B. (1978). Novel pathways in drug metabolism. *Xenobiotica* **8,** 1–25.
14. Testa, B., and Jenner, P. (1978). Novel drug metabolites produced by functionalization reactions: Chemistry and toxicology. *Drug Metab. Rev.* **7,** 325–369.
15. Testa, B., and Jenner, P. (1976). "Drug Metabolism: Chemical and Biochemical Aspects." Dekker, New York.
16. Brown, R. R., Miller, J. A., and Miller, E. C. (1954). The metabolism of methylated amino azo dyes. IV. Dietary factors enhancing demethylation *in vitro*. *J. Biol. Chem.* **209,** 211–222.
17. Coon, M. J., Vermilion, J. L., Vatsis, K. P., French, J. S., Dean, W. L., and Haugen, D. A. (1977). Biochemical studies on drug metabolism: Isolation of multiple forms of liver microsomal cytochrome *P*-450. *In* "Drug Metabolism Concepts" (D. M. Jerina, ed.), pp. 46–71. Am. Chem. Soc., Washington, D.C.
18. Mannering, G. J. (1971). Properties of cytochrome *P*-450 as affected by environmental factors: Qualitative changes due to administration of polycyclic hydrocarbons. *Metab., Clin. Exp.* **20,** 228–245.
19. Levin, W. (1977). Purification of liver microsomal cytochrome *P*-450: Hopes and promises. *In* "Microsomes and Drug Oxidations" (V. Ullrich, I. Roots, A. Hildebrandt, R. W. Estabrook, and A. H. Conney, eds.), pp. 735–747. Pergamon, Oxford.
20. Lu, A. Y. H. (1976). Liver microsomal drug-metabolizing enzyme system: Functional components and their properties. *Fed. Proc., Fed. Am. Soc. Exp. Biol.* **35,** 2460–2463.
21. Johnson, E. F. (1979). Multiple forms of cytochrome *P*-450: Criteria and significance. *Rev. Biochem. Toxicol.* **1,** 1–26.
22. Leibman, K. C. (1966). Effects of metyrapone on liver microsomal drug oxidations. *Fed. Proc., Fed. Am. Soc. Exp. Biol.* **25,** 417.

23. Leibman, K. C. (1969). Effects of metyrapone on liver microsomal drug oxidations. *Mol. Pharmacol.* **5**, 1–9.
24. Wiebel, F. J., Leutz, J. C., Diamond, L., and Gelboin, H. V. (1971). Aryl hydrocarbon hydroxylase in microsomes from rat tissues: Differential inhibition and stimulation by benzoflavones and organic solvents. *Arch. Biochem. Biophys.* **144**, 78–86.
25. Lu, A. Y. H., and West, S. B. (1972). Reconstituted liver microsomal system that hydroxylates drugs, other foreign compounds and endogenous substrates. III. Properties of the reconstituted 3,4-benzpyrene hydroxylase system. *Mol. Pharmacol.* **8**, 490–500.
26. Kawalek, J. C., and Lu, A. Y. H. (1975). Reconstituted liver microsomal system that hydroxylates drugs, other foreign compounds, and endogenous substrates. VII. Different catalytic activities of rabbit and rat cytochromes *P*-448. *Mol. Pharmacol.* **11**, 201–210.
27. Thomas, P. E., Lu, A. Y. H., West, S. B., Ryan, D., Miwa, G. T., and Levin, W. (1977). Accessibility of cytochrome *P*-450 in microsomal membranes: Inhibition of metabolism by antibodies to cytochrome *P*-450. *Mol. Pharmacol.* **13**, 819–831.
28. Dean, W. L., and Coon, M. J. (1977). Immunochemical studies on two electrophoretically homogeneous forms of rabbit liver microsomal cytochrome *P*-450: P-450$_{LM_2}$ and P-450$_{LM_4}$. *J. Biol. Chem.* **252**, 3255–3261.
29. Guengerich, F. P., and Mason, P. S. (1979). Immunological comparison of hepatic and extrahepatic cytochromes *P*-450. *Mol. Pharmacol.* **15**, 154–164.
30. Imai, Y., and Sato, R. (1966). Substrate interaction with hydroxylase system in liver microsomes. *Biochem. Biophys. Res. Commun.* **22**, 620–626.
31. Schenkman, J. B., Remmer, H., and Estabrook, R. W. (1967). Spectral studies of drug interaction with hepatic microsomal cytochrome. *Mol. Pharmacol.* **3**, 113–123.
32. Schenkman, J. B., Cinti, D. L., Moldeus, P. W., and Orrenius, S. (1973). Newer aspects of substrate binding to cytochrome *P*-450. *Drug Metab. Dispos.* **1**, 111–120.
33. Franklin, M. R. (1977). Inhibition of mixed-function oxidations by substrates forming reduced cytochrome *P*-450 metabolic-intermediate complexes. *Pharmacol. & Ther., Part A* **2**, 227–245.
34. Lu, A. Y. H., Junk, K. W., and Coon, M. J. (1969). Resolution of the cytochrome *P*-450-containing ω-hydroxylation system of liver microsomes into three components. *J. Biol. Chem.* **244**, 3714–3721.
35. Lu, A. Y. H., and Coon, M. J. (1968). Role of hemoprotein *P*-450 in fatty acid ω-hydroxylation in a soluble enzyme system from liver microsomes. *J. Biol. Chem.* **243**, 1331–1332.
36. Das, M. L., Orrenius, S., and Ernster, L. (1968). On the fatty acid and hydrocarbon hydroxylation in rat liver microsomes. *Eur. J. Biochem.* **4**, 519–523.
37. Lu, A. Y. H., Strobel, H. W., and Coon, M. J. (1970). Properties of a solubilized form of the cytochrome *P*-450-containing mixed-function oxidase of liver microsomes. *Mol. Pharmacol.* **6**, 213–220.
38. Frommer, U., Ullrich, V., Staudinger, H., and Orrenius, S. (1972). The monooxygenation of *n*-heptane by rat liver microsomes. *Biochim. Biophys. Acta* **280**, 487–494.
39. Lu, A. Y. H., Levin, W., West, S. B., Jacobson, M., Ryan, D., Kuntzman, R., and Conney, A. H. (1973). Reconstituted liver microsomal enzyme system that hydroxylates drugs, other foreign compounds and endogenous substrates. VI. Different substrate specificities of the cytochrome *P*-450 fractions from control and phenobarbital-treated rats. *J. Biol. Chem.* **248**, 456–460.
40. Haugen, D. A., van der Hoeven, T. A., and Coon, M. J. (1975). Purified liver micro-

somal cytochrome P-450. Separation and characterization of multiple forms. J. Biol. Chem. **250**, 3567–3570.

41. Huang, M. T., West, S. B., and Lu, A. Y. H. (1976). Separation, purification and properties of multiple forms of cytochrome P-450 from the liver microsomes of phenobarbital-treated mice. J. Biol. Chem. **251**, 4659–4665.

42. Ryan, D. E., Thomas, P. E., Korzeniowski, D., and Levin, W. (1978). Separation and characterization of highly purified forms of liver microsomal cytochrome P-450 from rats treated with polychlorinated biphenyls, phenobarbital and 3-methylcholanthrene. J. Biol. Chem. **254**, 1365–1374.

43. Groves, J. R., McClusky, G. A., White, R. E., and Coon, M. J. (1978). Aliphatic hydroxylation by highly purified liver microsomal cytochrome P-450. Evidence for a carbon radical intermediate. Biochem. Biophys. Res. Commun. **81**, 154–160.

44. Hjelmeland, L. M., Aronow, L., and Trudell, J. R. (1977). Intramolecular determination of primary kinetic isotope effects in hydroxylations catalyzed by cytochrome P-450. Biochem. Biophys. Res. Commun. **76**, 541–549.

45. Groves, J. T., and Van Der Puy, M. (1974). Stereospecific aliphatic hydroxylation by an iron-based oxidant. J. Am. Chem. Soc. **96**, 5274–5275.

46. Guroff, G., Daly, J. W., Jerina, D. M., Renson, J., Witkop, B., and Udenfriend, S. (1967). Hydroxylation-induced migration: The NIH shift. Science **157**, 1524–1530.

47. Daly, J., Jerina, D., and Witkop, B. (1968). Migration of deuterium during hydroxylation of aromatic substrates by liver microsomes. I. Influence of ring substituents. Arch. Biochem. Biophys. **128**, 517–527.

48. Daly, J. W., Jerina, D. M., and Witkop, B. (1972). Arene oxides and the NIH shift: The metabolism, toxicity and carcinogenicity of aromatic compounds. Experientia **28**, 1129–1149.

49. Jerina, D. M., and Daly, J. W. (1974). Arene oxides: A new aspect of drug metabolism. Science **185**, 573–582.

50. Rahimtula, A. D., O'Brien, P. J., Seifried, H. E., and Jerina, D. M. (1978). The mechanism of action of cytochrome P-450. Occurrence of the "NIH shift" during hydroperoxide dependent aromatic hydroxylation. Eur. J. Biochem. **89**, 133–141.

51. Billings, R. E., and McMahon, R. (1978). Microsomal biphenyl hydroxylation: The formation of 3-hydroxybiphenyl and biphenyl catechol. Mol. Pharmacol. **14**, 145–154.

52. Selander, H. G., Jerina, D. M., and Daly, J. W. (1975). Metabolism of chlorobenzene with hepatic microsomes and solubilized cytochrome P-450 systems. Arch. Biochem. Biophys. **168**, 309–321.

53. Tomaszewski, J. E., Jerina, D. M., and Daly, J. W. (1975). Deuterium isotope effects during formation of phenols by hepatic monooxygenases. Evidence for an alternative to the arene oxide pathway. Biochemistry **14**, 2024–2031.

54. Jollow, D. J., Mitchell, J. R., Zampaglione, N., and Gillette, J. R. (1974). Bromobenzene-induced liver necrosis. Protective role of glutathione and evidence for 3,4-bromobenzene oxide as the hepatotoxic metabolite. Pharmacology **11**, 151–169.

55. Reid, W. D., Christie, B., Krishna, G., Mitchell, J. R., Moskowitz, J., and Brodie, B. B. (1971). Bromobenzene metabolism and hepatic necrosis. Pharmacology **6**, 41–55.

56. Reid, W. D., Christie, B., Eichelbaum, M., and Krishna, G. (1971). 3-Methylcholanthrene blocks hepatic necrosis induced by administration of bromobenzene and carbon tetrachloride. Exp. Mol. Pathol. **15**, 363–372.

57. Zampaglione, N., Jollow, D. J., Mitchell, J. R., Stripp, B., Hamrick, M., and Gillette, J. R. (1973). Role of detoxifying enzymes in bromobenzene-induced liver necrosis. J. Pharmacol. Exp. Ther. **187**, 218–227.

58. Cooper, D. Y., Schleyer, H., Rosenthal, O., Levin, W., Lu, A. Y. H., Kuntzman, R., and Conney, A. H. (1977). Inhibition by CO of hepatic benzo[a]pyrene hydroxylation and its reversal by monochromatic light. *Eur. J. Biochem.* **74**, 69–75.

59. Holder, G., Yagi, H., Dansette, P., Jerina, D. M., Levin, W., Lu, A. Y H., and Conney, A. H. (1974). Effects of inducers and epoxide hydrase on the metabolism of benzo[a]pyrene by liver microsomes and a reconstituted system: Analysis by high pressure liquid chromatography. *Proc. Natl. Acad. Sci. U.S.A.* **71**, 4356–4360.

60. Yang, S. K., Selkirk, J. K., Plotkin, E. V., and Gelboin, H. V. (1975). Kinetic analysis of the metabolism of benzo[a]pyrene to phenols, dihydrodiols, and quinones by high-pressure liquid chromatography compared to analysis by aryl hydrocarbon hydroxylase assay, and the effect of enzyme induction. *Cancer Res.* **35**, 3642–3650.

61. Wiebel, F. J., Selkirk, J. K., Gelboin, H. V., Haugen, D. A., van der Hoeven, T. A., and Coon, M. J. (1975). Position-specific oxygenation of benzo[a]pyrene by different forms of purified cytochrome *P*-450 from rabbit liver. *Proc. Natl. Acad. Sci. U.S.A.* **72**, 3917–3920.

62. Thakker, D. R., Yagi, H., Levin, W., Lu, A. Y. H., Conney, A. H., and Jerina, D. M. (1977). Stereospecificity of microsomal and purified epoxide hydrase from rat liver. Hydration of arene oxides of polycyclic hydrocarbons. *J. Biol. Chem.* **252**, 6328–6334.

63. Yang, S. K., and Gelboin, H. V. (1976). Microsomal mixed-function oxidases and epoxide hydratase convert benzo[a]pyrene stereospecifically to optically active dihydroxydihydrobenzo[a]pyrenes. *Biochem. Pharmacol.* **25**, 2221–2225.

64. Yang, S. K., McCourt, D. W., Leutz, J. C., and Gelboin, H. V. (1977). Benzo[a]pyrene diol-epoxides: Mechanism of enzymatic formation and optically active intermediates. *Science* **196**, 1199–1201.

65. Yang, S. K., Roller, P. P., and Gelboin, H. V. (1977). Enzymatic mechanism of benzo[a]pyrene conversion to diols and phenols and an improved high-pressure liquid chromatographic separation of benzo[a]pyrene derivatives. *Biochemistry* **16**, 3680–3687.

66. Thakker, D. R., Yagi, H., Akagi, H., Koreeda, M., Lu, A. Y. H., Levin, W., Wood, A. W., Conney, A. H., and Jerina, D. M. (1977). Metabolism of benzo[a]pyrene. VI. Stereoselective metabolism of benzo[a]pyrene and benzo[a]pyrene 7,8-dihydrodiol to diol epoxides. *Chem.-Biol. Interact.* **16**, 281–300.

67. Yang, S. K., McCourt, D. W., Roller, P. P., and Gelboin, H. V. (1976). Enzymatic conversion of benzo[a]pyrene leading predominantly to the diol-epoxide r-7-t-8-dihydroxy-t-9,10-oxy-7,8,9,10-tetrahydrobenzo[a]pyrene through a single enantiomer of r-7-t-8-dihydroxy-7,8-dihydrobenzo[a]pyrene. *Proc. Natl. Acad. Sci. U.S.A.* **73**, 2594–2598.

68. Deutsch, J., Leutz, J., Yang, S. K., Gelboin, H. V., Chiang, Y. L., Vatsis, K. P., and Coon, M. J. (1978). Regio- and stereoselectivity of various forms of purified cytochrome *P*-450 in the metabolism of benzo[a]pyrene and (−)-trans-7,8-dihydroxy-7,8-dihydrobenzo[a]pyrene as shown by product formation and binding to DNA. *Proc. Natl. Acad. Sci. U.S.A.* **75**, 3123–3127.

69. Buening, M. K., Wislocki, P. G., Levin, W., Yagi, H., Thakker, D. R., Akagi, H., Koreeda, M., Jerina, D. M., and Conney, A. H. (1978). Turmogenicity of the optical enantiomers of the diastereomeric benzo[a]pyrene-7,8-diol-9,10-epoxides in newborn mice: Exceptional activity of (+)-7β,8α-dihydroxy-9α, 10α-epoxy-7,8,9,10-tetrahydrobenzo[a]pyrene. *Proc. Natl. Acad. Sci. U.S.A.* **75**, 5358–5361.

70. Wogan, G. N. (1975). Mycotoxins. *Annu. Rev. Pharmacol.* **15**, 437–451.

71. Garner, R. C., Miller, E. C., and Miller, J. A. (1972). Liver microsomal metabolism of aflatoxin B_1 to a reactive derivative toxic to *Salmonella typhimurium* TA-1530. *Cancer Res.* **32**, 2058–2066.

72. Wong, J. J., and Hsieh, D. P. H. (1976). Mutagenicity of aflatoxins related to their metabolism and carcinogenic potential. *Proc. Natl. Acad. Sci. U.S.A.* **73**, 2241–2244.

73. Swenson, D. H., Miller, J. A., and Miller, E. C. (1973). 2,3-Dihydro-2,3-dihydroxy-aflatoxin B₁: An acid hydrolysis product of an RNA-aflatoxin B₁ adduct formed by hamster and rat liver microsomes *in vitro*. *Biochem. Biophys. Res. Commun.* **53**, 1260–1267.

74. Swenson, D. H., Lin, J. K., Miller, E. C., and Miller, J. A. (1977). Aflatoxin B₁-2,3-oxide as a probable intermediate in the covalent binding of aflatoxins B₁ and B₂ to rat liver DNA and ribosomal RNA *in vivo*. *Cancer Res.* **37**, 172–181.

75. Essigmann, J. M., Croy, R. G., Nadzan, A. M., Busby, W. F., Jr., Reinhold, V. N., Büchi, G., and Wogan, G. N. (1977). Structural identification of the major DNA adduct formed by aflatoxin B *in vitro*. *Proc. Natl. Acad. Sci. U.S.A.* **74**, 1870–1874.

76. Martin, C. N., and Garner, R. C. (1977). Aflatoxin B₁-oxide generated by chemical or enzymic oxidation of aflatoxin B₁ causes guanine substitution in nucleic acids. *Nature (London)* **267**, 863–865.

77. Gurtoo, H. L., and Dave, C. V. (1975). *In vitro* metabolic conversion of aflatoxins and benzo[a]pyrene to nucleic acid-binding metabolites. *Cancer Res.* **35**, 382–389.

78. Alexandrov, K., and Frayssinet, C. (1974). Microsome-dependent binding of benzo(a)pyrene and aflatoxin B₁ to DNA, and benzo[a]pyrene binding to aflatoxin-conjugated DNA. *Cancer Res.* **34**, 3289–3295.

79. Raj, H. G., Santhanam, K., Gupta, R. P., and Venkitasubramanian, T. A. (1975). Oxidative metabolism of aflatoxin B₁: Observations on the formation of epoxide-glutathione conjugate. *Chem.-Biol. Interact.* **11**, 301–305.

80. Levin, W., Wood, A. W., Wislocki, P. G., Chang, R. L., Kapitulinik, J., Mah, H. D., Yagi, H., Jerina, D. M., and Conney, A. H. (1978). Mutagenicity and carcinogenicity of benzo[a]pyrene and benzo[a]pyrene derivatives. *In* "Polycyclic Hydrocarbons and Cancer" (H. V. Gelboin and P. O. P. Ts'o, eds.), Vol. 1, pp. 189–202. Academic Press, New York.

81. Yang, S. K., Deutsch, J., and Gelboin, H. V. (1978). Benzo[a]pyrene metabolism: Activation and detoxification. *In* "Polycyclic Hydrocarbons and Cancer" (H. V. Gelboin and P. O. P. Ts'o, eds.), Vol. 1, pp. 205–213. Academic Press, New York.

82. Jerina, D. M., Lehr, R. E., Yagi, H., Hernandez, O., Dansette, P. M., Wislocki, P. G., Wood, A. W., Chang, R. L., Levin, W., and Conney, A. H. (1976). Mutagenicity of benzo[a]pyrene derivatives and the description of a quantum mechanical model which predicts the ease of carbonium ion formation from diol epoxides. *In* "*In Vitro* Metabolic Activation in Mutagenesis Testing" (F. J. de Serres, J. R. Fouts, J. R. Bend, and R. M. Philpot, eds.), pp. 159–177. Elsevier/North-Holland Biomedical Press, Amsterdam.

83. Bartsch, H., and Montesano, R. (1975). Mutagenic and carcinogenic effects of vinyl chloride. *Mutat. Res.* **32**, 93–114.

84. Nakatsugawa, T., Ishida, M., and Dahm, P. A. (1965). Microsomal epoxidation of cyclodiene insecticides. *Biochem. Pharmacol.* **14**, 1853–1865.

85. DeMatteis, F. (1973). Drug-induced destruction of cytochrome *P*-450. *Drug Metab. Dispos.* **1**, 267–274.

86. Levin, W., Sernatinger, E., Jacobson, M., and Kuntzman, R. (1972). Destruction of cytochrome *P*-450 by secobarbital and other barbiturates containing allyl groups. *Science* **176**, 1341–1343.

87. Glatt, H. R., Metzler, M., Newmann, H. G., and Oesch, F. (1976). *Biochem. Biophys. Res. Commun.* **73**, 1025–1029.

88. Butler, T. C., and Bush, M. T. (1939). The metabolic fate of *N*-methylbarbituric acids. *J. Pharmacol. Exp. Ther.* **65**, 205–213.

89. McMahon, R. E. (1966). Microsomal dealkylation of drugs. Substrate specificity and mechanism. *J. Pharm. Sci.* **55**, 457–466.
90. Kamm, J. J., and Szuna, A. (1973). Studies on an unusual N-dealkylation reaction. II. Characteristics of the enzyme system and a proposed pathway for the reaction. *J. Pharmacol. Exp. Ther.* **184**, 729–738.
91. Bickel, M. H. (1969). The pharmacology and biochemistry of *N*-oxides. *Pharmacol. Rev.* **21**, 325–355.
92. Polonovski, M., and Polonovski, M. (1927). Sur les aminoxydes des alcaloides. III. Action des anhydrides et chlorules d'acides organiques. Preparations des bases nor. *Bull. Soc. Chim. Fr.* **41**, 1190–1208.
93. Oae, S., Kitao, T., and Kawamura, S. (1963). Model pathways for enzymatic oxidative demethylation. II. Polonovski reaction of *N,N*-dimethylaniline *N*-oxide, Pummerer reactions of dimethyl *n*-butylmethyl and methionine sulfoxide with acetylating agents and their implications in enzymatic demethylation. *Tetrahedron* **19**, 1783–1788.
94. Craig, J. C., Dwyer, F. P., Glazer, A. N., and Horning, E. C. (1961). Tertiary amine oxide rearrangements. I. Mechanism. *J. Am. Chem. Soc.* **83**, 1871–1878.
95. Ferris, J. P., Gerwe, R. D., and Gapski, G. R. (1968). Detoxication mechanisms. III. The scope and mechanism of the iron-catalyzed dealkylation of tertiary amine oxides. *J. Org. Chem.* **33**, 3493–3498.
96. Iwasaki, K., Noguchi, H., Kato, R., Imai, Y., and Sato, R. (1977). Reduction of tertiary amine *N*-oxide by purified cytochrome *P*-450. *Biochem. Biophys. Res. Commun.* **77**, 1143–1149.
97. Breck, G. D., and Trager, W. F. (1971). Oxidative N-dealkylation. A Mannich intermediate in the formation of a new metabolite of lidocaine in man. *Science* **173**, 544–545.
98. Schraven, E., Koss, F. W., Keck, J., Beisenherz, G., Bücheler, A., and Lindner, W. (1967). Excretion, isolation and identification of the metabolites of Bisolvon. *Eur. J. Pharmacol.* **1**, 445–451.
99. Hucker, H. B., Gillette, J. R., and Brodie, B. B. (1960). Enzymatic pathway for the formation of cotinine, a major metabolite of nicotine in rabbit liver. *J. Pharmacol. Exp. Ther.* **129**, 94–100.
100. Swartz, M. A., and Kolis, S. J. (1972). Pathways of medazepam metabolism. *J. Pharmacol. Exp. Ther.* **180**, 180–188.
101. Dorough, H. W., and Cassida, J. E. (1964). Nature of certain carbamate metabolites of the insecticide sevin. *J. Agric. Food Chem.* **12**, 294–304.
102. Gorrod, J. W., and Temple, D. J. (1976). The formation of an *N*-hydroxymethyl intermediate in the N-demethylation of *N*-methylcarbazole *in vivo* and *in vitro*. *Xenobiotica* **6**, 265–274.
103. Thompson, J. A., and Holtzman, J. L. (1974). Deuterium and tritium isotope effects on the microsomal N-demethylation of ethylmorphine. *Drug Metab. Dispos.* **2**, 577–582.
104. Nelson, S. D., Pohl, L. R., and Trager, W. F. (1975). Primary and β-secondary deuterium isotope effects in N-deethylation reactions. *J. Med. Chem.* **18**, 1062–1065.
105. Rosenblatt, D. H., Hull, L. A., De Luca, D. C., Davis, G. T., Weglein, R. C., and Williams, H. K. R. (1967). Oxidation of amines. II. Substituent effects in chlorine dioxide oxidations. *J. Am. Chem. Soc.* **89**, 1158–1163.
106. Northrop, D. B. (1975). Steady-state analysis of kinetic isotope effects in enzymic reactions. *Biochemistry* **14**, 2644–2651.
107. Westheimer, F. H. (1961). The magnitude of the primary kinetic isotope effect for compounds of hydrogen and deuterium. *Chem. Rev.* **61**, 265–273.
108. McMahon, R. E., Culp, H. W., and Occolowitz, J. C. (1969). Studies on the hepatic

microsomal N-dealkylation reaction. Molecular oxygen as the source of the oxygen atom. *J. Am. Chem. Soc.* **91**, 3389–3390.

109. Parli, C. J., and McMahon, R. E. (1973). The mechanism of microsomal deamination: Heavy isotope studies. *Drug Metab. Dispos.* **1**, 337–341.

110. Axelrod, J. (1955). Enzymatic deamination of amphetamine. *J. Biol. Chem.* **214**, 753–763.

111. Cho, A. K., and Wright, J. (1978). Pathways of metabolism of amphetamine and related compounds. *Life Sci.* **22**, 363–372.

112. Brodie, B. B., Gillette, J. R., and La Du, B. M. (1958). Enzymatic metabolism of drugs and other foreign compounds. *Annu. Rev. Biochem.* **27**, 427–454.

113. Parli, C. J., Wang, N., and McMahon, R. E. (1971). The mechanism of the oxidation of *d*-amphetamine by rabbit liver oxygenase. Oxygen-18 studies. *Biochem. Biophys. Res. Commun.* **43**, 1204–1209.

114. Beckett, A. H., and Al-Sarraj, S. (1972). The mechanism of oxidation of amphetamine enantiomorphs by liver microsomal preparations from different species. *J. Pharm. Pharmacol.* **24**, 174–176.

115. Lindeke, B., Cho, A. K., Thomas, T. L., and Michelson, L. (1973). Microsomal N-hydroxylation of phenylalkylamines. Identification of N-hydroxylated phenylalkylamines as their trimethylsilyl derivatives by GC/MC. *Acta Pharm. Suec.* **10**, 493–506.

116. Hucker, H. B., Michniewicz, B. M., and Rhodes, R. E. (1971). Phenylacetone oxime. An intermediate in the oxidative deamination of amphetamine. *Biochem. Pharmacol.* **20**, 2123–2128.

117. Wright, J., Cho, A. K., and Gal, J. (1977). The role of *N*-hydroxyamphetamine in the metabolic deamination of amphetamine. *Life Sci.* **20**, 467–474.

118. Coutts, R. T., Dawe, R., Dawson, G. W., and Kovach, S. H. (1976). *In vitro* metabolism of 1-phenyl-2-propanone oxime in rat liver homogenates. *Drug Metab. Dispos.* **4**, 35–39.

119. Beckett, A. H., and Jones, G. R. (1977). Metabolic oxidation of aralkyl oximes to nitro compounds by fortified 9000g liver supernatants from various species. *J. Pharm. Pharmacol.* **29**, 416–421.

120. Montesano, R. and Bartsch, H. (1976). Mutagenic and carcinogenic N-nitroso compounds: Possible environmental hazards. *Mutat. Res.* **32**, 179–228.

121. Bartsch, H., Malaveille, C., and Montesano, R. (1975). *In vitro* metabolism and microsome-mediated mutagenicity of dialkylnitrosamines in rat, hamster, and mouse tissues. *Cancer Res.* **35**, 644–651.

122. Czygan, P., Greim, H., Garro, A. J., Hutterer, F., Schaffner, F., Popper, H., Rosenthal, O., and Cooper, D. Y. (1973). Microsomal metabolism of dimethylnitrosoamine and the cytochrome P-450 dependency of its activation to a mutagen. *Cancer Res.* **33**, 2983–2986.

123. Guttenplan, J. B., Hutterer, F., and Garro, A. J. (1976). Effects of cytochrome P-448 and P-450 inducers on microsomal dimethylnitrosamine demethylase activity and the capacity of isolated microsomes to activate dimethylnitrosamine to a mutagen. *Mutat. Res.* **35**, 415–422.

124. Lotlikar, P. O., Baldy, W. J., Jr., and Dwyer, E. N. (1975). Dimethylnitrosamine demethylation by reconstituted liver microsomal cytochrome P-450 enzyme system. *Biochem. J.* **152**, 705–708.

125. Lijinsky, W., Loo, J., and Ross, A. E. (1968). Mechanism of alkylation of nucleic acids by nitrosodimethylamine. *Nature (London)* **218**, 1174–1175.

126. Magee, P. N., and Hultin, T. (1962). Toxic liver injury and carcinogenesis. Methylation of proteins of rat-liver slices by dimethylnitrosamine *in vitro*. *Biochem. J.* **83,** 106–114.
127. Magee, P. N., and Farber, E. (1962). Toxic liver injury and carcinogenesis. Methylation of rat-liver nucleic acids by dimethylnitrosamine *in vivo*. *Biochem. J.* **83,** 114–124.
128. Dagani, D., and Archer, M. C. (1976). Deuterium isotope effect in microsomal metabolism of dimethylnitrosamine, *J. Natl. Cancer Inst.* **57,** 955–957.
129. Elespuru, R. K. (1978). Deuterium isotope effects in mutagenesis by nitroso compounds. *Mutat. Res.* **54,** 265–270.
130. Keefer, L. K., Lijinsky, W., and Garcia, H. (1973). Deuterium isotope effect on the carcinogenicity of dimethylnitrosamine in rat liver. *J. Natl. Cancer Inst.* **51,** 299–302.
131. Duggan, D. E., Hogans, A. F., Kwan, K. C., and McMahon, F. G. (1972). The metabolism of indomethacin in man. *J. Pharmacol. Exp. Ther.* **181,** 563–575.
132. Parke, D. V. (1968). ''Biochemistry of Foreign Compounds.'' Pergamon, Oxford.
133. Mitoma, C., Yasuda, D. M., Tagg, J., and Tanabe, M. (1967). Effect of deuteration on the OCH_3 group on the enzymic demethylation of *o*-nitroanisole. *Biochim. Biophys. Acta* **136,** 566–567.
134. Foster, A. B., Jarman, M., Stevens, J. D., Thomas, P., and Westwood, J. H. (1974). Isotope effects in O- and N-demethylations mediated by rat liver microsomes. Application of direct insertion electron impact mass spectrometry. *Chem.-Biol. Interact.* **9,** 327–340.
135. Al-Gailany, K. A. S., Bridges, J. W., and Netter, K. J. (1975). The dealkylation of some *p*-nitrophenylalkylethers and their α-deuterated analogues by rat liver microsomes. *Biochem. Pharmacol.* **24,** 867–870.
136. Garland, W. A., Nelson, S. D., and Sasame, H. A. (1976). Primary and β-secondary isotope effects in the O-deethylation of phenacetin. *Biochem. Biophys. Res. Commun.* **72,** 539–545.
137. Thurman, R. G., Marazzo, D. P., Jones, L. S., and Kauffman, F. C. (1977). The continuous kinetic determination of *p*-nitroaniline O-demethylation in hemoglobin-free perfused rat liver, *J. Pharmacol. Exp. Ther.* **201,** 498–506.
138. Ullrich, V., Frommer, U., and Weber, P. (1973). Differences in the O-dealkylation of 7-ethoxycoumarin after pretreatment with phenobarbital and 3-methylcholanthrene. *Hoppe-Seyler's Z. Physiol. Chem.* **354,** 514–520.
139. Burke, M. D., and Mayer, R. T. (1975). Inherent specificities of purified cytochromes *P*-450 and *P*-448 toward biphenyl hydroxylation and ethoxyresorufin deethylation. *Drug. Metab. Dispos.* **3,** 245–253.
140. Burke, M. D., Prough, R. A., and Mayer, R. T. (1977). Characteristics of a microsomal cytochrome *P*-448-mediated reaction. Ethoxyresorufin O-deethylation. *Drug Metab. Dispos.* **5,** 1–8.
141. Dybing, E., Aune, T., and Nelson, S. D. (1979). Metabolic activation of 2,4-diaminoanisole, a hair-dye component. II. Role of cytochrome *P*-450 metabolism in irreversible binding *in vitro*. *Biochem. Pharmacol.* **28,** 43–50.
142. Gutman, H. R., and Bell, P. (1977). N-hydroxylation of arylamides by the rat and guinea pig. Evidence for substrate specificity and participation of cytochrome P_1-450. *Biochim. Biophys. Acta* **498,** 229–243.
143. Lotlikar, P. D., and Zaleski, K. (1974). Inhibitory effect of carbon monoxide on the N- and ring-hydroxylation of 2-acetamidofluorene by hamster hepatic microsomal preparations. *Biochem. J.* **144,** 427–430.
144. Thorgeirsson, S. S., Jollow, D. J., Sasame, H. A., Green, I., and Mitchell, J. R. (1973). The role of cytochrome *P*-450 in N-hydroxylation of 2-acetylaminofluorene. *Mol. Pharmacol.* **9,** 398–404.

145. Lotlikar, P. D., Wertman, K., and Luha, L. (1973). Role of mixed-function amine oxidase in N-hydroxylation of 2-acetaminofluorene by hamster liver microsomal preparations. *Biochem. J.* **136**, 1137–1140.

146. Lotlikar, P. D., and Zaleski, K. (1975). Ring and N-hydroxylation of 2-acetamidofluorene by rat liver reconstituted cytochrome P-450 enzyme system. *Biochem. J.* **150**, 561–564.

147. Lotlikar, P. D., Luha, L., and Zaleski, K. (1974). Reconstituted hamster liver microsomal enzyme system for N-hydroxylation of the carcinogen 2-acetylaminofluorene. *Biochem. Biophys. Res. Commun.* **59**, 1349–1355.

148. Miller, J. A., and Miller, E. C. (1969). The metabolic activation of carcinogenic aromatic amines and amides. *Prog. Exp. Tumor Res.* **11**, 273–301.

149. Hinson, J. A., Nelson, S. D., and Mitchell, J. R. (1977). Studies on the microsomal formation of arylating metabolites of acetaminophen and phenacetin. *Mol. Pharmacol.* **13**, 625–633.

150. Hinson, J. A., Pohl, L. R., and Gillette, J. R. (1979). N-Hydroxyacetaminophen: A microsomal metabolite of N-hydroxyphenacetin but apparently not of acetaminophen. *Fed. Proc., Fed. Am. Soc. Exp. Biol.* **38**, 426.

151. Hinson, J. A., and Mitchell, J. R. (1976). N-Hydroxylation of phenacetin by hamster liver microsomes. *Drug Metab. Dispos.* **4**, 430–435.

152. Cho, A. K., Lindeke, B., and Sum, C. Y. (1974). The N-hydroxylation of phentermine (2-methyl-2-amino-1-phenylpropane). Properties of the enzyme system. *Drug Metab. Dispos.* **2**, 1–8.

153. Cho, A. K., Sum, C. Y., Jonsson, J., and Lindeke, B. (1978). The role of N-hydroxylation in the metabolism of phentermine and amphetamine by liver preparations. *In* "Biological Oxidation of Nitrogen" (J. W. Gorrod, ed.), pp. 15–23. Elsevier/North-Holland Biomedical Press, Amsterdam.

154. Gorrod, J. W. (1973). Differentiation of various types of biological oxidation of nitrogen in organic compounds. *Chem.-Biol. Interact.* **7**, 289–303.

155. Beckett, A. H., and Belanger, P. M. (1976). The microsomal N-oxidation of phentermine. *J. Pharm. Pharmacol.* **28**, 692–699.

156. Coutts, R. T., Dawson, G. W., and Beckett, A. H. (1976). *In vitro* metabolism of 1-phenyl-2-(n-propylamino)propane (*N*-propylamphetamine) by rat liver homogenates. *J. Pharm. Pharmacol.* **28**, 815–821.

157. Beckett, A. H., and Gibson, G. G. (1975). Microsomal N-hydroxylation of dibenzylamine. *Xenobiotica* **5**, 677–686.

158. Franklin, M. R. (1974). The formation of a 455 nm complex during cytochrome P-450-dependent *N*-hydroxyamphetamine metabolism. *Mol. Pharmacol.* **10**, 975–985.

159. Franklin, M. R. (1974). Complexes of metabolites of amphetamines with hepatic cytochrome P-450. *Xenobiotica* **4**, 133–142.

160. Kawalek, J. C., Levin, W., Ryan, D., and Lu, A. Y. H. (1976). Reconstituted liver microsomal enzyme system that hydroxylates drugs, other foreign compounds, and endogenous substrates. IX. The formation of a 455 nm metabolite–cytochrome P-450 complex. *Drug Metab. Dispos.* **4**, 190–194.

161. Mansuy, D., Beaune, P., Chottard, J. C., Bartoli, J. F., and Gans, P. (1976). The nature of the "455 nm absorbing complex" formed during the cytochrome P-450 dependent oxidative metabolism of amphetamine. *Biochem. Pharmacol.* **25**, 609–612.

162. Hirata, M., Lindeke, B., and Orrenius, S. (1979). Cytochrome P-450 product complexes and glutathione consumption produced in isolated hepatocytes by norbenzphetamine and its N-oxidized congeners. *Biochem. Pharmacol.* **28**, 479–484.

163. Jonsson, J., and Lindeke, B. (1976). On the formation of cytochrome P-450 product

complexes during the metabolism of phenylalkylamine. *Acta Pharm. Suec.* **13,** 313–320.

164. Lindeke, B., Jonsson, J., Hallström, G., and Paulsen, U. (1978). The effect of α-carbon substitution on the 455 nm complex formed by phenethylamines and their N-oxidized congeners. *In* "Biological Oxidation of Nitrogen" (J. W. Gorrod, ed.), pp. 47–52. Elsevier/North-Holland Biomedical Press, Amsterdam.

165. Neal, R. A., Kamataki, T., Hunter, A. L., and Catignani, G. (1977). Monooxygenase catalyzed activation of thiono-sulfur containing compounds to reactive intermediates. *In* "Microsomes and Drug Oxidations" (V. Ullrich ed.), pp. 467–475. Pergamon, Oxford.

166. Poulsen, L. L., Hyslop, R. M., and Ziegler, D. M. (1974). S-oxidation of thioureylene catalyzed by a microsomal flavoprotein mixed function oxidase. *Biochem. Pharmacol.* **23,** 3431–3440.

167. Norman, B. J., Poore, R. E., and Neal, R. A. (1974). Studies of the binding of sulfur released in the mixed-function oxidase-catalyzed metabolism of diethyl *p*-nitrophenylphosphorothionate (parathion) to diethyl *p*-nitrophenyl phosphate (paraoxon). *Biochem. Pharmacol.* **23,** 1733–1744.

168. Ptashne, K. A., Wolcott, R. M., and Neal, R. A. (1971). Oxygen-18 studies on the chemical mechanisms of the mixed function oxidase catalyzed desulfuration and dearylation reactions of parathion. *J. Pharmacol. Exp. Ther.* **179,** 380–385.

169. Kamataki, T., Belcher, D. H., and Neal, R. A. (1976). Studies of the metabolism of diethyl *p*-nitrophenyl phosphorothionate (parathion) and benzphetamine using an apparently homogeneous preparation of rat liver cytochrome *P*-450: Effect of a cytochrome *P*-450 antibody preparation. *Mol. Pharmacol.* **12,** 921–932.

170. Dalvi, R. R., Poore, R. E., and Neal, R. A. (1974). Studies of the metabolism of carbon disulfide by rat liver microsomes. *Life Sci.* **14,** 1785–1796.

171. Ptashne, K. A., and Neal, R. A. (1972). Reaction of parathion and malathion with peroxytrifluoroacetic acid, a model system for the mixed function oxidases. *Biochemistry* **11,** 3224–3228.

172. Lee, P. W., and Neal, R. A. (1978). Metabolism of methimazole by rat liver cytochrome *P*-450-containing monooxygenases. *Drug Metab. Dispos.* **6,** 591–600.

173. Boyd, M. R., and Neal, R. A. (1976). Studies on the mechanism of toxicity and of development of tolerance to the pulmonary toxin, α-naphthylthiourea (ANTU). *Drug Metab. Dispos.* **4,** 314–322.

174. Dalvi, R. R., Hunter, A. L., and Neal, R. A. (1975). Toxicological implications of the mixed function oxidase catalyzed metabolism of carbon disulfide. *Chem.-Biol. Interact.* **10,** 349–361.

175. De Matteis, F. (1974). Covalent binding of sulfur to microsomes and loss of cytochrome *P*-450 during the oxidative desulfuration of several chemicals. *Mol. Pharmacol.* **10,** 849–854.

176. Catignani, G. L., and Neal, R. A. (1975). Evidence for the formation of a protein bound hydrodisulfide resulting from the microsomal mixed function oxidase catalyzed desulfuration of carbon disulfide. *Biochem. Biophys. Res. Commun.* **65,** 629–636.

177. Kamataki, T., and Neal, R. A. (1976). Metabolism of diethyl *p*-nitrophenyl phosphorothionate (parathion) by a reconstituted mixed-function oxidase enzyme system. Studies of the covalent binding of the sulfur atom. *Mol. Pharmacol.* **12,** 933–944.

178. Gillette, J. R. (1969). Significance of mixed oxygenases and nitroreductase in drug metabolism. *Ann. N. Y. Acad. Sci.* **160,** 558–570.

179. Coccia, P. F., and Westerfeld, W. W. (1967). The metabolism of chlorpromazine by liver microsomal enzyme systems. *J. Pharmacol. Exp. Ther.* **157,** 446–458.

180. Van Dyke, R. A., and Gandolfi, A. J. (1975). Characteristics of a microsomal dechlorination system. *Mol. Pharmacol.* **11**, 809–817.
181. Van Dyke, R. A., and Wineman, C. G. (1971). Enzymatic dechlorination. Dechlorination of chloroethanes and propanes *in vitro*. *Biochem. Pharmacol.* **20**, 463–470.
182. Gandolfi, A. J., and Van Dyke, R. A. (1973). Dechlorination of chloroethane with a reconstituted liver microsomal system. *Biochem. Biophys. Res. Commun.* **53**, 687–692.
183. Loew, G., Trudell, J., and Motulsky, H. (1973). Quantum chemical studies of the metabolism of a series of chlorinated ethane anesthetics. *Mol. Pharmacol.* **9**, 152–162.
184. Kubic, V. L., and Anders, M. W. (1975). Metabolism of dihalomethanes to carbon monoxide. II. *In vitro* studies. *Drug Metab. Dispos.* **3**, 104–112.
185. Kubic, V. L., and Anders, M. W. (1978). Metabolism of dihalomethanes to carbon monoxide. III. Studies on the mechanism of the reaction. *Biochem. Pharmacol.* **27**, 2349–2355.
186. Anders, M. W., Kubic, V. L., and Ahmed, R. E. (1977). Metabolism of halogenated methanes and molecular binding. *J. Environ. Pathol. Toxicol.* **1**, 117–124.
187. Pohl, L. R., and Krishna, G. (1978). Study of the mechanism of metabolic activation of chloramphenicol by rat liver microsomes. *Biochem. Pharmacol.* **27**, 335–341.
188. Pohl, L. R., Nelson, S. D., and Krishna, G. (1978). Investigation of the mechanism of the metabolic activation of chloramphenicol by rat liver microsomes. Identification of a new metabolite. *Biochem. Pharmacol.* **27**, 491–496.
189. Pohl, L. R., Bhooshan, B., Whittaker, N. F., and Krishna, G. (1977). Phosgene: A metabolite of chloroform. *Biochem. Biophys. Res. Commun.* **79**, 684–691.
190. Pohl, L. R., and Krishna, G. (1978). Deuterium isotope effect in bioactivation and hepatotoxicity of chloroform. *Life Sci.* **23**, 1067–1072.
191. Pohl, L. R., George, J. W., Martin, J. L., and Krishna, G. (1979). Deuterium isotope effect in *in vivo* bioactivation of chloroform to phosgene. *Biochem. Pharmacol.* **28**, 561–563.
192. Ullrich, V., Hermann, G., and Weber, P. (1978). Nitrate formation from 2-nitropropane by microsomal monooxygenases. *Biochem. Pharmacol.* **27**, 2301–2304.
193. Jonsson, J., Kammerer, R. C., and Cho, A. K. (1977). Metabolism of 2-nitro-1-phenylpropane to phenylacetone by rabbit liver microsomes. *Res. Commun. Chem. Pathol. Pharmacol.* **18**, 75–82.
194. Fouts, J. R., and Brodie, B. B. (1957). The enzymatic reduction of chloramphenicol, *p*-nitrobenzoic acid and other aromatic nitro compounds in mammals. *J. Pharmacol. Exp. Ther.* **119**, 197–207.
195. Gillette, J. R., Kamm, J. J., and Sasame, H. A. (1968). Mechanism of *p*-nitrobenzoate reduction in liver: The possible role of cytochrome *P*-450 in liver microsomes. *Mol. Pharmacol.* **4**, 541–548.
196. Feller, D. R., Morita, M., and Gillette, J. R. (1971). Enzymatic reduction of niridazole by rat liver microsomes. *Biochem. Pharmacol.* **20**, 203–215.
197. Yoshida, Y., and Kumaoka, H. (1970). Studies on mechanisms of nitro- and azo-reducing systems of rat liver microsomes. *Proc. Symp. Drug Metab. Action, 1st, 1969* Vol. 1, pp. 57–65.
198. Kato, R., Oshima, T., and Takanaka, A. (1969). Studies on the mechanism of nitro reduction by rat liver. *Mol. Pharmacol.* **5**, 487–498.
199. Mason, R. P., and Holtzman, J. L. (1975). The mechanism of microsomal and mitochondrial nitroreductase. Electron spin resonance evidence for nitroaromatic free radical intermediates. *Biochemistry* **14**, 1626–1632.
200. Mason, R. P., and Holtzman, J. L. (1975). The role of catalytic superoxide formation in the O_2 inhibition of nitroreductase. *Biochem. Biophys. Res. Commun.* **67**, 1267–1274.

201. Sum, C. Y., and Cho, A. K. (1976). Properties of microsomal enzyme systems that reduce N-hydroxyphentermine. *Drug Metab. Dispos.* **4**, 436–441.
202. Hernandez, P. H., Mazel, P., and Gillette, J. R. (1967). Studies on the mechanisms of action of mammalian hepatic azoreductase. II. The effects of phenobarbital and 3-methylcholanthrene on carbon monoxide sensitive and insensitive azoreductase activities. *Biochem. Pharmacol.* **16**, 1877–1888.
203. Hernandez, P. H., Gillette, J. R., and Mazel, P. (1967). Studies on the mechanism of action of mammalian hepatic azoreductase. I. Azoreductase activity of reduced nicotinamide adenine dinucleotide phosphate-cytochrome *c* reductase. *Biochem. Pharmacol.* **16**, 1859–1875.
204. Shargel, L., and Mazel, P. (1972). Influence of 2,4-dichloro-6-phenoxyethylamine (DPEA) and β-diethylaminoethyl diphenylpropylacetate (SKF 525-A) on hepatic microsomal azoreductase activity from phenobarbital or 3-methylcholanthrene induced rats. *Biochem. Pharmacol.* **21**, 69–75.
205. Fujita, S., and Peisach, J. (1978). Liver microsomal cytochrome *P*-450 and azoreductase activity. *J. Biol. Chem.* **253**, 4512–4513.
206. Iwasaki, K., Noguchi, H., Kato, R., Imai, Y., and Sato, R. (1977). Reduction of tertiary amine N-oxide by purified cytochrome P-450. *Biochem. Biophys. Res. Commun.* **77**, 1143–1149.
207. Kato, R., Iwasaki, K., and Noguchi, H. (1978). Reduction of tertiary amine N-oxides by cytochrome *P*-450. Mechanisms of the stimulatory effect of flavins and methyl viologen. *Mol. Pharmacol.* **14**, 654–664.
208. Sugiura, M., Iwasaki, K., and Kato, R. (1976). Reduction of tertiary amine N-oxides by liver microsomal cytochrome *P*-450. *Mol. Pharmacol.* **12**, 322–334.
209. Kato, R., Iwasaki, K., and Noguchi, H. (1976). Stimulatory effect of FMN and methyl viologen on cytochrome *P*-450 dependent reduction of tertiary amine N-oxide. *Biochem. Biophys. Res. Commun.* **72**, 267–274.
210. Booth, J., Hewer, A., Keysell, G. R., and Sims, P. (1975). Enzymic reduction of aromatic hydrocarbon epoxides by the microsomal fraction of rat liver. *Xenobiotica* **5**, 197–203.
211. Kato, R., Iwasaki, K., Shiraga, T., and Noguchi, H. (1976). Evidence for the involvement of cytochrome *P*-450 in reduction of benzo[a]pyrene 4,5-oxide by rat liver microsomes. *Biochem. Biophys. Res. Commun.* **70**, 681–687.
212. Uehleke, H., Hellmer, K. H., and Tabarelli, S. (1973). Binding of ^{14}C-carbon tetrachloride to microsomal proteins *in vitro* and formation of $CHCl_3$ by reduced liver microsomes. *Xenobiotica* **3**, 1–11.
213. Van Dyke, R. A., and Gandolfi, A. J. (1976). Anaerobic release of fluoride from halothane. Relationship to the binding of halothane to hepatic cellular constituents. *Drug Metab. Dispos.* **4**, 40–44.
214. Mansuy, D., Nastainczyk, W., and Ullrich, V. (1974). The mechanism of halothane binding to microsomal cytochrome *P*-450. *Naunyn-Schmiedeberg's Arch. Pharmacol.* **285**, 315–324.
215. Kato, R. (1974). Sex-related differences in drug metabolism. *Drug Metab. Rev.* **3**, 1–32.

Chapter 8

The Role of
NADPH-Cytochrome *c* (*P*-450)
Reductase in Detoxication

BETTIE SUE SILER MASTERS

I. INTRODUCTION

Historically, NADPH-cytochrome *c* reductase (EC 1.6.2.4) has been known as a distinct entity since 1950 when Horecker[1] isolated it from whole liver acetone powder and purified and characterized it as a flavoprotein. Since its subcellular localization was not known, a possible role in mitochondrial electron transport could not be discounted at this time.

In 1962, first, Williams and Kamin[2] and, then, Phillips and Langdon[3] reported that NADPH-cytochrome *c* reductase was localized in the endoplasmic reticulum of the cell within vesicles isolated in the ultracentrifuge as microsomes. Williams and Kamin[2] compared their steapsin-

ENZYMATIC BASIS OF DETOXICATION, VOL. I

solubilized preparation with Horecker's on the basis of the identity of the prosthetic groups, similar specific activities, comparable molecular weights, reactivation of the apoenzyme with either FAD or FMN, comparable pH optima, and competitive inhibition by NADP+; Phillips and Langdon[3] suggested that, where comparisons were possible, their trypsin-solubilized NADPH-cytochrome c reductase was similar to the whole liver preparation of Horecker. Both groups recognized the possibility that this flavoprotein could be involved in aromatic and steroid hydroxylations and drug demethylations.

La Du et al.,[4] in 1955, and Gillette et al.,[5] in 1957, had reported the inhibition of the NADPH-mediated microsomal demethylation of monomethyl-4-aminoantipyrine by cytochrome c and Krisch and Staudinger,[6] in 1961, demonstrated the inibition of acetanilide hydroxylation in liver microsomes by cytochrome c.

The final and definitive proof for participation of NADPH-cytochrome c reductase in microsomal drug metabolism reactions resulted from its purification to homogeneity. The proteolytically solubilized preparations, which have been shown to be missing that portion of the enzyme which anchors it to the membrane but which contain both the FAD and FMN prosthetic groups, have been used to produce antibodies. In 1969, these inhibitory, precipitating antibodies were first shown by Omura[7] to inhibit aniline hydroxylation catalyzed by liver microsomes and by Kuriyama et al.[8] to inhibit both the uninduced and phenobarbital-induced forms of NADPH-cytochrome c reductase. During the same year, using the ionic detergent, sodium deoxycholate, in the presence of dithiothreitol and glycerol, Lu et al.[9] were able to reconstitute the metabolism of various drug substrates with partially purified NADPH-cytochrome c reductase and cytochrome P-450 from liver microsomes. These studies also demonstrated the requirement for this flavoprotein in cytochrome P-450-mediated reactions.

II. MICROSOMAL ELECTRON TRANSPORT SYSTEMS

A. Hepatic Metabolism

The implication of NADPH-cytochrome c (P-450) reductase in microsomal electron transport reactions, particularly those mediated by cytochrome P-450, began with the discovery in 1958 by Klingenberg[10] and Garfinkel,[11] simultaneously and independently, of a carbon monoxide binding pigment in liver microsomes. Klingenberg[10] reported that G. R. Williams actually observed such a pigment first, in 1955, using the double-beam recording spectrophotometer in the laboratory of Britton Chance. It was not until 1962, when Omura and Sato[12,13] reported the purification of a heme protein from liver microsomes, which bound carbon monoxide in its reduced form and exhibited a unique absorption spectrum indicating that it was a new b-type cytochrome, that its identity with the membrane-bound, carbon monoxide-binding pigment was established.

Subsequently, Estabrook, Cooper, and Rosenthal[14] established that cytochrome P-450 was the terminal oxygenase in the C-21-hydroxylation reaction catalyzed by adrenal cortical microsomes, using the principle of the photochemical action spectrum developed by Warburg[15] for the involvement of cytochrome oxidase in mitochondrial electron transport. Cooper et al.[16] then utilized this experimental approach to demonstrate that cytochrome P-450 was required in the oxidative metabolism of various drugs and steroids by liver microsomes.

Thus, the evidence that cytochrome P-450 played a vital role in the metabolism of a variety of both endogenous substrates (steroids) and exogenous compounds (drugs) was firmly established. Although the early studies of La Du,[4] Gillette,[5] and Omura[7] strongly suggested that NADPH-cytochrome c reductase was the source of electrons for this process, the definitive proof came with the demonstration by Masters et al.,[17,18] that the antibody to purified NADPH-cytochrome c reductase inhibited aminopyrine and ethylmorphine demethylation concomitantly with the inhibition of cytochrome c and cytochrome P-450 reduction catalyzed by either pig[17] or human liver[18] microsomes. Glazer et al.[19] also measured the inhibition of cytochrome P-450 reduction by anti-NADPH-cytochrome c reductase globulin, but not concomitantly with cytochrome c reduction. Nevertheless, both groups[17-19] established unequivocally that this flavoprotein supplies the electrons from NADPH for the oxidative demethylation of aminopyrine by liver microsomes from rat, pig, and human.

The variety of cytochrome P-450-mediated monooxygenase reactions catalyzed by liver microsomes has been well described, but the nature of the specificity of these reactions remains a subject of controversy and the object of much research effort. A recent comprehensive review of these reactions and the role of cytochrome P-450 has been presented by Sato and Omura.[20] The exhaustive list of substrates includes laurate, hexane, aniline, benzo[a]pyrene, cholesterol, testosterone, aminopyrine, ethylmorphine, codeine, 7-ethoxycoumarin, amphetamines, chloropromazine, phenothiazine, fluoraniline, aldrin, and many others. The reaction types involve aliphatic, alicyclic, and aromatic hydroxylations; steroid hydroxylations; N-hydroxylations; oxidative dealkylations of N-, O-, and

S-alkyl compounds; oxidative deaminations of primary amines; oxidation of thioethers to sulfoxides; and epoxidation of hydrocarbons and halogenated hydrocarbons. The reader is referred to Omura *et al.*[20] and to Chapters 6 and 7 in this volume for a discussion of these reactions.

B. Extrahepatic Metabolism

1. *Adrenal*

From a historical viewpoint, the adrenal cortical microsomal C-21 hydroxylation of steroids was the first cytochrome *P*-450-mediated reaction to be identified.[14] This system is principally concerned with the metabolism of endogenous steroid substrates, such as 17-hydroxyprogesterone. Since the cytochromes *P*-450 of both adrenal cortical microsomes and mitochondria are highly substrate specific, their capacity for the oxidation of exogenous compounds is virtually nonexistent, except for the weak benzo[a]pyrene hydroxylase activity reported by Zampaglione and Mannering.[21] Although carbon monoxide inhibited benzo[a]pyrene hydroxylase activity in adrenal microsomes, SKF-525A, aniline, ethylmorphine, and 3-methyl-4-aminoazobenzene had no effect.

2. *Lung*

The metabolism of foreign compounds has been demonstrated in lung microsomes from rat,[22] rabbit,[22] mouse,[22] hamster,[22] guinea pig,[22] and human[23] and in reconstituted systems from rabbit.[24,25] Aniline and biphenyl hydroxylation and aminopyrine demethylation activities were demonstrable in lung microsomes from all of the animal species studied[22] and trace amounts of laurate hydroxylation and measurable benzo[a]-pyrene hydroxylation were found in human lung microsomes.[23] Using highly purified cytochrome *P*-450 and NADPH-cytochrome *P*-450 reductase from rabbit lung microsomes, Guengerich was able to reconstitute, in the presence of phospholipid, benzphetamine demethylation, benzo[a]-pyrene hydroxylation, 7-ethoxycoumarin O-dealkylation, and a variety of other activities.[24] These studies substantiate that lung microsomes are capable of metabolizing exogenous compounds to some degree. It is interesting to note that the ratio of cytochrome *P*-450 content in lung compared to liver varies from 0.07–0.15 in the various animal species studied.[22]

3. *Kidney*

In 1966, kidney microsomes were shown by Kato[26] to contain cytochrome *P*-450, and Wada *et al.*[27] and Ichihara *et al.*[28] demonstrated a significant amount of lauric acid ω-hydroxylation activity in kidney. Ichihara et al.[28] reported ω-hydroxylation of decanoate and dodecanoate (laurate) in a variety of tissues from five different species. More recently, kidney cortex microsomes have been shown to catalyze the monooxygenation of aminopyrine,[22,23,29] aniline,[22] and benzo[a]pyrene.[30,31] The studies of Jakobsson and Cinti[29] and Prough et al.,[31] included both rodent and human microsomal systems. In addition, Jakobsson et al.[23] found measurable amounts of aminopyrine demethylase, laurate hydroxylase, and benzo[a]pyrene hydroxylase in kidney cortex microsomes from ten human cases from which livers, lungs, and kidneys were obtained at autopsy. The ratio of cytochrome P-450 content in kidney compared to liver ranges from 0.13–0.33 in a variety of species.[22] Thus, it would appear that kidney microsomes possess a capacity for the metabolism of xenobiotics and that, in the case of benzo[a]pyrene metabolism in rat kidney, aryl hydrocarbon administration induces this activity.[30]

4. Spleen

Spleen microsomes were shown by Tenhunen et al.,[32] in 1968, to catalyze the oxidation of heme at the α-methene bridge to form biliverdin and carbon monoxide. The reaction was shown to require NADPH and oxygen and the incorporation of ^{18}O-labeled molecular oxygen into bilirubin and CO was demonstrated by mass spectrometry.[33] Schacter et al.[34] showed that heme oxygenase activity catalyzed by either rat liver or spleen microsomes could be inhibited by an antibody to NADPH-cytochrome c reductase, confirming the requirement for this microsomal flavoprotein as a source of electrons. The role of cytochrome P-450 in this process, however, was equivocal since inducing agents, such as phenobarbital, and inhibitors, such as metyrapone and SKF-525A, of the cytochrome P-450-mediated monooxygenases were without effect on heme oxygenase activity.[34] The role of spleen microsomes in the metabolism of exogenous compounds has not been established.

5. Intestine

The capacity of the intestinal mucosa to metabolize benzo[a]pyrene has been known for some time,[21,35] and supports the contention of Wattenberg and Leong[36] that benzo[a]pyrene hydroxylase activity occurs in organs which serve as portals of entry to the body. A recent report by Fang and Strobel[37] demonstrated that rat colon mucosa can metabolize precarcinogens to mutagenic metabolites without the mediation of intestinal bacteria using the Ames test as a criterion for mutagenicity. These investigators had published data previously documenting the presence of a drug and carcinogen metabolism system in microsomes isolated from the mucosal layer of the colon.[38] Stohs *et al.*[39] demonstrated the metabolism of benzo[*a*]pyrene, 7-ethoxycoumarin, 7-ethoxyresorufin, and biphenyl in microsomes prepared from the mucosal lining of the small intestine. The rates of metabolism of both drugs and carcinogens are increased by pretreatment of animals with inducing agents.[37]

III. PURIFICATION OF NADPH-CYTOCHROME c (P-450) REDUCTASE

Although NADPH-cytochrome *c* was purified as early as 1950 by Horecker[1] using trypsin digestion of a whole pig liver acetone powder, the microsomal localization of this flavoprotein was not known until 1962, when Williams and Kamin[2] and Phillips and Langdon[3] independently published the purification of the enzyme from pig liver microsomes using proteolytic digestion to release the enzyme from its membrane-bound milieu. Kinetic parameters were determined with these preparations and studies of the effects of pH, ionic strength, and inhibitors were performed.[2,3] Homogeneous preparations of the reductase were obtained by Masters *et al.*,[40,41] using a preparation procedure employing pancreatic steapsin as a solubilizing agent. These studies[40,42] led to the description of a unique catalytic mechanism for flavoproteins, which will be discussed later, and determined that the enzyme contained 2 flavins per mole. At this time, NADPH-cytochrome *c* reductase was thought to contain only FAD as a prosthetic group,[1-3,43-45] although several laboratories had reported that the apoenzyme of NADPH-cytochrome *c* reductase was reactivated more efficiently with FMN than with FAD.[1-3] However, Iyanagi and Mason[46] showed in 1973 that the trypsin-solubilized microsomal flavoprotein contained both FAD and FMN as prosthetic groups and this observation was confirmed by Masters *et al.*,[47] Vermilion and Coon,[48,49] Dignam and Strobel,[50,51] and Yasukochi and Masters[52] on both steapsin-solubilized and detergent-solubilized preparations.

Since Coon's laboratory[53] reported in 1969 that preparations of proteolytically solubilized NADPH-cytochrome *c* reductase obtained from Kamin's laboratory were not capable of reconstituting microsomal cytochrome *P*-450-mediated reactions, efforts have been directed toward its purification from detergent-solubilized microsomes. Various procedures have been employed for the detergent solubilization and subsequent purification of NADPH-cytochrome *P*-450 reductase,[48-52,54] and several of these have resulted in homogeneous preparations. Yasukochi and Masters[52] introduced a biospecific affinity chromatography procedure for the highly selective purification of NADPH-cytochrome *P*-450 reductase in high yield (30–50%) from microsomes. The specific affinity of the reductase for the 2′-phosphate group on NADPH and its lower binding affinity for 2′-AMP permitted the separation of the reductase from other solubilized microsomal constituents on 2′,5′-ADP-Sepharose 4B [Sepharose 4B-bound N^6-(6-aminohexyl)adenosine 2′,5′-bisphosphate]. Dignam and Strobel[51] have described an alternative procedure using an affinity column of NADP-Sepharose 4B.

The procedure of Yasukochi and Masters[52,54] involves the solubilization of microsomes with a combination of nonionic (Renex 690 or Emulgen 913) and ionic (sodium deoxycholate) detergents, batch chromatography on DEAE-cellulose, affinity chromatography on 2′,5′-ADP-Sepharose 4B (Pharmacia), gel filtration on Ultrogel AcA 34, treatment with potassium ferricyanide (to oxidize the flavins), charcoal (to remove nucleotides), and hydroxylapatite chromatography (to remove detergent). If a spectrally pure reductase preparation is not required, a homogeneous preparation is obtained after Ultrogel AcA 34 gel filtration.[52] The specific activities of these preparations vary between 49 and 63 μmoles of cytochrome c reduced per min/mg in 0.3 M potassium phosphate, pH 7.7, 0.3 mM EDTA, at 30°. The turnover numbers, based on flavin concentration, are between 2380 and 2840/min. The molecular weights of these preparations are approximately 78,000. NADPH-cytochrome c (P-450) reductases have been purified from variety of sources, including lung,[24,25,55] kidney,[56,57] spleen,[58] and intestine.[59] Only recently has it been possible to obtain preparations of reductase from extrahepatic organs which were reconstitutively active in various monooxygenase reactions in the presence of purified cytochromes P-450 from the same or different organs and a phospholipid.[24,25,57,59] The properties of these various reductase preparations will be discussed in Section VI.

IV. PROPERTIES OF NADPH-CYTOCHROME c (P-450) REDUCTASE FROM VARIOUS ORGANS AND SPECIES

The purification of NADPH-cytochrome c reductases and/or NADPH-cytochrome P-450 reductases from extrahepatic tissues has led to the conclusion that these flavoproteins are quite similar kinetically and physicochemically to the liver reductase.[24,55–58] Furthermore, all experiments which involve the use of reductase fractions from one organ with cytochromes P-450 from another organ indicate that the reductases from various organs can function interchangeably.[24,25,57] Table I shows the properties of NADPH-cytochrome c (P-450) reductases from several organs, the higher molecular weight forms being active in reconstituting

TABLE I

Physicochemical, Kinetic, and Immunochemical Properties of Purified NADPH-/Cytochrome C (P-450) Reductases from Various Organs

Organ		Molecular weight	Flavin content 1:1:1 FAD/FMN/Apoenzyme	K_m NADPH ($\times 10^6 M$)	K_m Cytochrome c ($\times 10^6 M$)	Specific activity[b] and/or turnover number	Immunochemical characteristics
Liver							
Rabbit	Fp-T	68,000 (46) SDS-PAGE	Fp-T (46)	3.2	4.3 (43)	Fp-D	35 μmoles min^{-1} mg^{-1} (75) 0.35 M, 25°	
	Fp-D[a]	79,000 (46) Sed. equil.	Fp-D (68,75)				59.8 μmoles min^{-1} mg^{-1} (68) 0.3 M, 30°	
		74,000 (46) SDS-PAGE				Fp-T	38–41 μmoles min^{-1} mg^{-1} (46) 0.1 M, 25°	
		74,000 (68) SDS-PAGE						
Rat	Fp-D	79,000 (48) SDS-PAGE	Fp-D (48,50,52)			Fp-D	40.9 μmoles min^{-1} mg^{-1} (48) 0.3 M, 30°	Ab cross-reacts with pig reductase and microsomes[a]
		80,000 (50) SDS-PAGE					43.8 μmoles min^{-1} mg^{-1} (52) 0.3 M, 30°	
		78,000 (52) SDS-PAGE					48.8 μmoles min^{-1} mg^{-1} (50) 0.1 M, 30°	
	Fp-B	71,000 (55) SDS-PAGE	Fp-S (47)			Fp-T	41.2 μmoles min^{-1} mg^{-1} (45) 0.1 M, 25°	
						Fp-B	51.0 μmoles min^{-1} mg^{-1} (55)	
Pig	Fp-S	68,000 (1) Flavin:protein	Fp-S (47)	2.6 (41)	5.6 (40,41)	Fp-S	1200–1400 min^{-1} (40) 0.05 M, 25°	Ab cross-reacts with rat and human reductases and microsomes (17,63)
		68,000 (60) Sed. equil.					37.4 μmoles min^{-1} mg^{-1} (52) 0.3 M, 30°	
	Fp-D	79,000[c]	Fp-D (52)			Fp-D	2856 min^{-1} (52) 0.3 M, 30°	
							49–63 μmoles min^{-1} mg^{-1} (54) 0.3 M, 30°	
							2380–2840 min^{-1} (54) 0.3 M, 30°	

Organ	Purified form, MW (method)	Specific activity[b]		Turnover number[b]	Antibody cross-reactivity
Human[a]				Fp-S 850 min^{-1}	Ab cross-reacts with rat and pig reductases and microsomes
Kidney	Fp-S (56) 71,000 (56) SDS-PAGE, 68,000 (56) Seph. G-200; Fp-T (58) 66,000 (58) Seph. G-100; Fp-D 75,000 (78) SDS-PAGE	10.0 (56); 9.5 (58); 29 (78)	4.6 (56); 3.3 (58); 20 (78)	Fp-S 1245 min^{-1} (56) 0.05 M, 25°; Fp-D 61.0 μmoles min^{-1} mg (78) 0.3 M, 30°	Ab to liver reductase cross-reacts with pig kidney reductase and microsomes (56, 78)
Lung — Rabbit	Fp-D 79,000 (64) SDS-PAGE			Fp-D 47.7 μmoles min^{-1} mg^{-1} (64)	Ab to liver reductase cross-reacts with lung reductase (55)
Lung — Rat	Fp-D 79,000 (55) SDS-PAGE; Fp-B' 71,000 (55) SDS-PAGE			Fp-B 31.0 μmoles min^{-1} mg^{-1} (55)	
Spleen — Pig	Fp-T (58) 66,000 (58) Seph. G-100	6.0 (58)	4.7 (58)		Ab to pig liver reductase inhibits pig and rat spleen microsomal reductase activity (34)

[a] The notations Fp-D, Fp-T, Fp-S, and Fp-B represent detergent-, trypsin-, pancreatic steapsin-, and bromelain-solubilized microsomal reductases, respectively, from the various organs indicated.

[b] Specific activities and turnover numbers have been determined under a variety of conditions including different ionic strengths of phosphate buffer and different temperatures. All determinations are performed in potassium phosphate buffer, pH 7.5–7.7, at the ionic strengths and temperatures indicated in the table.

[c] Y. Yasukochi, R. T. Okita, and B. S. S. Masters, unpublished observations.

[d] B. S. S. Masters, L. Dean Coe, N. I. Mock, and R. Burgess, unpublished observations.

monooxygenase activities in the presence of cytochrome P-450 and a phospholipid. The lower molecular weight forms of the reductase have been obtained by proteolytic digestion procedures involving the use of trypsin,[45] pancreatic steapsin,[41] or bromelain[55] and the molecular weights vary between 66,000 and 71,000 daltons[1,2,43,55,56,58,60] from various species and organs. On the other hand, the detergent-solubilized reductase preparations exhibited molecular weights ranging between 74,000 and 80,000[24,49,57,61,62] and these preparations, contrary to the proteolytically solubilized forms of the reductase,[53] are capable of reconstituting monooxygenation activities with a variety of substrates in the presence of cytochrome P-450.

Immunochemical studies have been performed in a number of laboratories showing that the NADPH-cytochrome P-450 reductases from various organs within the same species are quite similar, if not identical.[55-58] There appear to be some differences, however, among reductases from different species as determined by immunochemical[17,18,63] and peptide mapping techniques.[64] Comparisons of molecular weights and amino acid content determined in different laboratories on liver reductases from the same species (rat) reveal somewhat different values. Knapp *et al.*,[61] reported a molecular weight of 79,000–80,000 daltons as determined by sodium dodecyl sulfate–polyacrylamide disc gel electrophoresis[65] while estimations of molecular weight by sedimentation equilibrium gave a value of 76,500 daltons.[61] On the other hand, Vermilion and Coon[62] reported a molecular weight of 76,000, determined by sodium dodecyl sulfate–polyacrylamide slab gel electrophoresis using the discontinuous buffer system of Laemmli.[66] The remarkable differences between the preparations from these laboratories[61,62] were related to the amino acid content. There were nine tryptophan, five tyrosine, six serine, and one cysteine residues *more* and nine threonine, eight methionine, five proline, four glycine, and four leucine residues *less* in the Vermilion and Coon preparation[62] than in the purified reductase of Knapp *et al.*[61] These differences cannot be due simply to a difference in molecular weight, since there are more of some residues and less of others although tryptophan can be differentially lost during analysis. Recently, however, Gum and Strobel[67] have reexamined both the detergent-solubilized rat liver NADPH-cytochrome P-450 reductase and a preparation of rat liver reductase obtained after incubation of phospholipid vesicles containing reductase with steapsin. The amino acid content of the detergent-solubilized reductase of Gum and Strobel[67] more nearly resembles that of Vermilion and Coon[62] with notable differences in the number of glycine, methionine, isoleucine, leucine, phenylalanine, lysine, and tryptophan residues (in excess of three residues more in the Vermilion and Coon[62] preparation for these amino acids except for methionine of which there are 19 in the Gum and Strobel[67] preparation and 16 in the Vermilion and Coon[62] sample). In order to compare subtle differences among reductase preparations from different organs of the same species or from the same organs from different species, these discrepancies must be resolved.

V. SPECIFICITY OF REDUCTASE-CYTOCHROME P-450 INTERACTION IN VARIOUS MONOOXYGENASE REACTIONS; OTHER METABOLIC FUNCTIONS

The cross-organ specificities of preparations of cytochrome P-450 have been examined by several laboratories in reconstituted systems in order to ascertain the basis for specificity. The studies of Ichihara *et al.*[28] in 1973 suggested that the specificity for the electron-donating flavoprotein for reconstitution of fatty acid ω-hydroxylation with a partially purified cytochrome P-450 from pig kidney cortex microsomes was not absolute; a partially purified NADPH-cytochrome c reductase, solubilized from pig liver microsomes by Triton X-100, was found capable of reconstituting the ω-hydroxylation of lauric acid.[28] As purification techniques have improved, recent studies[57] have led to the conclusion that the minimal components required for such a reconstitution experiment include NADPH-cytochrome c (P-450) reductase, cytochrome P-450, and a lipid component (either with added phospholipid or in phospholipid-containing vesicles). The studies of Yasukochi *et al.*,[57] permit the conclusion that either pig liver or pig kidney microsomal NADPH-cytochrome P-450 reductase (purified by detergent solubilization) can reconstitute the ω- and ω-1-hydroxylation of laurate catalyzed by pig kidney microsomal cytochrome P-450 or the oxidative demethylation of benzphetamine catalyzed by pig liver microsomal cytochrome P-450. Although the liver and kidney microsomal cytochromes P-450 are immunochemically distinct, the reductases from the two organs are immunochemically similar and cross-reactive.[56,57]

The studies of Guengerich[24] on the reconstitution of benzphetamine N-demethylation with preparations of cytochrome P-450 from either rabbit liver or lung microsomes revealed a similar type of cross-reactivity. The liver and lung reductases were equally competent in reconstituting benzphetamine demethylation with either liver or lung cytochrome P-450, indicating that the cytochrome P-450 component dictates the specificity of the system and that only subtle differences exist among the reductase preparations.

It is therefore unreasonable to assume that different forms of NADPH-cytochrome P-450 reductase reside in the same organ. Such a conclusion was reached by Coon et al.,[68] and Vermilion and Coon[62] in their studies on purified reductases from rat and rabbit liver microsomes in which they reported that rat liver reductases of molecular weights 78,000 and 76,000 and rabbit liver reductases of molecular weights 74,000 and 68,000 were capable of reconstituting benzphetamine-enhanced NADPH-oxidation in reconstituted systems with purified cytochromes P-450.[68] Experiments performed in the author's laboratory on pig, rat, and rabbit systems do not support the existence of multiple forms of the reductase within an organ but rather indicate a proteolytic degradation process which varies from organ to organ and species to species. This proteolytic process can be prevented in the preparation of rat liver NADPH-cytochrome P-450 reductase by including the specific lysosomal protease inhibitors,[69] leupeptin, and pepstatin.[70]

Finally, it must be mentioned that NADPH-cytochrome P-450 reductase may function in other physiologically important reactions, not involving cytochrome P-450. One of these systems is the heme oxygenase reaction which occurs in a variety of organs but primarily in liver and spleen. This system was shown by Yoshida and Kikuchi[71,72] to involve NADPH-cytochrome c (P-450) reductase and an apoprotein, which does not contain heme until the heme substrate is provided, effective in catalyzing the oxidative cleavage of the α-methene bridge to biliverdin and carbon monoxide. Since this enzymatic process can be considered part of a detoxification mechanism for an important endogenous metabolite it bears mentioning in this review. However, the mechanism of the interaction of NADPH-cytochrome c (P-450) reductase with heme oxygenase is not understood and its cellular localization has not yet been defined.

The mechanism by which NADPH-cytochrome c (P-450) reductase interacts with cytochrome P-450 is being addressed by a number of different laboratories[54,57,73-75] and the status of our present knowledge has been reviewed recently.[76,77] It is sufficient to say that the hydroxylation cycle requires two electrons to be introduced into the cytochrome P-450 molecule during substrate binding, oxygen binding, and concomitant intramolecular oxidation–reduction steps leading to the final product of the reaction. It is known that NADPH-cytochrome c (P-450) reductase is necessary for this process, but it is not known whether it is sufficient in the microsomal system for supplying both electrons. The oxidation–reduction states which NADPH-cytochrome c (P-450) reductase undergoes during catalytic turnover has been the subject of much discussion,[46,54,75-77] since the studies of Masters et al.[40,42] first suggested the involvement of an air-stable, free radical form of the reductase as the oxidized partner of the catalytic cycle. While this aspect of the catalytic mechanism has been preserved, the actual role of other oxidation–reduction states of the flavoprotein has not been defined. Recent experiments of Yasukochi *et al.*,[54] have demonstrated the rapid formation of fully reduced flavin (FlH$_2$) prior to the formation of flavin semiquinone (FIH·) and, although this work[54] did not identify the flavins involved in these reductive steps, it established the mechanism of reduction by NADPH as a 2-electron process. The subsequent formation of the 1-electron-reduced, air-stable semiquinone results, in these experiments[54] from the reaction of the flavoprotein with oxygen to produce superoxide anion (O$_2^-$) although earlier studies showed that a variety of electron acceptors oxidize the enzyme to this state during turnover.[40,42] It is of great interest to determine how the FAD and FMN prosthetic groups cooperate in the process of donating *two* electrons, *one* at a time, to cytochrome *P*-450 during hydroxylation reactions catalyzed by reconstituted systems containing only the reductase, cytochrome *P*-450, and phospholipid(s). However, it remains to be demonstrated if *both* electrons are necessarily donated through the NADPH-cytochrome *P*-450 reductase in microsome-catalyzed monooxygenation reactions.

ACKNOWLEDGMENT

Supported in part by USPHS Grant No. HLBI 13619 and GM 16488 and by Grant No. I-453 from The Robert A. Welch Foundation.

REFERENCES

1. Horecker, B. L. (1950). Triphosphopyridine nucleotide-cytochrome *c* reductase in liver. *J. Biol. Chem.* **183**, 593–605.
2. Williams, C. H., Jr., and Kamin, H. (1962). Microsomal triphosphopyridine nucleotide-cytochrome *c* reductase of liver. *J. Biol. Chem.* **237**, 587–595.
3. Phillips, A. H., and Langdon, R. G. (1962). Hepatic triphosphopyridine nucleotide-cytochrome *c* reductase: Isolation, characterization, and kinetic studies. *J. Biol. Chem.* **237**, 2652–2660.
4. La Du, B. N., Gaudette, L., Trousof, N., and Brodie, B. B. (1955). Enzymatic dealkylation of aminopyrine (Pyramidon) and other alkylamines. *J. Biol. Chem.* **214**, 741–752.
5. Gillette, J. R., Brodie, B. B., and La Du, B. N. (1957). The oxidation of drugs by liver microsomes: On the role of TPNH and oxidase. *J. Pharmacol. Exp. Ther.* **119**, 532–540.
6. Krisch, K., and Staudinger, H. (1961). Untersuchungen zur enzymatischen hydroxylierung: Hydroxylierung von acetanilid und deren beziehungen zur mikrosomalen pyridinucleotidoxydation. *Biochem. Z.* **334**, 312–327.

7. Omura, T. (1969). *In* "Microsomes and Drug Oxidations" (J. R. Gillette, A. H. Conney, G. J. Cosmides, R. W. Estabrook, J. R. Fouts, and G. J. Mannering, eds.), pp. 160–161. Academic Press, New York.

8. Kuriyama, Y., Omura, T., Siekevitz, P., and Palade, G. E. (1969). Effects of phenobarbital on the synthesis and degradation of the protein components of rat liver microsomal membranes. *J. Biol. Chem.* **244**, 2017–2026.

9. Lu, A. Y. H., Strobel, H. W., and Coon, M. J. (1969). Hydroxylation of benzphetamine and other drugs by a solubilized form of cytochrome *P*-450 from liver microsomes: Lipid requirement for drug demethylation. *Biochem. Biophys. Res. Commun.* **36**, 545–551.

10. Klingenberg, M. (1958). Pigments of rat liver microsomes. *Arch. Biochem. Biophys.* **75**, 379–386.

11. Garfinkel, D. (1958). Studies on pig liver microsomes. I. Enzymic and pigment composition of different microsomal fractions. *Arch. Biochem. Biophys.* **77**, 493–509.

12. Omura, T., and Sato, R. (1964). The carbon monoxide-binding pigment of liver microsomes. I. Evidence for its hemoprotein nature. *J. Biol. Chem.* **239**, 2370–2378.

13. Omura, T., and Sato, R. (1964). The carbon monoxide-binding pigment of liver microsomes. II. Solubilization, purification, and properties. *J. Biol. Chem.* **239**, 2379–2385.

14. Estabrook, R. W., Cooper, D. Y., and Rosenthal, O. (1963). The light reversible carbon monoxide inhibition of the steroid C-21-hydroxylase system of adrenal cortex. *Biochem. Z.* **338**, 741–755.

15. Warburg, O. (1949). "Heavy Metal Prosthetic Groups and Enzyme Action." Oxford Univ. Press, London and New York.

16. Cooper, D. Y., Levin, S., Narasimhulu, S., Rosenthal, O., and Estabrook, R. W. (1964). Photochemical action spectrum of the terminal oxidase of mixed function oxidase systems. *Science* **147**, 400–402.

17. Masters, B. S. S., Baron, J., Taylor, W. R., Isaacson, E. L., and LoSpalluto, J. (1971). Immunochemical studies on electron transport chains involving cytochrome *P*-450. I. Effects of antibodies to pig liver microsomal reduced triphosphopyridine nucleotide-cytochrome *c* reductase and the non-heme iron protein from bovine adrenocortical mitochondria. *J. Biol. Chem.* **246**, 4143–4150.

18. Masters, B. S. S., Nelson, E. B., Ziegler, D. M., Baron, J., Raj, P. P., and Isaacson, E. L. (1971). Immunochemical studies utilizing antibody to NADPH-cytochrome *c* reductase as a specific inhibitor of microsomal electron transport. *Chem.-Biol. Interact.* **3**, 296–299.

19. Glazer, R. I., Schenkman, J. B., and Sartorelli, A. C. (1971). Immunochemical studies on the role of reduced nicotinamide adenine dinucleotide phosphate-cytochrome *c* (*P*-450) reductase in drug oxidation. *Mol. Pharmacol.* **7**, 683–688.

20. Omura, T., Takemori, S., Suhara, K., Katagiri, M., Yoshida, Y., and Tagawa, K. (1978). Hepatic microsomal systems. *In* "Cytochrome *P*-450" (R. Sato and T. Omura, eds.), pp. 138–163. Academic Press, New York.

21. Zampaglione, N. G., and Mannering, G. J. (1973). Properties of benzpyrene hydroxylase in the liver, intestinal mucosa, and adrenal of untreated and 3-methylcholanthrene-treated rats. *J. Pharmacol. Exp. Ther.* **185**, 676–685.

22. Litterst, C. L., Mimnaugh, E. G., Reagan, R. L., and Gram, T. E. (1975). Comparison of *in vitro* drug metabolism by lung, liver, and kidney of several common laboratory species. *Drug Metab. Dispos.* **3**, 259–265.

23. Jakobsson, S. V., Okita, R. T., Prough, R. A., Mock, N. I., Buja, L. M., Graham, J. W., Petty, C. A., and Masters, B. S. S. (1977). Studies on the cytochrome *P*-450-dependent monooxygenase systems of human liver, lung and kidney microsomes. *Excerpta Med. Int. Cong. Ser.* **440**, 71–73.

24. Guengerich, F. P. (1977). Preparation and properties of highly purified cytochrome P-450 and NADPH-cytochrome P-450 reductase from pulmonary microsomes of untreated rabbits. *Mol. Pharmacol.* **13**, 911–923.
25. Wolf, C. R., Smith, B. R., Ball, L. M., Serabjit-Singh, C., Bend, J. R., and Philpot, R. M. (1979). *J. Biol. Chem.* **254**, 3658–3663.
26. Kato, R. (1966). Possible role of cytochrome P-450 in the oxidation of drugs in liver microsomes. *J. Biochem. (Tokyo)* **59**, 574–583.
27. Wada, F., Shibata, H., Goto, M., and Sakamoto, Y. (1968). Participation of the microsomal electron transport system involving cytochrome P-450 in ω-oxidation of fatty acids. *Biochim. Biophys. Acta* **162**, 518–524.
28. Ichihara, K., Kusunose, E., and Kusunose, M. (1973). Some properties of NADPH-cytochrome c reductase reconstitutively active in fatty acid ω-hydroxylation. *Eur. J. Biochem.* **38**, 463–472.
29. Jakobsson, S. V., and Cinti, D. L. (1973). Studies on the cytochrome P-450 containing monooxygenase system in human kidney cortex microsomes. *J. Pharmacol. Exp. Ther.* **185**, 226–234.
30. Grundin, R., Jakobsson, S., and Cinti, D. L. (1973). Induction of microsomal aryl hydrocarbon (3,4-benzo[a]pyrene) hydroxylase and cytochrome P-450$_K$ in rat kidney cortex. *Arch. Biochem. Biophys.* **158**, 544–555.
31. Prough, R. A., Patrizi, V. W., Okita, R. T., Masters, B. S. S., and Jakobsson, S. W. (1979). Characteristics of benzo[a]pyrene metabolism by kidney, liver, and lung microsomal fractions from rodents and humans. *Cancer Res.* **39**, 1199–1206.
32. Tenhunen, R., Marver, H. S., and Schmid, R. (1968). The enzymatic conversion of heme to bilirubin by microsomal heme oxygenase. *Proc. Natl. Acad. Sci. U.S.A.* **61**, 748–755.
33. Tenhunen, R., Marver, H. S., and Schmid, R. (1969). Microsomal heme oxygenase: Characterization of the enzyme. *J. Biol. Chem.* **244**, 6388–6394.
34. Schacter, B. A., Nelson, E. B., Marver, H. S., and Masters, B. S. S. (1972). Immunochemical evidence for an association of heme oxygenase with the microsomal electron transport system. *J. Biol. Chem.* **247**, 3601–3607.
35. Hietanen, E., and Vainio, A. (1973). Interspecies variations in small intestinal and hepatic drug hydroxylation and glucuronidation. *Acta Pharmacol. Toxicol.* **33**, 57–64.
36. Wattenberg, L. W., and Leong, J. L. (1971). Tissue distribution studies of polycyclic hydrocarbon hydroxylase activity. *In* "Handbook of Experimental Pharmacology" (B. B. Brodie, J. R. Gillette, and H. S. Ackerman, eds.), Vol. 28, Part 2, pp. 422–430. Springer-Verlag, Berlin and New York.
37. Fang, W.-F., and Strobel, H. W. (1978). Activation of carcinogens and mutagens by rat colon mucosa. *Cancer Res.* **38**, 2939–2944.
38. Fang, W.-F., and Strobel, H. W. (1978). The drug and carcinogen metabolism system of rat colon microsomes. *Arch. Biochem. Biophys.* **186**, 128–138.
39. Stohs, S. J., Grafstrom, R. C., Burke, M. D., and Orrenius, S. (1976). Xenobiotic metabolism and enzyme induction in isolated rat intestinal microsomes. *Drug Metab. Dispos.* **4**, 517–521.
40. Masters, B. S. S., Kamin, H., Gibson, Q. H., and Williams, C. H., Jr. (1965). Studies on the mechanism of microsomal triphosphopyridine nucleotide-cytochrome c reductase. *J. Biol. Chem.* **240**, 921–931.
41. Masters, B. S. S., Williams, C. H., Jr., and Kamin, H. (1967). The preparation and properties of microsomal TPNH-cytochrome c reductase from pig liver. *In* "Methods in Enzymology" (R. W. Estabrook and M. E. Pullman, eds.), Vol. 10, pp. 565–573. Academic Press, New York.

42. Masters, B. S. S., Bilimoria, M. H., Kamin, H., and Gibson, Q. H. (1965). The mechanism of 1- and 2-electron transfers catalyzed by reduced triphosphopyridine nucleotide-cytochrome c reductase. *J. Biol. Chem.* **240**, 4081–4088.

43. Ichikawa, Y., and Yamano, T. (1969). Studies on the microsomal nicotinamide adenine dinucleotide phosphate-cytochrome c reductase of rabbit liver. *J. Biochem. (Tokyo)* **66**, 351–360.

44. Nishibayashi-Yamashita, H., and Sato, R. (1970). Vitamin K_3-dependent NADPH oxidase of liver microsomes: Purification, properties, and identity with microsomal NADPH-cytochrome c reductase. *J. Biochem. (Tokyo)* **67**, 199–210.

45. Omura, T., and Takesue, S. (1970). A new method for simultaneous purification of cytochrome b_5 and NADPH-cytochrome c reductase from rat liver microsomes. *J. Biochem. (Tokyo)* **67**, 249–257.

46. Iyanagi, T., and Mason, H. S. (1973). Some properties of hepatic reduced nicotinamide adenine dinucleotide phosphate-cytochrome c reductase. *Biochemistry* **12**, 2297–2308.

47. Masters, B. S. S., Prough, R. A., and Kamin, H. (1975). Properties of the stable aerobic and anaerobic half-reduced states of NADPH-cytochrome c reductase. *Biochemistry* **14**, 607–613.

48. Vermilion, J. L., and Coon, M. J. (1974). Highly purified detergent-solubilized NADPH-cytochrome P-450 reductase from phenobarbital-induced rat liver microsomes. *Biochem. Biophys. Res. Commun.* **60**, 1315–1322.

49. Vermilion, J. L., and Coon, M. J. (1976). Properties of highly purified detergent-solubilized NADPH-cytochrome P-450 reductase from liver microsomes. *In* "Flavins and Flavoproteins" (T. P. Singer, ed.), pp. 674–678. Elsevier, Amsterdam.

50. Dignam, J. D., and Strobel, H. W. (1975). Preparation of homogeneous NADPH-cytochrome P-450 reductase from rat liver. *Biochem. Biophys. Res. Commun.* **63**, 845–852.

51. Dignam, J. D., and Strobel, H. W. (1977). NADPH-cytochrome P-450 reductase from rat liver: Purification by affinity chromatography and characterization. *Biochemistry* **16**, 1116–1123.

52. Yasukochi, Y., and Masters, B. S. S. (1976). Some properties of a detergent-solubilized NADPH-cytochrome c (cytochrome P-450) reductase purified by biospecific affinity chromatography. *J. Biol. Chem.* **251**, 5337–5344.

53. Lu, A. Y. H., Junk, K. W., and Coon, M. J. (1969). Resolution of the cytochrome P-450-containing ω-hydroxylation system of liver microsomes into three components. *J. Biol. Chem.* **244**, 3714–3721.

54. Yasukochi, Y., Peterson, J. A., and Masters, B. S. S. (1979). NADPH-cytochrome c (P-450) reductase: Spectrophotometric and stopped-flow kinetic studies on the formation of reduced flavoprotein intermediates. *J. Biol. Chem.* **254**, 7097–7104.

55. Buege, J. A., and Aust, S. D. (1975). Comparative studies of rat liver and lung NADPH-cytochrome c reductase. *Biochim. Biophys. Acta* **385**, 371–379.

56. Fan, L. L., and Masters, B. S. S. (1974). Properties of purified kidney microsomal NADPH-cytochrome c reductase. *Arch. Biochem. Biophys.* **165**, 665–671.

57. Yasukochi, Y., Okita, R. T., and Masters, B. S. S. (1979). Reaction cross-specificities between the NADPH-cytochrome c (P-450) reductases and cytochromes P-450 of pig liver and kidney microsomes. *In* "Flavins and Flavoproteins" (T. Yamano and K. Yagi, eds.). pp. 703–710. Japan Scientific Societies Press, Tokyo.

58. Iyanagi, T. (1974). Some properties of kidney cortex and splenic microsomal NADPH-cytochrome c reductase. *FEBS Lett.* **46**, 51–54.

59. Oshinsky, R. J., and Strobel, H. W. (1979). Reconstitution of resolved mixed function

oxidase activity from rat colon mucosa microsomal components. *Fed. Proc., Fed. Am. Soc. Exp. Biol.* **38**, 2436.

60. Masters, B. S. S., and Ziegler, D. M. (1971). The distinct nature and function of NADPH-cytochrome *c* reductase and the NADPH-dependent mixed-function amine oxidase of porcine liver microsomes. *Arch. Biochem. Biophys.* **145**, 358–364.
61. Knapp, J. A., Dignam, J. D., and Strobel, H. W. (1977). NADPH-cytochrome *P*-450 reductase: Circular dichroism and physical studies. *J. Biol. Chem.* **252**, 437–443.
62. Vermilion, J. L., and Coon, M. J. (1978). Purified liver microsomal NADPH-cytochrome *P*-450 reductase: Spectral characterization of oxidation-reduction states. *J. Biol. Chem.* **253**, 2694–2704.
63. Nelson, E. B., Raj, P. P., Belfi, K. J., and Masters, B. S. S. (1971). Oxidative drug metabolism in human liver microsomes. *J. Pharmacol. Exp. Ther.* **178**, 580–588.
64. Guengerich, F. P. (1978). Comparison of highly purified microsomal cytochromes *P*-450 and NADPH-cytochrome *P*-450 reductases by peptide mapping. *Biochem. Biophys. Res. Commun.* **82**, 820–827.
65. Weber, K., and Osborn, M. (1969). The reliability of molecular weight determinations by dodecyl sulfate-polyacrylamide gel electrophoresis. *J. Biol. Chem.* **244**, 4406–4412.
66. Laemmli, U. K. (1970). Cleavage of structural proteins during the assembly of the head of bacteriophage T4. *Nature (London)* **227**, 680–685.
67. Gum, J. R., and Strobel, H. W. (1979). Purified NADPH-cytochrome *P*-450 reductase: Interaction with hepatic microsomes and phospholipid vesicles. *J. Biol. Chem.* **254**, 4177–4185.
68. Coon, M. J., Vermilion, J. L., Vatsis, K. P., French, J. S., Dean, W. L., and Haugen, D. A. (1977). Biochemical studies on drug metabolism: Isolation of multiple forms of liver microsomal cytochrome *P*-450. *In* "Concepts in Drug Metabolism (D. M. Jerina, ed.), pp. 46–71. Am. Chem. Soc. Symposium Series **44**, Washington, D.C.
69. Omura, T., Noshiro, M., Harada, N., and Masters, B. S. S., unpublished observations.
70. Umezawa, H., and Aoyagi, T. (1977). Activities of proteinase inhibitors. *In* "Proteinases in Mammalian Cells and Tissues" (A. J. Barrett, ed.), pp. 637–662. North-Holland Publ., New York.
71. Yoshida, T., and Kikuchi, G. (1978). Purification and properties of heme oxygenase from pig spleen microsomes. *J. Biol. Chem.* **253**, 4224–4229.
72. Yoshida, T., and Kikuchi, G. (1978). Features of the reaction of heme degradation catalyzed by the reconstituted microsomal heme oxygenase system. *J. Biol. Chem.* **253**, 4230–4236.
73. Vermilion, J. L., and Coon, M. J. (1978). Identification of the high and low potential flavins of liver microsomal NADPH-cytochrome *P*-450 reductase. *J. Biol. Chem.* **253**, 8812–8819.
74. Taniguchi, H., Imai, Y., Iyanagi, T., and Sato, R. (1979). Interaction between NADPH-cytochrome *P*-450 reductase and cytochrome *P*-450 in the membrane of phosphatidylcholine vesicles. *Biochim. Biophys. Acta* **550**, 341–356.
75. Iyanagi, T., Anan, F. K., Imai, Y., and Mason, H. S. (1978). Studies on the microsomal mixed function oxidase system: Redox properties of detergent-solubilized NADPH-cytochrome *P*-450 reductase. *Biochemistry* **17**, 2224–2230.
76. Masters, B. S. S., and Okita, R. T. (1979). The history, properties, and function of NADPH-cytochrome *P*-450 reductase. *In* "Encyclopedia of Pharmacology and Therapeutics. Mixed Function Oxidases" (J. Schenkman and D. Kupfer, eds.). Pergamon, Oxford (in press).
77. Estabrook, R. W., Werringloer, J., Masters, B. S. S., and Peterson, J. A. (1979). The

microsomal electron transport system revisited: A new look at cytochrome *P*-450 function. *In* "Oxidases and Related Redox Systems" (T. E. King, H. S. Mason, and M. Morrison, eds.). Univ. Park Press, Baltimore, Maryland (in press).

78. Yasukochi, Y., Okita, R. T., and Masters, B. S. S. (1980). "Comparison of the Properties of Detergent-Solubilized NADPH-Cytochrome *P*-450 Reductases from Pig Liver and Kidney: Immunochemical Kinetic, and Reconstitutive Properties". *Arch. Biochem. Biophys.,* in press.

Chapter 9

Microsomal Flavin-Containing Monooxygenase: Oxygenation of Nucleophilic Nitrogen and Sulfur Compounds

DANIEL M. ZIEGLER

I. INTRODUCTION

Foreign compounds containing nucleophilic nitrogen or sulfur atoms are frequently metabolized by an oxidative attack on the heteroatom. The

201

ENZYMATIC BASIS OF DETOXICATION, VOL. I

literature dealing with N-oxygenation of arylamines, described in detail in a recent review,[1] is particularly extensive. Interest in this pathway was stimulated by the early studies of the Millers[2] and by Kiese[3] and associates on the metabolic N-oxygenation of arylamines and related compounds, to form more reactive intermediates. Since most of these studies were restricted to known toxic or carcinogenic compounds, it is not surprising that this work led to the frequently expressed, but erroneous, view that N-oxygenation usually leads to more toxic metabolites. While this statement holds for a limited number of compounds, the vast majority of medicinal tertiary aliphatic amines and naturally occurring alkaloids yield pharmacologically less toxic derivatives upon N-oxygenation. However, as with any group, there are exceptions, and metabolites from a specific compound have to be evaluated on their own merits.

Depending upon the nature of the parent compound and reaction conditions, oxidative attacks on nucleophilic centers of organic nitrogen or sulfur compounds can produce a variety of products. A detailed description of the number and properties of oxygenated products of different organic nitrogen or sulfur compounds is beyond the scope of this chapter. Only the structure, nomenclature, and basic chemical and biological properties of some major products formed metabolically will be considered here.

A. Amines

N-Oxygenation of tertiary amines yields products usually referred to as amine oxides (Fig. 1). In an aqueous environment the oxygen is protonated, and Smith[4] suggests that this form is more accurately described as a quaternary derivative of hydroxylamine. Amine oxides are stable compounds and form crystalline N,N,N-trisubstituted hydroxylammonium salts with acids. These compounds are quite polar, and amine oxides of low molecular weight are readily excreted.

The role of N-oxygenation in the metabolism of tertiary amine drugs has been extensively reviewed[5-8] and will be discussed only briefly here. N-Oxides of many aliphatic and alicyclic tertiary amine drugs have been detected in urine of several species after administration of the parent compound.[8] Although the rate of N-oxygenation of lipophilic aliphatic amines can be quite rapid in vivo, Bickel[5] points out that the amount of amine oxide excreted represents only that fraction not reduced or metabolically converted to other products. Large species differences have been reported,[6] and the highest rates are usually observed with hepatic tissue from hogs, dogs, and guinea pigs. The rate of N-oxygenation is also exceptionally high in the human[9] and is a major route for the metabolism

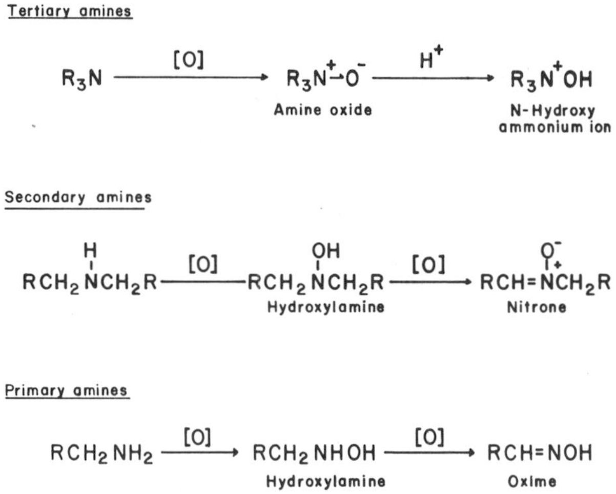

$$R_3N \xrightarrow{\text{[O]}} R_3\overset{+}{N}\!\!-\!\!O^- \xrightarrow{H^+} R_3\overset{+}{N}OH$$

Amine oxide   N-Hydroxy
                                ammonium ion

Secondary amines

$$\underset{\text{Hydroxylamine}}{RCH_2\overset{\overset{\displaystyle H}{|}}{N}CH_2R} \xrightarrow{\text{[O]}} \underset{\text{Hydroxylamine}}{RCH_2\overset{\overset{\displaystyle OH}{|}}{N}CH_2R} \xrightarrow{\text{[O]}} \underset{\text{Nitrone}}{RCH=\overset{\overset{\displaystyle O^-}{|}}{\overset{+}{N}}CH_2R}$$

Primary amines

$$RCH_2NH_2 \xrightarrow{\text{[O]}} \underset{\text{Hydroxylamine}}{RCH_2NHOH} \xrightarrow{\text{[O]}} \underset{\text{Oxime}}{RCH=NOH}$$

Fig. 1. N-Oxygenated products of amines.

and disposition of trimethylamine and a number of medicinal tertiary amines.[6]

N-Oxygenation of secondary amines yields the corresponding hydroxylamines. The pK_a of most secondary hydroxylamines is below 6, and, at physiological pH, they are predominantly present as neutral molecules. Hydroxylamines are fairly reactive compounds, particularly susceptible to further oxidation to nitrones. Nitrones are N-substituted oximes and exhibit many of the physical and chemical properties of the latter group, but are frequently more susceptible to nucleophilic attack. This is especially true of aliphatic nitrones that readily decompose in aqueous solution as indicated in the following reaction.

$$\underset{RCH=\overset{\overset{\displaystyle O^-}{|}}{N}\overset{+}{\;}R+\;H_2O\;\rightarrow\;RCHO\;+\;RNH_2OH}{} \tag{1}$$

Nitrones with an aryl substituent on the α-carbon are far less susceptible to nucleophilic substitution and are fairly stable in neutral aqueous solutions. A few α-phenylnitrones have been isolated and characterized from reactions catalyzed by hepatic microsomes.[10]

In a recent comprehensive review, Coutts and Beckett[11] suggest that N-oxygenation may be a significant route for the metabolism of many medicinal secondary amines. This conclusion is based largely on the relatively rapid N-oxygenation of lipophilic amines *in vitro*, but secondary *N*-hydroxyamines and their further oxidation products have also been detected *in vivo*. Tissue concentrations of secondary hydroxylamines were quite low, but this would be expected since hepatic microsomes also contain a NADH-dependent hydroxylamine reductase that catalyzes rapid reduction of primary and secondary hydroxylamines.[10] Rapid reduction would also limit further oxidation. The contribution of N-oxygenation to the disposition of secondary amine drugs has not been adequately resolved.

B. Organic Sulfur Compounds

Foreign compounds bearing nucleophilic divalent sulfur atoms that have received the most attention fall into four categories: sulfides, thiols, disulfides, and thiones. Depending upon reaction conditions, oxidation of the latter three classes can produce an extraordinarily large number of products, many of which arise from secondary reactions of initial oxygenated derivatives. Since many of the secondary products formed *in vitro* have little physiological significance, only the major initial oxygenated products are listed in Fig. 2. More detailed descriptions are given in reviews.[12-14]

Alkyl and aryl sulfides are readily oxygenated to sulfoxides and sulfones and higher oxidation products are rarely obtained with mild oxidizing agents. Sulfoxides and sulfones are more polar than the parent sulfide and a number are readily excreted in the urine. S-Oxygenation of lipophilic sulfides is a common pathway for their metabolism and disposition[15] and is apparently an important route for detoxication of sulfides.

Oxidation of aryl or alkyl thiols yields extremely reactive sulfenic acids that react almost instantly with thiols (cf. Fig. 2), yielding disulfides as the first stable oxidation products. Further oxidation of disulfides can generate a host of intermediate products formed by sequential S-oxygenation or by S-oxygenations followed by hydrolytic cleavage of the sulfur–sulfur bond. However, oxidation of foreign thiols past the disulfide probably would not occur *in vivo*. Furthermore, alkyl or aryl sulfenic acids generated metabolically would preferentially react with glutathione, yielding mixed disulfides, the reduction of which by GSH would regenerate the foreign thiol [Eq. (2)]. Oxygenation followed by reduction leads to a cycle

$$RSOH + GSH \longrightarrow RSSG \xrightarrow{GSH} RSH + GSSG \qquad (2)$$

that could deplete tissues of glutathione. While these reactions may account for toxicity of some foreign thiols and disulfides, they do not produce more polar metabolites. S-Methylation (see Vol. II, Chapter 7) followed by sulfoxidation may be the major route for disposition of foreign thiols and disulfides.[16]

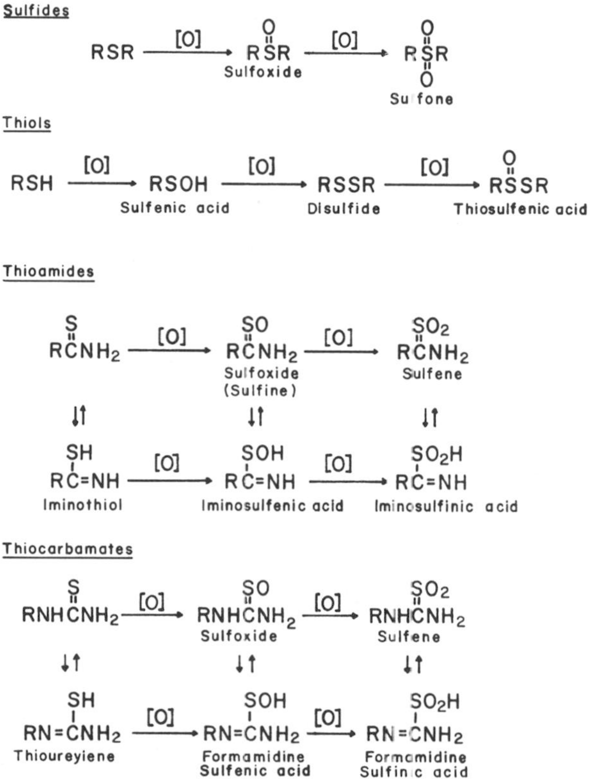

Sulfides

$$RSR \xrightarrow{[O]} R\overset{O}{\underset{}{\overset{\|}{S}}R} \xrightarrow{[O]} R\overset{O}{\underset{\underset{O}{\|}}{\overset{\|}{S}}}R$$

Sulfoxide   Sulfone

Thiols

$$RSH \xrightarrow{[O]} RSOH \xrightarrow{[O]} RSSR \xrightarrow{[O]} R\overset{O}{\overset{\|}{S}}SR$$

Sulfenic acid   Disulfide   Thiosulfenic acid

Thioamides

$$R\overset{S}{\overset{\|}{C}}NH_2 \xrightarrow{[O]} R\overset{SO}{\overset{\|}{C}}NH_2 \xrightarrow{[O]} R\overset{SO_2}{\overset{\|}{C}}NH_2$$

Sulfoxide (Sulfine)   Sulfene

$$\updownarrow \qquad\qquad \updownarrow \qquad\qquad \updownarrow$$

$$R\overset{SH}{\overset{|}{C}}=NH \xrightarrow{[O]} R\overset{SOH}{\overset{|}{C}}=NH \xrightarrow{[O]} R\overset{SO_2H}{\overset{|}{C}}=NH$$

Iminothiol   Iminosulfenic acid   Iminosulfinic acid

Thiocarbamates

$$RNH\overset{S}{\overset{\|}{C}}NH_2 \xrightarrow{[O]} RNH\overset{SO}{\overset{\|}{C}}NH_2 \xrightarrow{[O]} RNH\overset{SO_2}{\overset{\|}{C}}NH_2$$

Sulfoxide   Sulfene

$$\updownarrow \qquad\qquad \updownarrow \qquad\qquad \updownarrow$$

$$RN=\overset{SH}{\overset{|}{C}}NH_2 \xrightarrow{[O]} RN=\overset{SOH}{\overset{|}{C}}NH_2 \xrightarrow{[O]} RN=\overset{SO_2H}{\overset{|}{C}}NH_2$$

Thioureyiene   Formamidine Sulfenic acid   Formamidine Sulfinic acid

Fig. 2. S-Oxygenated products of organic sulfur compounds.

Oxygenation of the thione of thioamides produces the relatively stable S-oxide (sulfine) but, in N-aryl substituted thioamides, significant amounts of the S-oxide can exist as the far less stable iminosulfenic acid. The chemical properties of the two tautomers are quite different and both have to be considered in the metabolism of N-substituted thioamides. Further S-oxygenation of unsubstituted thioamide sulfoxides yields the extremely reactive dioxygenated derivatives that can be considered as either sulfenes or iminosulfinic acids. The toxicity of thioamides is usually attributed to their metabolic oxidation to mono- and/or di-S-oxygenated metabolites. For example, thioacetamide is metabolized to the more hepatotoxic S-oxide both *in vivo*[17] and *in vitro*.[18]

Compounds bearing a thiocarbamide moiety can exist in both the thiol and thione form. The amount of each tautomer depends upon substituents on the nitrogen, and (except for 1,1,3,3-tetrasubstituted thioureas) both forms are usually present. Formamididinesulfenic acids are the predominant initial products obtained by an oxidative attack on the sulfur of thiourea and most mono- and di-1,1- or di-1,3-substituted thioureas. While form-amidinesulfenic acids are quite reactive, they are more stable than alkyl sulfenic acids and, according to Walter and Randau,[19] possess some of the properties of sulfoxides (sulfenes). Their further oxidation produces the more stable dioxygenated derivatives that are best described as formam-idinesulfinic acids. Thiocarbamides, like thioamides, are toxic com-pounds, and metabolic S-oxygenation may be responsible for their toxic-ity. Like alkyl sulfenic acids [Eq. (2)], formamidine sulfenic acids are reduced by GSH and can form relatively stable mixed disulfides with protein thiols. The mixed disulfides, however, are readily reduced by GSH and *in vivo* protein-formamidine mixed disulfides are probably not significant except at abnormally low concentrations of glutathione. How-ever, protein-formamidine mixed disulfide formation may be responsible for the observed binding of oxygenated thiourea metabolites *in vitro*.[20] In the intact animal, covalent binding of desulfurated thiourea metabolites reported by Boyd and Neal[20] is more likely due to the reaction of form-amidinesulfinic acids with amino groups of macromolecules. Form-amidinesulfinic acids readily react with nucleophilic amines to form N-substituted guanidines [Eq. (3)].

$$R_1—N{=}C—NH_2 + R_2NH_2 \rightarrow R_1N{=}C—NH_2 + \tfrac{1}{2}H_2S_2O_3 + \tfrac{1}{2}H_2O \overset{\overset{\displaystyle SO_2H}{|}}{} \qquad \overset{\overset{\displaystyle NHR_2}{|}}{} \tag{3}$$

Low molecular weight thiols such as glutathione (GSH) accelerate the decomposition of formamidine sulfinic acids to ureas and cyanamides by an unknown mechanism. This GSH-dependent decomposition of form-amidine sulfinic acids may be a significant route for detoxication of S-oxygenated thiocarbamide metabolites.

C. Multiplicity of Microsomal Monooxygenases

Prior to 1960 it was generally assumed that NADPH-dependent oxyge-nations catalyzed by hepatic microsomes were mediated through cyto-chrome *P*-450-dependent monooxygenases. However, subsequent studies from several laboratories demonstrated that the enzyme catalyzing oxygenation of nucleophilic tertiary and secondary amines differed in properties from the cytochrome *P*-450 system. The early studies on mul-tiplicity of microsomal monooxygenases are described in detail in several reviews[7,21,22] and were unambiguously verified by the purification to homogeneity of a microsomal flavin-containing monooxygenase free from cytochromes that catalyzes oxygenation of nucleophilic organic nitrogen or sulfur compounds. The properties, substrate specificity, and species and tissue distribution of the flavin-containing monooxygenase (EC 1.14.13.8) [the enzyme is presently listed in *Enzyme Nomenclature* as dimethylaniline monooxygenase (*N*-oxide forming)] are summarized in the following sections.

II. HOG LIVER MICROSOMAL FLAVIN-CONTAINING MONOOXYGENASE

A. Assay Methods

The activity of the liver enzyme is readily measured in whole homogenates and all subfractions by following methimazole (*N*-methyl-2-mercaptoimidazole)-dependent oxygen uptake in the presence of NADPH and glutathione. The monooxygenase catalyzes S-oxygenation of methimazole to the imidazole-sulfenic acid and GSH immediately reduces the intermediate, regenerating the parent substrate as illustrated in Fig. 3. Only reaction a in Fig. 3 is enzymatic. At GSH concentrations above 1 mM the nonenzymatic reduction of the sulfenic acid (reactions b and c in Fig. 3) is much faster than the catalyzed step. Since methimazole saturates the monooxygenase at less than 200 μM, high concentrations of substrate are not required because it is constantly regenerated. Furthermore, GSH prevents accumulation of the reactive sulfenic acid that can produce slow, progressive inactivation of highly purified preparations of the monooxygenase. At concentrations less than 2 mM, methimazole is S-oxygenated at detectable rates only by the flavin-containing monooxygenase. Interference by other microsomal oxygenases in crude preparations is also minimized by carrying out measurements at pH 8.4–8.5 in the presence of *n*-octylamine or another lipophilic primary alkylamine. These compounds not only stimulate methimazole-dependent

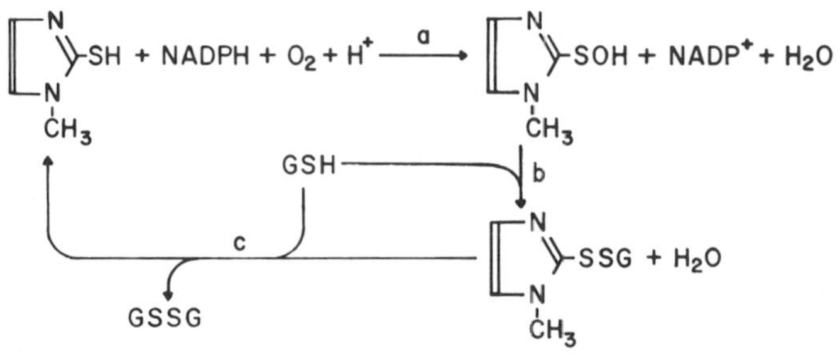

Fig. 3. Monooxygenase catalyzed oxygenation of methimazole in the presence of glutathione.

oxygen uptake but suppress oxygen reduction by other microsomal NADPH oxidases. Any of a large number of thiocarbamide, thioamide, or tertiary amine substrates can be used in place of methimazole, but none have distinct advantages over methimazole.

Activity of the flavin-containing monooxygenase of microsomes and subfractions can also be assessed by following oxygenatable substrate-dependent oxidation of NADPH spectrophotometrically or by following generation of tertiary amine oxides.[23]

B. Enzyme Isolation

Steps in tissue selection, preparation of hog liver microsomes and detergents used to extract the monooxygenase have been described in some detail.[24] However, the necessity for rapid chilling of the tissue cannot be overemphasized. The monooxygenase is unusually sensitive to thermal inactivation, and little or no activity can be detected in liver that is not removed and chilled in less than 10 min after exsanguination. The temperature of internal organs of large animals rises rapidly during the first few minutes after death; the temperature of hog liver can reach 42°–43° if not removed within 15 min (unpublished observations, this laboratory). The rate of monooxygenase inactivation increases sharply with increase in temperature; above 45° it is completely inactivated within 1 or 2 min. Furthermore, unknown endogenous inhibitors that affect yield and stability of monooxygenase also rapidly increase in amount at temperatures above 20°, and it is essential that all steps in tissue processing and enzyme isolation be carried out at as close to 0° as practical.

The monooxygenase, extracted from microsomes with detergents, is purified by initial fractionation with polyethylene glycol (PEG) followed by column chromatography on ion-exchange cellulose. The average liver from an adult female hog contains slightly more than 2 g monooxygenase, about 10% of which is recovered in the final fraction. Some of the original activity is inactivated during purification, but most is lost to other fractions. The monooxygenase present in other fractions is, however, not different from homogeneous preparations with regard to thermal stability or basic catalytic properties. There is no evidence that liver from adult hogs contains more than one form of this monooxygenase.

The homogeneous enzyme can be stored at 2° as a suspension in 13% PEG for several weeks with little or no loss in activity. In solution at pH 7.2–7.4 the monooxygenase is somewhat less stable and rapid loss of activity (probably due to microbial contamination) is sometimes encountered after 2 weeks at 2°. Preparations frozen and lyophilized from 50 mM phosphate at pH 7.2 lose from 10 to 30% of the original activity. However, the dried powders are quite stable, and further loss in activity has not been observed in preparations shipped long distances without refrigeration to other laboratories.

C. Composition and Physical Properties

The purified hog liver monooxygenase contains 15.1 to 15.3 nmole FAD per mg protein and is free from hemoproteins and other chromophores that absorb in the visible spectrum. The preparations are also free from copper, iron, molybdenum, and cobalt, but traces of zinc are usually present. The concentration of zinc is usually less than one-tenth that of flavin, and its participation in the catalytic cycle is unlikely. The most highly purified preparations contain small but variable amounts of lipid. Lipid is apparently not required for activity, but effects of lipids on enzyme stabliity or on kinetic constants for lipophilic substrates have not been adequately explored.

The minimum molecular weight based on mass of amino acids per mole flavin is 65,000. This value is somewhat higher than monomer mass based on migration rates on polyacrylamide gels containing sodium dodecyl sulfate. However, the monooxygenase, like other membrane-bound proteins, binds unusually high amounts of the detergent, and its mobility would be faster than soluble protein standards of equal mass.

Gel filtration and sedimentation velocity patterns indicate that the catalytically active monooxygenase exists predominantly as aggregates of four or more monomeric units, but the degree of aggregation required for activity has not been determined.

D. Spectral Characteristics and Kinetic Mechanism

Four different forms of the monooxygenase can be distinguished by changes in flavin absorbance between 300 and 500 nm (Fig. 4). The spectrum of the fully oxidized enzyme is essentially identical with that of other pure flavoproteins. Neither oxygen nor hydroxylatable substrate perturb the spectrum of the oxidized flavoprotein, but NADP shifts both peaks to shorter wavelengths. The absorbance of the shorter wavelength peak and of the shoulder at 485 nm are also decreased by NADP.

Anaerobically NADPH fully reduces the flavin. Reduction apparently occurs via a 2-electron transfer, since spectra characteristic of flavin semiquinones are not detectable. Reoxidation of the fully reduced flavoprotein by oxygen yields a species with a single absorbance band at 366 nm in the presence of high (0.1 mM) NADP. This band shifts to longer wavelengths with decreasing NADP, but its basic shape remains un-

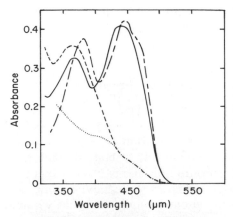

Fig. 4. Spectral form of the FAD-containing monooxygenase. Fully oxidized enzyme (– –); oxidized plus NADP⁺ (——); reduced with NADPH (·····); and reduced plus oxygen (- - - -).

changed. The absorption spectra of the oxygenated flavoprotein is similar to that of the stable peroxy·FMN–luciferin complex[25] and of per-oxy·FAD complexes of hydroxybenzoate hydroxylase.[26] The peroxyflavin form of the microsomal enzyme is stable for 1–2 h at 2°, but above 25° it decomposes within minutes yielding the oxidized flavoprotein and H_2O_2. Addition of an oxygenatable substrate also immediately abolishes the spectrum of the intermediate, suggesting that the peroxyflavin is the oxygenating species that reacts with the oxygenatable substrate.

This interpretation is also consistent with kinetic studies carried out with nucleophilic nitrogen and sulfur substrates.[27] The rate equation consistent with initial velocity measurements suggests an ordered Ter-Bi mechanism with an irreversible step between the second and third substrate. NADPH is added first, followed by oxygen, and the oxygenatable substrate added last; oxygenated product is released before NADP. The essential features of this mechanism are illustrated in Fig. 5. Although this basic mechanism is supported by spectral and kinetic experiments, a number of details (i.e., position of protons and rate limiting step) have not been resolved.

E. Catalytic Properties

1. pH Optimum and Turnover

At saturating NADPH and oxygen, oxygenations catalyzed by the monooxygenase exhibit a distinct pH optima at 8.3–8.4. The pH profile is essentially the same for all substrates and the velocity at pH 7.4 is

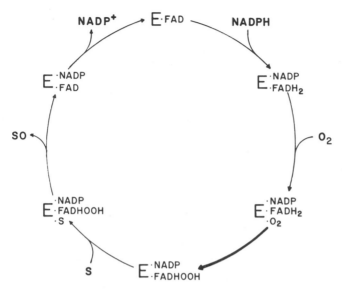

Fig. 5. Probable mechanism of the FAD-containing monooxygenase. E, enzyme; S, oxygenatable substrate; SO, oxygenated product. The heavy arrow indicates probable position of the irreversible step.

approximately 50 to 55% of that at pH 8.4. The turnover per mole of flavin (k_{cat}) with simple tertiary amine or sulfur-containing substrates at pH 7.4 is relatively constant at 34–35 min^{-1}. Positive effectors, i.e., octyl-guanidine or n-octylamine, nearly double k_{cat} at all hydrogen ion concentrations to a maximum 130 min^{-1} at pH 8.4. The turnover with more lipophilic amines, i.e., phenothiazine drugs, is twice that of simple amines and is only marginally affected by positive effectors. Complex lipophilic tertiary amine drugs, like positive effectors, apparently modify a rate-limiting step during catalysis.

Divalent cations are not required for activity. Neither Mg^{2+} nor Mn^{2+} affect catalytic properties at any concentration tested, but Ba^{2+} and Ca^{2+} at concentrations above 10 mM inhibit oxygenation of some substrates.

The concentration of NADPH (5 μM) required for half-maximal activity is not affected by chemical properties of the oxygenatable substrate and appears to be independent of pH between pH 7.0 and 8.5. However, concentrations of oxygen required for half-maximal activity change markedly with pH.[27] At pH 7.4, 21–25 μM oxygen half saturates the enzyme, but this value increases to 210 μM at pH 8.4. The concentration of oxygen in buffer equilibrated with air is slightly less than this, and at the pH optimum reaction velocities are usually limited by oxygen. Increasing oxygen concentration above ambient increases reaction rates, but the reaction is not linear with time since oxygen much above 0.4 mM begins to inactivate the enzyme.

2. Stability

The monooxygenase is extremely unstable above 30° in the absence of NADP or NADPH. Kitchell *et al.*[28] have shown that the flavoprotein undergoes irreversible structural changes above 30° as measured by fluorescence and circular dichroism. The structural changes correlate with changes in oxidative activity toward methimazole and diethylaniline. In the presence of NADPH, preparations completely free from endogenous inhibitors and contaminating proteases retain 95–98% of their original activity after 20 min incubation at 37°. However, above 45° irreversible denaturation occurs in a few seconds. The unusual thermal lability appears to be an intrinsic property of the microsomal flavin-containing monooxygenase and has been used to differentiate reactions catalyzed by the flavin-containing enzyme from other monooxygenases in microsomes from several species.[21]

3. Substrate Control of Overall Oxidation

In the absence of oxygenatable substrate, preparations of the purified monooxygenase catalyze slow NADPH-dependent reduction of oxygen to hydrogen peroxide. This rate is frequently less than 2–3% of the oxygenatable substrate-dependent rate and probably reflects modification of the enzyme during purification. In partially purified fractions, oxygenatable substrate control of overall oxidation is almost complete. However, oxygenatable substrate-independent NADPH oxidase activity of the homogeneous flavoprotein or partially purified fractions can be induced by exposure to nonionic detergents at 0° for prolonged periods, by brief exposure to a pH below 5 or above 9, or by incubation with 10 mM calcium or barium ions at pH 7.8. Structural changes induced by these agents are not known.

Rauckman *et al.*[29] have shown that the purified monooxygenase also produces superoxide anions (Chapter 15). The rate of this reaction, approximately 0.04 mole/min/mole FAD at 37°, is not related to the rate of hydrogen peroxide generation and may be an intrinsic property of the flavoprotein.

4. Activators and Inhibitors

The purified flavoprotein apparently has at least one modifier site capable of binding with positively charged lipophilic compounds.[30] The modifier site can be saturated by lipophilic primary alkylamines or alkyl guanidines. While these compounds are not oxidized, they consistently double k_{cat}; activation by primary alkylamines is one of the most characteristic properties of the monooxygenase from hog liver.

An inhibitor specific for the FAD-containing monooxygenase has not been described. Highly purified preparations are quite sensitive to anionic soaps and detergents at pH 8.4[24] and concentrations of stearate or sodium dodecyl sulfate only 20 times greater than enzyme inactivate the monooxygenase. The membrane-bound enzyme is orders of magnitude less sensitive to exogenous anionic detergents, which limits their use as selective inhibitors.

F. Oxygenatable Substrates

1. Nitrogen Containing

The purified monooxygenase catalyzes N-oxygenations of diverse types of nucleophilic organic nitrogen compounds. Products formed and kinetic constants for representative compounds are listed in the following sections. The constants for trimethylamine were calculated from velocities measured at variable NADPH, oxygen, and hydroxylatable substrate,[27] but all others listed are based only on changes in initial velocity as a function of amine concentration at constant NADPH and oxygen. Since most published values were obtained from initial velocity measurements at pH 8.4 in buffer equilibrated with air, oxygen was less than saturating. Furthermore, kinetic constants for some of the better substrates are based on rates measured only at substrate concentrations above their apparent K_m and the true K_m for these substrates are undoubtedly lower than the values listed. Despite these limitations, apparent constants determined under the same conditions reflect real differences and can serve for a qualitative assessment of substrate specificity.

     a. Tertiary amines. Tertiary amines are oxidized to the relatively stable and polar amine oxides as illustrated in the following reaction.

$$R_3N + NADPH + O_2 + H^+ \rightarrow R_3N^+{\rightarrow}O^- + NADP^+ + H_2O \qquad (4)$$

Only one position is N-oxygenated in substrates containing more than one tertiary amine group.[31] Oxygenations of asymmetric tertiary amines are stereo-specific[32] as expected for an enzyme catalyzed reaction. Virtually all N,N-disubstituted alkyl- and arylamines free from negatively charged groups tested are N-oxygenated, but concentrations required for half-maximal activity vary considerably, as illustrated by a few examples listed in Table I. Changes in k_{cat}/k_m are primarily due to changes in K_m. At saturation all tertiary amines with lipophilic side chains longer than seven carbons are oxidized at the same rate. Compounds with shorter side

TABLE I

Nitrogen-Containing Substrates for the Microsomal FAD-Containing
Monooxygenase

Compound	K_m (μM)	$\dfrac{k_{cat}}{K_m}$ $\dfrac{\text{liter}}{\text{mol min}}$ ($\times 10^{-4}$)	Reference
Tertiary amines			
Dimethylaniline	3	2090	27
Chlorpromazine	9	1390	31
Trifluoperazine	13	970	31
Imipramine	22	570	32
Amitryptyline	100	125	[a]
Brompheniramine	200	63	32
Benzphetamine	128	49	[a]
Guanethidine	170	37	[a]
Trimethylamine	4200	2	27
Secondary Amines			
Demethyltrifluoperazine	9	1390	[a]
N-Benzylamphetamine	120	104	[a]
Desipramine	250	25	[a]
N-Methylaniline	320	20	32
Nortryptyline	500	13	[a]
Secondary Hydroxylamines			
N-Hydroxydemethyltrifluoperazine	7	1610	[a]
N-Benzyl-N-hydroxyamphetamine	220	30	[a]
N-Methyl-N-hydroxybenzylamine	1500	4	[a]

[a] Unpublished measurements, this laboratory.

chains are oxidized at one-half this rate, but addition of a positive effector
increases the rate to that observed with the more lipophilic tertiary
amines. In addition to lipophilicity, steric factors also have marked effects
on apparent K_m. In general, methyl substituents are preferred over other
alkyl groups and amines with lipophilic side chains branched 3–4 carbons
from the nitrogen are among the best substrates. However, there are
exceptions, e.g., dimethylaniline, and it is evident that steric factors
determining substrate binding are not well-defined.

   b. Secondary amines. The purified monooxygenase catalyzes rapid
N-oxygenation of secondary amines and kinetic constants for a few are
listed in Table I. All secondary amine analogues of tertiary amine substrates are oxidized, but usually exhibit a somewhat higher apparent K_m

and lower k_{cat}. Some compounds containing a heterocyclic ring are exceptions. For example, the K_m for demethyltrifluoperazine is lower than that for trifluoperazine.

In contrast to N-oxygenation of a tertiary amine where only a single product is formed, N-oxygenation of a secondary amine is far more complex and produces several different products (Fig. 6). The products

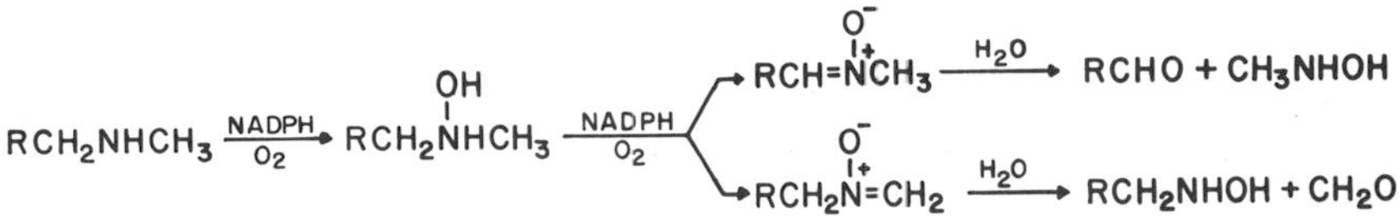

$$RCH_2NHCH_3 \xrightarrow[O_2]{NADPH} RCH_2\overset{\underset{\displaystyle OH}{|}}{N}HCH_3 \xrightarrow[O_2]{NADPH} \begin{cases} RCH=\overset{\underset{\displaystyle O^-}{|+}}{N}CH_3 \xrightarrow{H_2O} RCHO + CH_3NHOH \\ RCH_2\overset{\underset{\displaystyle O^-}{|+}}{N}=CH_2 \xrightarrow{H_2O} RCH_2NHOH + CH_2O \end{cases}$$

Fig. 6. Products formed by sequential enzymatic oxygenation of secondary alkylamines.

are formed by sequential enzymatic oxygenation of the amine, followed by hydrolysis of the nitrones, as indicated in the Introduction [Eq. (1)].

The reaction sequence and products formed have been established for oxidation of *N*-methylbenzylamine in the reactions catalyzed by the purified monooxygenase,[33] and formation of formaldehyde by this mechanism during oxidation of secondary *N*-methylamines by hepatic microsome has also been described.[34] The studies of Beckett and associates (see ref. 11 for reviews) suggest that this may be a significant route for metabolic transformations of medicinal secondary amines. Kinetic data on the few secondary *N*-hydroxyamines listed in Table I support this conclusion. At concentrations of substrate less than one-tenth K_m, oxygenations catalyzed by this enzyme are directly proportional to the magnitude of k_{cat}/K_m. For example, at concentrations below K_m, demethyltrifluoperazine is N-oxygenated about 1,000-fold faster than trimethylamine. Since N-oxygenation is considered a major route for metabolism of trimethylamine, rapid N-oxygenation of demethyltrifluoperazine and other secondary amines must also occur *in vivo*.

c. Hydrazines. The monooxygenase also catalyzes rapid oxygenation of 1,1-disubstituted hydrazines (R_2N—NH_2) and N-substituted aziridines

$$\begin{array}{c} RN\text{---}NH \\ \diagdown\diagup \\ CH_2 \end{array}$$

However, monosubstituted (RNH—NH$_2$) and 1,2-disubstituted (R$_1$NH—NHR$_2$) hydrazines are usually poor substrates.[35] By analogy with amines, the oxidative attack on 1,1-disubstituted hydrazines probably occurs on the tertiary nitrogen, but this has not been established.

Oxygenated hydrazines are extremely unstable, reactive compounds and few secondary products have been identified. Prough[35] has shown that enzymatic N-oxygenation of methylhydrazine produces some methane, suggesting that one or more of the intermediates are extremely reactive. N-oxygenation catalyzed by this enzyme may be one of the metabolic reactions responsible for toxicity of 1,1-disubstituted hydrazines and some monosubstituted hydrazines.

2. Sulfur-Containing Substrates

Specificity of the purified monooxygenase for nucleophilic sulfur compounds is quite broad and thiocarbamates, thioamides, thiols, sulfides, and disulfides are among the better substrates. The enzyme is a better S-oxygenase than a N-oxygenase since compounds bearing both nucleophilic sulfur and nitrogen atoms are preferentially S-oxygenated.

Kinetic constants and types of products formed from a few representative sulfur compounds oxygenated by the purified monooxygenase are listed in the following sections. The facile nonenzymatic oxidation of thiols in slightly alkaline solutions complicates measurements at the pH optimum, and all constants listed were calculated from initial rate measurements at pH 7.4.

a. Thiocarbamides and Thioamides. Thioureas and N-substituted thioureas are S-oxygenated to formamidinesulfinic acids by sequential oxidative attacks as illustrated in Fig. 1. The formamidinesulfinic acids are formed by a second enzymatic oxygenation of the intermediate formamidine sulfenic acid.[36] Kinetic constants for a number of thioureas and formamidinesulfenic acids demonstrate that concentrations required to half saturate the monooxygenase are in the micromolar range (Table II). S-Oxygenation of thioureas is undoubtedly a physiologically significant route for their metabolism and may be responsible for their activation to toxic metabolites.

The thiocarbamide groups of 2-mercaptoimidazoles and 2-mercaptopyrimidines are structurally related to the thiourea thiol-tautomer and most of these compounds are di-S-oxygenated[37] by the mechanism illustrated for thioureas (Fig. 7). However, the reactivity and stability of aromatic heterocyclic sulfinic acids are quite different from formamidinesulfinic acids, although the sulfenic acid derivatives of both classes are readily reduced by thiols [cf. Eq. (2) and Fig. 3]. Differences in toxicity of thiocarbamide containing compounds may be due to differences in stability and reactivity of S-oxygenated metabolites rather than to differences in routes of metabolism.

Thioamides are among the better substrates for the purified mono-

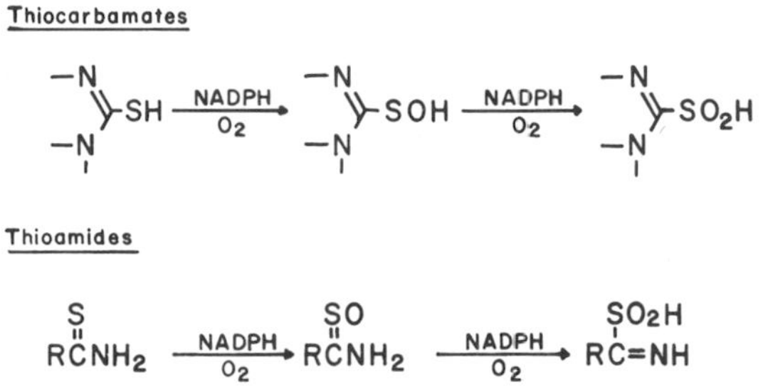

Thiocarbamates

Thioamides

Fig. 7. Sequential enzymatic S-oxygenation of thiocarbamates and thioamides.

oxygenase, but only a few have been tested (Table II). They are all rapidly oxygenated to S-oxides and then converted to di-S-oxygenated derivatives (Fig. 7). The rate of the second reaction is quite slow and products formed from the second oxygenation have not been identified. By analogy with thioureas the second oxidative attack probably occurs on the sulfur, but this has not been demonstrated unequivocally.

b. Sulfides, Disulfides, and Thiols.   The monooxygenase catalyzes oxygenation of alkyl sulfides to sulfoxides, as illustrated in Eq. (5). Sulfides saturate the monooxygenase at very low concentrations and the kinetic constants listed in Table III are based only on initial velocity measurements above apparent K_m; the true values may be considerably lower. Alkyl sulfoxides may be oxygenated to sulfones by this monooxygenase, but the concentration of the only one tested, dimethyl sulfoxide, required to saturate the enzyme is quite high. The kinetic constants suggest that oxygenation of alkyl sulfides to sulfoxides as catalyzed by the flavin-containing monooxygenase may be a major route for sulfoxide formation but their further oxidation to sulfones by this enzyme *in vivo* is uncertain.

$$\text{R—S—R + NADPH + H}^+ + O_2 \rightarrow \text{R—}\overset{\displaystyle O}{\underset{\displaystyle \|}{\text{S}}}\text{—R + NADP}^+ + \text{H}_2\text{O} \qquad (5)$$

Alkyl thiols and disulfides are also excellent substrates for the monooxygenase. The reaction mechanism has not been studied in detail, but the oxidation of alkyl thiols and disulfides yields alkyl sulfenic acids. The final product may arise by sequential oxygenations, as illustrated in Fig. 8.

The disulfide formed by reaction of thiol with the enzymatically generated sulfenic acid is oxygenated to the thiosulfenate (thiosulfoxide), which can hydrolyze to the sulfinic acid and thiol. However, the possible formation of alkyl sulfinates by hydrolysis of enzymatically generated thio-

TABLE II

Thiocarbamide and Thioamide Substrates for the Microsomal FAD-Containing Monooxygenase

	Apparent kinetic constants		
Compound	K_m (μM)	$\frac{k_{cat}}{K_m} \cdot \frac{\text{liter}}{\text{mol min}}$ ($\times 10^{-4}$)	Reference
Thiocarbamides			
Naphthyl-2-thiourea	4.0	1570	37
Phenylthiourea	4.1	1530	36
Methimazole	5.2	1200	37
Thiocarbanilide	6.7	940	36
Thiourea	23	270	36
Ethylenethiourea	34	190	36
Propylthiouracil	950	7	36
Sulfenic Acids			
1,3-Diphenylformamidinesulfenic acid	6.9	910	36
N-Methylimidazole-2-sulfenic acid	22	290	36
Phenylformamidinesulfenic acid	34	180	36
Formamidinesulfenic acid	140	45	36
Ethylene formamidinesulfenic acid	190	33	36
Thioamides			
Thionamide	25	250	[a]
Thioacetamide	65	97	37
Thionicotinamide	460	14	[a]

[a] Unpublished measurements, this laboratory.

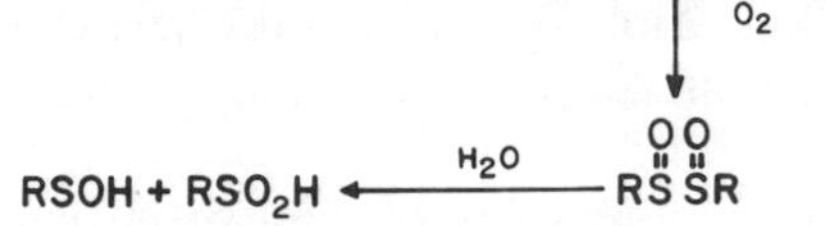

Fig. 8. Alternate routes for oxygenation of alkyl thiols or disulfides to alkyl sulfinates.

TABLE III

Sulfide, Thiol, and Disulfide Substrates for the Microsomal FAD-Containing Monooxygenase

Compound	K_m (μM)	$k_{cat} \cdot \dfrac{\text{liter}}{K_m} \dfrac{}{\text{mol min}}$ ($\times 10^{-4}$)	Reference
Sulfides			
Dimethyl sulfide	8	816	a
Methylphenyl sulfide	13	483	a
Ethylene sulfide	60	105	a
Diphenyl sulfide	74	85	a
Thiols			
n-Heptylmercaptan	15	419	a
n-Butylmercaptan	33	190	a
Thiophenol	33	190	a
Benzylmercaptan	55	114	a
Cysteamine	120	52	a
Dithiothreotol	465	14	37
Thioglycerol	721	9	37
β-Mercaptoethanol	2800	2	37
Disulfides			
o-Dithiane	< 3	< 2100	a
Dibenzyldisulfide	45	140	a
Dihydroxy-o-dithiane	739	9	37

Apparent kinetic constants

[a] Unpublished experiments, this laboratory.

sulfinate cannot be excluded. While the generation of alkylsulfinic acids from thiols and disulfides by this mechanism can be demonstrated *in vitro*, it is unlikely that oxidation of thiols would proceed past the disulfide *in vivo*. Thioltransferase (Chapter 12, Volume II) catalyzed reduction of disulfide by glutathione would prevent significant accumulation of steady state concentrations of disulfide. Disposition of xenobiotic thiols, and of thiols produced by reduction of disulfides, probably occurs by the route discussed in the Introduction.

3. Physiological Substrates

Cysteamine is the only physiological compound tested that is oxidized at a significant rate by the purified monooxygenase. This aminothiol is oxidized only to the disulfide as shown in Eq. (6). In contrast to alkyl

$$2N^+H_3CH_2CH_2SH + NADPH + O_2 + H^+ \rightarrow$$
$$(N^+H_3CH_2CH_2S-)_2 + NADP^+ + 2H_2O \tag{6}$$

disulfides, further oxidation of the disulfide, cystamine, does not occur: the diamine character of cystamine prevents its interaction with the enzyme. Inability of the monooxygenase to interact with compounds containing two positive charges that are separated by four or more carbons has been known for some time. Configuration of groups near the catalytic site apparently prevents interaction of diamines with the enzyme. Kinetic studies on mechanism of cysteamine oxidation have not been completed, but oxidation of this substrate appears more complex than the mechanism illustrated in Fig. 5 for oxygenation of xenobiotic compounds.

The possible cellular role of the monooxygenase in generating disulfides required for synthesis of peptide disulfides has been described.[38]

G. Substrate Specificity

Though the flavin-containing monooxygenase catalyzes oxygenation of many different types of compounds, it is not a completely nonspecific monooxygenase. Oxygenation of primary alkylamines, N-alkylamides, N-alkylcarbamates or heterocyclic aromatic amines, e.g., pyridines, purines, pyrimidines, etc., cannot be detected. The monooxygenase also does not catalyze N-oxygenation of amines where the lone pair of electrons is delocalized by conjugation with an aromatic system, e.g., phenothiazine and diphenylamine. Primary arylamines that form imine tautomers, e.g., 2-naphthylamine, are oxygenated[39] but all others, including aniline and benzidine, are not attacked at measurable rates. N-Oxygenation of the latter arylamines, amides, and heterocyclic aromatic amines are catalyzed by microsomal monooxygenases other than the flavin-containing enzyme. Gorrod[7] has recently reviewed probable biochemical mechanisms for N-oxidation of different types of organic nitrogen compounds and suggests that the more basic compounds are oxygenated by the flavin-containing enzyme and the more acidic ones by cytochrome P-450. However, there are numerous exceptions and this generalization, although useful as a preliminary guide is limited in predicting the substrate specificity of microsomal monooxygenases.

Sulfur compounds as or less nucleophilic than thiophene are not substrates. Delocalization of electrons on sulfur through conjugation with an aromatic ring prevents enzymatic oxidation. For example, diphenylthiophene and methylene blue are not oxygenated at detectable rates. The enzyme also does not catalyze oxygenation of S-methylmercaptoimidazoles or of pseudothioureas. The reason for this is not apparent but it illustrates steric restrictions on substrate specificity that would be difficult to predict.

While substrate specificity of the monooxygenase has not been rigorously defined by the standards of enzymology, the available kinetic data do support a few general conclusions. As stated, the monooxygenase does not catalyze oxygenation of amides, carbamates, heterocyclic aromatic amines, or primary alkylamines; the low nucleophilicity of the first two classes prevents facile oxidation of the nitrogen atom, and this is probably the principal reason that they are not substrates. However, this is not true of the other two classes. Pyridine and primary alkylamines are more basic than some of the better substrates, e.g., dimethylaniline. Delocalization of the heteroatom electrons in aromatic amines may be a factor, but there is no satisfactory electronic basis for the lack of substrate activity of primary alkylamines.

Ionic charge is also an important factor. Neutral compounds free from polar groups or compounds bearing a single positive charge are the preferred substrates. Compounds with a negative charge anywhere on the molecule are not substrates. This is true even for compounds that do not contain a net negative charge, e.g., cysteine, methionine, methyl red, N,N-dimethyl-p-aminobenzoate. Apparently negative charges on the enzyme near the catalytic site prevent interaction of negatively charged molecules with the peroxyflavin. Formamidinesulfenic acids may be exceptions, but the exact structure of mono-S-oxygenated thiocarbamides is not known. Walter and Randau[19] suggest that the physical and chemical properties of formamidinesulfenic acids are more consistent with a sulfoxide structure, although this designation is probably not completely descriptive for all mono-S-oxygenated thioureas. Phenyl- and ethyleneformamidine sulfenic acids are less polar than their parent thioureas as judged by chromatography on silicic acid.[36] Formation of a stable hydrogen bond between oxygen and nitrogen apparently prevents ionization and these compounds exist as neutral molecules in aqueous solutions.

Within the limits discussed above, the monooxygenase catalyzes oxygenation of surprisingly diverse nucleophilic nitrogen or sulfur compounds. Structural features responsible for differences in affinity (based on k_{cat}/K_m differences) are as yet impossible to assess with precision. Lipophilicity appears to be a factor. In a homologous series, the more lipophilic substrates exhibit the lowest K_m but other equally important steric factors are much more difficult to define.

III. SPECIES AND TISSUE DISTRIBUTION

A. Activity Measurements and Specificity

The flavin-containing monooxygenase has been purified to homogeneity only from hog liver and in all other tissues or hepatic tissue from other species this enzyme is usually equated with an activity, frequently with the N-oxygenation of tertiary amines. Associating an activity in crude preparations with a specific enzyme is subject to error, both qualitative and quantitative. However, similarities in cofactor requirements, pH optima, thermal stability, subcellular localization, and effects of differential inhibitors suggest that N-oxygenations of tertiary alkylamines are catalysed by basically the same type of monooxygenase in tissues from all vertebrates.

Quantitative species differences in substrate specificity and kinetic properties would be expected and some have been described. For example, primary alkylamine stimulation of N-oxygenations catalyzed by hog, hamster, or guinea pig liver microsomes cannot be demonstrated with rat or rabbit liver microsomes.[37,40] Other species differences in properties of the flavin-containing monooxygenase will undoubtedly be encountered, but there is no evidence presently supporting marked differences in substrate specificity among different tissues or species.

Activity of the flavin-containing monooxygenase is most frequently determined by following NADPH- and oxygen-dependent N-oxidation of dimethylaniline. The product, dimethylaniline N-oxide, is readily measured by a sensitive and specific colorimetric procedure. As little as 10 μM N-oxide added to tissue homogenates can be recovered quantitatively and normal components of tissue homogenates do not interfere. Whereas the product formed is readily measured, the parent substrate has a number of disadvantages which are not widely recognized. Above pH 7.0 it exists almost entirely as the free base, and its solubility in neutral aqueous solutions is less than 1 mM. In open flasks, maintained at 37°, dimethylaniline rapidly escapes into the atmosphere, and the rate of loss is a function of mixing rate and exposed surface area. At concentrations slightly above 1 mM about one-half the substrate is lost during the first 5 min from media agitated at a rate sufficient to maintain constant oxygen. This unavoidable physical loss of substrate is probably the major reason for nonlinearity of dimethylaniline N-oxidation for periods longer than 5 to 10 min. Furthermore, in the absence of a lipophilic primary amine the rate of dimethylaniline N-oxide formation underestimates potential activity substantially in tissues from most species. Lipophilic alkylamines apparently bind with inhibitors that can inactivate the monooxygenase

above 20°–25°. As stated earlier, the nature of these inhibitors is not known, but lipolytic products formed during collection and processing of tissues are the most likely agents. Despite such limitations, N-oxidation of dimethylaniline is frequently used to measure activity of flavin-containing monooxygenases in tissue homogenates and isolated microsomes.

The specificity of this reaction has been questioned,[40] but the evidence cited supporting possible cytochrome *P*-450 catalyzed N-oxygenation of dimethylaniline is not consistent with this hypothesis. For example, inhibitors specific for cytochrome *P*-450 do not inhibit the reaction at any concentration of substrate measured experimentally. Furthermore, no correlation was found between rates of N-oxygenation and cytochrome *P*-450 concentration in different tissues or in the same tissue from different species. While some N-oxidation of dimethylaniline by lipid peroxides, or by hydrogen peroxide generated from reduced cytochrome *P*-450, cannot be entirely excluded under the conditions used,[40] this is a very slow reaction compared with that catalyzed by the flavin-containing monooxygenase.

All of our studies over the past 10 years indicate that N-oxygenation of dimethylaniline is a reliable "marker" activity for the flavin-containing monooxygenase. In all tissues examined, maximum rates are obtained at pH 8.3 to 8.5 and the reaction is not inhibited by agents specific for other microsomal monooxygenases.[34] In the absence of NADPH, the activity is also sensitive to thermal inactivation at temperatures above 30°, and about half is destroyed by thermal equilibration at 37° for as little as 6 min.[22] This unusual thermal instability is characteristic of the microsomal flavin-containing monooxygenase in all tissues examined and appears an intrinsic property of the enzyme. Since thermal equilibration of microsomes in the absence of NADPH is a common practice in many laboratories, it is not surprising that reactions catalyzed by this monooxygenase are frequently not detected.

B. Dimethylaniline N-Oxygenation

NADPH- and oxygen-dependent N-oxygenation of dimethylaniline has been detected in a variety of tissues from several species (Table IV). Although measurements were conducted under such different conditions that direct comparisons are impossible, some trends are evident. In adults of all species, liver contains the highest activity with lesser but detectable amounts in most other tissues except muscle. The activity of rodent lung microsomes is also quite high and, in some species, can equal or exceed that of liver microsomes. This is not true for the hog or human (unpub-

TABLE IV

Dimethylaniline N-Oxygenase Activity of Different Mammalian Tissues

Species	Organ	Preparation	Activity (nmoles/min/mg proteins)	Reference
Human	Liver	Whole homogenate	1.0–4.2	9
Hog	Liver	Whole homogenate	4.1	24
		Microsomes	1.2	42
			13	37
			50	a
	Lung	Microsomes	0.3	42
	Kidney	Microsomes	0.2	42
	Bladder mucosa	Microsomes	0.3	42
	Testicle	Microsomes	0.2	42
	Corpus luteum	Microsomes	4.1	42
			21	a
	Thyroid	Microsomes	0.02	42
	Thymus	Microsomes	0.01	42
	Adrenal	Microsomes	0.05	42
Rat (male)	Liver	Microsomes	10	37
			4.0	22
			1.9	43
	Lung	Whole homogenate	1.8	a
	Kidney	Whole homogenate	1.2	a
Rabbit	Liver	Microsomes	10	37
			5.3	22
	Lung	Microsomes	5.9	22
Hamster	Liver	Microsomes	16	37
	Lung	Microsomes	11	a
	Kidney	Microsomes	5.4	a
Guinea pig	Liver	Microsomes	9	37
			1.2	43
Mouse				
Male	Liver	Microsomes	4.0	44
			3.2	45
	Lung	Microsomes	7.9	44
	Kidney	Microsomes	5.1	44
Female	Liver	Microsomes	21	44
			4.4	45
	Lung	Microsomes	8.3	44
	Kidney	Microsomes	4.5	44

[a] Unpublished values, this laboratory.

lished experiments), where activity of lung tissue is never more than 20% that of liver.

In newborn rodents, hepatic dimethylaniline N-oxygenase develops rapidly and reaches twice adult levels just before weaning.[21] Whether this occurs in other species is not known, but activity is also high in human fetal liver and increases with age of the fetus.[41] Heinze et al.[42] reported that this activity also increases dramatically during maturation of corpora lutea in hogs. Activity, barely detectable at ovulation, increases substantially by the day 16–17 and then decreases sharply within 2 days. Activity of the monooxygenase in liver and corpora lutea appears to correlate with rapid cell growth, but whether this is a general phenomena or peculiar to these tissues is not known.

The flavin-containing monooxygenase is not induced by most drugs or other xenobiotic agents, although induction of rat hepatic dimethylaniline N-oxygenase by prednisolone has been reported.[43] This activity is also affected by steroid sex hormones in some tissues of mice[44]; the concentration (activity) of flavin-containing monooxygenase in liver and kidney of the mouse is apparently regulated by steroid hormones. Testosterone decreases and progesterone increases hepatic activity of both gonadectomized males and females. Activity of kidney is also affected by these hormones, although the changes are not as great; changes in activity of lung tissue were not detected. A laboratory animal in which enzyme concentration can be reproducibly manipulated should prove a useful model for studies on the physiological function of the flavin-containing monooxygenase and its role in metabolism of xenobiotic compounds.

REFERENCES

1. Weisburger, J. H., and Weisburger, E. K. (1973). Biochemical formation and pharmacological, toxicological, and pathological properties of hydroxylamines and hydroxyamic acids. *Pharmacol. Rev.* **25,** 1–66.
2. Cramer, J. W., Miller, J. A., and Miller, E. C. (1960). N-Hydroxylation: A new metabolic reaction observed in the rat with the carcinogen 2-acetylaminofluorene. *J. Biol. Chem.* **235,** 885–888.
3. Kiese, M. (1966). The biochemical production of ferrihemoglobin forming derivatives from aromatic amines and mechanisms of ferrihemoglobin formation. *Pharmacol. Rev.* **18,** 1091–1161.
4. Smith, P. A. S. (1966). "Open-Chain Nitrogen Compounds," Vol. II, pp. 21–36. Benjamin, New York.
5. Bickel, M. H. (1969). The pharmacology and biochemistry of N-oxides. *Pharmacol. Rev.* **21,** 325–355.
6. Bickel, M. H. (1971). N-oxide formation and related reactions in drug metabolism. *Xenobiotica* **1,** 313–319.
7. Gorrod, J. W. (1978). On the multiplicity of microsomal N-oxidase systems. *In* "Mechanisms of Oxidizing Enzyme" (T. P. Singer and R. N. Ondarza, eds.), pp. 189–197. Elsevier, Amsterdam.
8. Jenner, P. (1971). The role of nitrogen oxidation in the excretion of drugs and foreign compounds. *Xenobiotica* **1,** 399–418.
9. Gold, M., and Ziegler, D. M. (1973). Dimethylaniline N-oxidase and aminopyrine N-demethylase activities of human liver tissue. *Xenobiotica* **3,** 179–189.
10. Kadlubar, F. F., McKee, E. M., and Ziegler, D. M. (1973). Reduced pyridine

nucleotide-dependent *N*-hydroxyamine oxidase and reductase activities of hepatic microsomes. *Arch. Biochem. Biophys.* **156**, 46–57.

11. Coutts, R. T., and Beckett, A. H. (1977). Metabolic N-oxidation of primary and secondary medicinal amines. *Drug Metab. Rev.* **6**, 51–104.

12. Capuzzi, G., and Modena, G. (1974). Oxidation of thiols. *In* "The Chemistry of the Thiol Group" (S. Patai, ed.), pp. 785–840. Wiley, New York.

13. Loosmore, S. M., and McKinnon, D. M. (1976). Oxidation products of cyclic thiones. *Phosphorus Sulfur* **1**, 185–209.

14. Sterling, C. J. M. (1974). The sulfinic acids and their derivatives. *Int. J. Sulfur Chem.* **6**, 277–316.

15. Gillette, J. R., and Kamin, J. J. (1960). The enzymatic formation of sulfoxides: The oxidation of chlorpromazine and 4,4′-diaminodiphenyl sulfide by guinea pig liver microsomes. *J. Pharmacol. Exp. Ther.* **130**, 262–267.

16. Suzuki, Z., Murakami, K., Kikuchi, S., Nishimaga, K., and Numata, M. (1967). Urinary metabolites of thioamide tetrahydrofurfuryl disulfide in rats. *J. Pharmacol. Exp. Ther.* **158**, 353–364.

17. Ammon, R., Berninger, H., Haas, H. J., and Landsberg, I. (1967). Thioacetamidsulfoxide, ein stoffwechsel product des thioacetamids. *Arzneim.-Forsch.* **17**, 521–523.

18. Porter, W. R., and Neal, R. A. (1978). Metabolism of thioacetamide and thioacetamide *S*-oxide by rat liver microsomes. *Drug Metab. Dispos.* **6**, 379–388.

19. Walter, W., and Randau, G. (1969). Thioharstoff-*S*-monooxide. *Justus Liebigs Ann. Chem.* **722**, 52–79.

20. Boyd, M. R., and Neal, R. A. (1976). Studies on the mechanisms of toxicity and of development of tolerance to the pulmonary toxin, α-naphthylthiourea. *Drug Metab. Dispos.* **4**, 314–321.

21. Uehleke, H. (1971). N-Hydroxylation. *Xenobiotica* **1**, 327–340.

22. Uehleke, H. (1973). The role of cytochrome *P*-450 in the N-oxidation of individual amines. *Drug Metab. Dispos.* **1**, 299–313.

23. Mitchell, C. H., and Ziegler, D. M. (1969). A quantitative micro method for the estimation of amine oxides. *Anal. Biochem.* **28**, 261–268.

24. Ziegler, D. M., and Poulsen, L. L. (1978). Hepatic microsomal mixed-function amine oxidase. *In* "Methods in Enzymology" (S. Fleischer and L. Packer, eds.), Vol. 52, Part C, pp. 142–151. Academic Press, New York.

25. Ghisla, S., Hustings, J., Favaudon, V., and Lhoste, J. (1978). Structure of the oxygen adduct intermediate in the bacterial luciferase reaction: ^{13}C nuclear magnetic reasonance determination. *Proc. Natl. Acad. Sci. U.S.A.* **75**, 5860–5863.

26. Ghisla, S., Entsch, B., Massey, V., and Husian, M. (1977). On the structure of flavin-oxygen intermediates involved in enzymatic reactions. *Eur. J. Biochem.* **76**, 139–148.

27. Poulsen, L. L., and Ziegler, D. M. (1979). The microsomal flavin-containing monooxygenase: Spectral characterization and kinetic studies. *J. Biol. Chem.* **254**, 6449–6455.

28. Kitchell, B., Rauckman, E., and Rosen, G. M. (1978). The effect of temperature on mixed-function amine oxidase intrinsic fluorescence and oxidative activity. *Mol. Pharmacol.* **14**, 1092–1098.

29. Rauckman, E. J., Rosen, G. M., and Kitchell, B. B. (1979). Superoxide radical as an intermediate in the oxidation of hydroxylamines by mixed function amine oxidase. *Mol. Pharmacol.* **15**, 131–137.

30. Ziegler, D. M., McKee, E. M., and Poulsen, L. L. (1973). Microsomal-catalyzed N-oxidation of arylamines. *Drug Metab. Dispos.* **1**, 314–321.

31. Sofer, S. S., and Ziegler, D. M. (1978). Microsomal mixed-function amine oxidase:

Oxidation products of piperazine-substituted phenothiazine drugs. *Drug Metab. Dispos.* **6**, 232–239.

32. Ziegler, D. M., Jollow, D., and Cook, D. (1971). Properties of a purified microsomal mixed-function amine oxidase. *In* "Flavins and Flavoproteins" (H. Kamin, ed.), pp. 504–522. Univ. Park Press, Baltimore, Maryland.

33. Poulsen, L. L., Kadlubar, F. F., and Ziegler, D. M. (1974). Role of the microsomal mixed-function amine oxidase in the oxidation of N,N-disubstituted hydroxylamines. *Arch. Biochem. Biophys.* **164**, 774–775.

34. Prough, R. A., and Ziegler, D. M. (1977). The relative participation of liver microsomal amine oxidase and cytochrome *P*-450 in N-demethylation reactions. *Arch. Biochem. Biophys.* **180**, 363–373.

35. Prough, R. A. (1973). The N-oxidation of alkylhydrazines catalyzed by the microsomal mixed-function amine oxidase. *Arch. Biochem. Biophys.* **158**, 442–444.

36. Poulsen, L. L., Hyslop, R. M., and Ziegler, D. M. (1979). S-Oxygenation of N-substituted thioureas catalyzed by the liver microsomal FAD-containing monooxygenase. *Arch. Biochem. Biophys.* **198**, 78–88.

37. Poulsen, L. L., Hyslop, R., and Ziegler, D. M. (1974). S-Oxidation of thioureylenes catalyzed by a microsomal flavoprotein mixed-function oxidase. *Biochem. Pharmacol.* **23**, 3431–3440.

38. Ziegler, D. M., and Poulsen, L. L. (1977). Protein disulfide bond synthesis: A possible intracellular mechanism. *Trends Biochem. Sci.* **2**, 79–81.

39. Poulsen, L. L., Masters, B. S. S., and Ziegler, D. M. (1976). Mechanism of 2-naphthylamine oxidation catalyzed by pig liver microsomes. *Xenobiotica* **6**, 481–498.

40. Hlavica, P., and Kehl, M. (1977). The role of cytochrome *P*-450 and mixed-function amine oxidase in the N-oxidation of N,N-dimethylaniline. *Biochem. J.* **164**, 487–496.

41. Rane, A. (1973). N-oxidation of a tertiary amine (N,N-dimethylaniline) by human fetal liver microsomes. *Clin. Pharmacol. Ther.* **15**, 32–38.

42. Heinze, E., Hlavica, P., Kiese, M., and Lipowsky, G. (1970). N-Oxygenation of arylamines in microsomes prepared from corpora lutea of the cycle and other tissues of the pig. *Biochem. Pharmacol.* **19**, 641–649.

43. Arrhenius, E. (1968). Effects on hepatic microsomal N- and C-oxygenation of aromatic amines by *in vivo* corticosteroid or aminofluorene treatment, diet, or stress. *Cancer Res.* **28**, 264–273.

44. Duffel, M. W., Poulsen, L. L., and Ziegler, D. M. (1979). Sex hormone induced changes in activity of the microsomal flavin-containing monooxygenase of mouse tissues. *Fed. Proc., Fed. Am. Soc. Exp. Biol.* **38**, 732.

45. Wirth, P. J., and Thorgeirsson, S. S. (1978). Amine oxidase in mice—sex differences and developmental aspects. *Biochem. Pharmacol.* **27**, 601–603.

Part III

Other Oxidation–Reduction Systems

Chapter 10

Alcohol Dehydrogenase

WILLIAM F. BOSRON and TING-KAI LI

I. INTRODUCTION

The metabolic elimination of alcohols in experimental animals and man occurs primarily by two pathways: (1) oxidation first to the aldehyde and subsequently to the corresponding carboxylic acid, and (2) direct conjugation of the alcohol with glucuronic acid.[1] The principal enzymes involved in alcohol oxidation are the NAD^+-dependent alcohol (EC 1.1.1.1) [Eq. (1)] and aldehyde dehydrogenases (EC 1.2.1.3) (Chapter 12, this volume).

$$RCH_2OH + NAD^+ \rightleftharpoons RCHO + NADH + H^+ \qquad (1)$$

231

ENZYMATIC BASIS OF DETOXICATION, VOL. I
Copyright © 1980 by Academic Press, Inc.
ISBN 0-12-380001-3

Alcohols can also be oxidized by peroxide in a catalase-dependent reaction or by a high K_m, NADPH-dependent, microsomal ethanol-oxidizing system.[2,3] The extent to which any alcohol is oxidized or conjugated is dependent upon the structure of the alcohol. For example, an intraperitoneal dose of *tert*-amyl alcohol takes approximately 50 h to be eliminated from the rat, whereas, the primary alcohol, *n*-amyl alcohol, is eliminated in 3 to 6 h.[4] In rabbits, approximately 58% of *tert*-amyl alcohol appears as the glucuronide conjugate but only 6.7% of *n*-amyl alcohol is conjugated.[1] Such elimination specificity is entirely consistent with the relative reactivities of mammalian liver alcohol dehydrogenases toward these alcohols: *tert*-amyl alcohol is not oxidized by horse liver alcohol dehydrogenase, whereas the enzyme exhibits high activity and a low K_m for *n*-amyl alcohol.[5]

While dehydrogenase-dependent alcohol oxidation reactions represent one mechanism by which endogenous metabolites and toxic foreign compounds or drugs are *detoxified*, Mulder[6] observed that this same oxidative process alternatively may be responsible for *toxification* of other compounds. Such is the case for the dehydrogenase-dependent oxidation of methanol and ethylene glycol in man. The ultimate metabolic products, formate and oxalate, respectively, are more toxic than their alcohol precursors. Inhibition of ethylene glycol oxidation by administration of ethanol, a competitive substrate for the dehydrogenase[7] or inhibition of methanol oxidation by 4-methylpyrazole[8] has been suggested as a means by which the oxidative toxification of these alcohols may be prevented.

In man, the specificity and regulation of alcohol oxidation is even more complex than in experimental animals. The direct roles of catalase and the microsomal ethanol-oxidizing system in man remain a matter of debate.[9-11] Moreover, human livers contain multiple molecular forms or isoenzymes of alcohol dehydrogenase, some of which exhibit strikingly different substrate specificity and kinetic properties. Therefore, genetic variation in isoenzyme distribution or content might result in differences in alcohol elimination rate and/or physiologic sensitivity to alcohol among certain individuals. In fact, racial differences in ethanol elimination rate, blood acetaldehyde concentrations after ethanol ingestion, physiologic reaction to ethanol administration, and incidence of alcoholism have been observed.[10,12-16]

II. DISTRIBUTION OF ALCOHOL DEHYDROGENASE

Pyridine nucleotide-dependent alcohol dehydrogenases are ubiquitously distributed in plants and animals. The tissue distribution of NAD+-dependent alcohol dehydrogenase activity has been examined in man.[17]

The highest activity per gram of tissue occurs in liver, while stomach, lung, intestine, and kidney contain less than 5% of the activity in liver. The properties of the partially purified stomach enzyme have been studied.[18] Because it occurs in such low amounts, it was concluded that the enzyme in stomach contributes little to beverage ethanol oxidation. There is uncertainty concerning the content of alcohol dehydrogenase in brain.[19-20] Since ethanol oxidation has been postulated to alter biogenic amine metabolism,[21] definitive assessment of the capacity of brain to oxidize alcohol is a matter of considerable importance. From studies in the rat, Van Thiel et al.[22] postulated that competition between ethanol and retinol oxidation by alcohol dehydrogenase may disrupt retinol metabolism and be a contributing factor to the observed increase in frequency of sterility in chronic alcoholics. Small amounts of alcohol dehydrogenase activity have been detected in human testicular tissue (W. F. Bosron and T.-K. Li, unpublished observations).

All evidence indicates that liver is the principal site of ethanol oxidation in man. Approximately 75% of a dose of alcohol is metabolized by the liver as determined by direct hepatic vein catheterization studies[23,24] in which hepatic elimination capacity ranged from 0.87 to 2.29 mmole/min, while extrahepatic metabolism accounted for only 0.4 mmole/min.[25] Most experimental findings indicate that hepatic alcohol dehydrogenase is the principal and rate-limiting enzyme responsible for ethanol oxidation in man, particularly when the concentration of alcohol is less than 20 mM.[10,26]

III. MULTIPLE MOLECULAR FORMS OF HUMAN LIVER ALCOHOL DEHYDROGENASE

It has long been known that there are multiple molecular forms of human liver alcohol dehydrogenase.[27] This heterogeneity is best identified by electrophoresis on starch gels where as many as 16 different activity bands or molecular forms have been observed in different liver specimens. A genetic model for their formation as isoenzymes has been proposed. From analysis of activity patterns of over 300 liver autopsy specimens by starch gel electrophoresis, Smith et al.[28,29] postulated that there are 3 gene loci, ADH_1, ADH_2, and ADH_3 coding for α, β and γ subunits (Table I). Moreover, they proposed polymorphism at the ADH_2 and ADH_3 gene loci giving rise to β_1 or β_2 and γ_1 or γ_2 subunits, respectively. Since ADH is dimeric, a total of 5 different homodimeric and 10 different heterodimeric active isoenzymes can be formed by the random combination of such subunits (Table I).

In 1975, a previously unidentified, "anodic" molecular form of the

TABLE I

Molecular Forms of Human Liver Alcohol Dehydrogenase[a]

Gene locus	Homodimers	Heterodimers
? ———————————————— Π-ADH ————————————		
ADH_1	$\alpha\alpha$	
		$\alpha\gamma_2$
	$\gamma_2\gamma_2$	
		$\alpha\gamma_1$
ADH_3		$\alpha\beta_1, \gamma_1\gamma_2$
		$\alpha\beta_2$
		$\beta_1\gamma_2$
		$\beta_2\gamma_2$
	$\gamma_1\gamma_1$	
		$\beta_1\gamma_1$
		$\beta_2\gamma_1$
	$\beta_1\beta_1$	
ADH_2		$\beta_1\beta_2$
	$\beta_2\beta_2$	

[a] The genetic loci[28] and corresponding homodimeric and heterodimeric isoenzymes are listed according to their relative mobility on starch gel electrophoresis from the anode at the top. The genetic locus and subunit structure of Π-alcohol dehydrogenase have not been determined.

enzyme was identified by starch gel electrophoresis in high specific activity biopsy and autopsy liver specimens.[30] It was subsequently purified to homogeneity.[31] This molecular form was found to be considerably more labile than the other isoenzymes both *in vitro*[32] and *in situ*,[30] a property that apparently precluded its detection by previous investigators. Because of its relative insensitivity to inhibition by pyrazole and 4-methylpyrazole (K_i = 30 and 2 mM, respectively) as compared to the other isoenzymes, this molecular form was designated Π-ADH.[33] The genetic relationship between Π-ADH and the other alcohol dehydrogenase isoenzymes is presently unknown (Table I). However, Adinolfi *et al.*[34] recently reported that antibody prepared against the $\beta_1\beta_1$ form of the enzyme cross-reacted with all other isoenzymes except Π-ADH, suggesting that Π-ADH may be genetically distinct. Developmentally, only the $\alpha\alpha$ and Π-ADH forms are present in the early fetus. During fetal development, the $\alpha\beta$ isoenzyme becomes observable, and in the neonate Π, $\alpha\alpha$, $\alpha\beta$, and $\beta\beta$ forms are present. The ADH_3 gene, which codes for the γ forms is not expressed in liver until 3 to 6 months of age, thereby giving rise to the complex adult electrophoretic pattern described above.[28]

IV. PURIFICATION OF HUMAN LIVER ALCOHOL DEHYDROGENASE

A. Affinity Chromatography

The most important recent advance that has enabled the rapid and specific purification of these isoenzymes from human liver is the development of affinity chromatography techniques. An alcohol dehydrogenase-specific, double-ternary complex affinity chromatography procedure was devised by Lange and Vallee.[35] The method employs the 4-substituted pyrazole analogue, 4[3-(N-6-aminocaproyl)aminopropyl] pyrazole immobilized on Sepharose (CapGapp-Sepharose) as the affinity ligand. Pyrazole and its 4-substituted derivatives are potent inhibitors of horse liver alcohol dehydrogenase[36,37] and most molecular forms of the human enzyme,[38,39] with K_i values ranging from 0.1 to 7 μM. Theorell[40] reported that pyrazole binds to the enzyme·NAD⁺ binary complex and, therefore, is a competitive inhibitor with respect to ethanol. Lange and Vallee[35] found that the liver enzyme from various animal species, including the human, could be bound selectively to CapGapp-Sepharose in the presence of NAD⁺ as the ternary complex. Enzyme was subsequently eluted with substrate, i.e., ethanol, through the formation of a second, catalytically active, ternary complex. In their studies, crude extracts from human, rat, horse, and rabbit livers were first partially purified in batch on DEAE-cellulose and then chromatographed on CapGapp-Sepharose. The yield of enzyme purified in this manner was 65–90% leading to a single protein band on sodium dodecyl sulfate-gel electrophoresis. Hence, this technique represents a rapid and specific method for the purification of large quantities of alcohol dehydrogenase.

Recent studies have indicated that CapGapp-Sepharose can also be utilized to *separate* molecular forms of the human liver enzyme, on the basis of their differences in affinity for pyrazole compounds. When partially purified homogenate supernatants from certain high specific activity livers were chromatographed on CapGapp-Sepharose, II-ADH or the "anodic" molecular form did not bind but eluted in the column void volume.[31,33] This behavior is consistent with the high K_i of this molecular form for 4-methylpyrazole (2 mM), which is as much as 10,000 times the value of the other molecular forms (*vide infra*). The molecular and catalytic properties of II-ADH, purified to homogeneity, have been reported.[32]

Another approach to the purification of alcohol dehydrogenase by affinity chromatography is to bind enzyme in a binary complex to an immobilized coenzyme analogue. Toward this end, NAD⁺ and AMP resins have been synthesized.[41] While these resins are not specific for alcohol dehydrogenase per se, purification of the horse liver enzyme,[42,43] human liver Π-ADH[31,33] and the human stomach enzyme[18] has been achieved by employing specific conditions for enzyme elution such as NAD^+ or NADH gradients. Recent studies by Adinolfi and Hopkinson[44] demonstrated that the human liver enzyme can also be adsorbed onto the general affinity resin, Blue-Dextran Sepharose. Owing to the differential affinity of the isoenzymes for the resin and/or NAD^+, they can be partially separated by elution with increasing concentrations of NAD^+. The sequential use of both specific ligand affinity chromatography (CapGapp-Sepharose) and general ligand affinity chromatography (AMP-agarose and Blue-Dextran Sepharose), has yielded a highly efficient scheme for purification of large quantities of human liver alcohol dehydrogenase isoenzymes.[31]

B. Ion-Exchange Chromatography

Conventional protein isolation procedures have also been utilized for the purification of the human liver enzyme and separation of isoenzymes. The isoenzymes have been partially resolved or purified by ion-exchange chromatography on carboxymethyl cellulose (CM-cellulose) or DEAE-cellulose.[27,45-49] In addition, the isoenzymes have been partially separated by column isoelectric focusing.[30] A procedure for enzyme crystallization has also been reported.[45]

C. Stability of Purified Alcohol Dehydrogenase

Certain molecular forms of the human liver enzyme are relatively unstable both in the crude and purified state. "Storage bands" have been observed on electrophoresis after repeated freezing and thawing of liver tissue[28] or after prolonged storage of the purified enzyme.[18,46,48] It has been postulated that such electrophoretic changes may arise from limited degradation[48] or disulfide bond formation.[46] Π-Alcohol dehydrogenase is particularly labile both *in situ*[30] and in the purified state.[32] Although the basis of this instability is presently unknown, purified Π-ADH can be effectively stabilized for several weeks by storage at 4° in the presence of 10 mM ethanol. Other preparations of partially purified isoenzymes are reported to be stable for several weeks if stored in buffer at 4°.[35,48]

V.   MOLECULAR PROPERTIES OF HUMAN LIVER ALCOHOL DEHYDROGENASE

All mammalian liver alcohol dehydrogenases examined thus far, from man, horse, or rat, appear to be remarkably similar with respect to molecular weight, subunit structure, amino acid composition, and zinc content.[50] Analysis of the enzyme from human liver by ultracentrifugation has yielded molecular weight values ranging from 78,000 to 87,000.[32,49,51-53] Its mobility on sodium dodecyl sulfate-gel electrophoresis[31,53] is identical to that of the horse liver enzyme, which has been shown to have a subunit molecular weight of approximately 40,000.[54] These results indicate that, as with horse liver alcohol dehydrogenase, the human enzyme is composed of two subunits of equal molecular weight.

Amino acid analysis of samples containing unspecified mixtures of isoenzymes of human liver alcohol dehydrogenase[45,53] as well as for the $\beta\beta$[52] and the Π-ADH[32] molecular forms indicate that the total number of residues per subunit range from 365 to 412. Recent preparations have been found to contain also 4 g-atoms of tightly bound zinc per mole of enzyme.[32,53] Sequence analyses of tryptic peptides have provided evidence for the existence of heterodimeric forms of isoenzymes.[52,55] However, detailed correlation between catalytic and structural features of the individual isoenzymes must await further refinement of separation and purification techniques.

VI. CATALYTIC PROPERTIES OF HUMAN LIVER ALCOHOL DEHYDROGENASE

A. Assay of Activity

Pyridine nucleotide-dependent dehydrogenases represent one of the easiest classes of enzymes to assay. The production or utilization of the nucleotide substrate, NADH, is readily monitored spectrophotometrically at 340 nm,[56] fluorometrically[57] or by bioluminescence with bacterial luciferase.[58] Protein concentration of human liver alcohol dehydrogenase can be determined spectrophotometrically by using a value of $A_{280\ nm}^{mg/ml} = 0.58$.[53] Enzyme active sites can be quantified by titration with NADH in the presence of isobutyramide[48]; under these circumstances a highly fluorescent ternary complex is formed. The dissociation constant for NADH in the presence of isobutyramide is 2 to 7 nM at pH 7.0.[59]

B. pH Optimum

The ethanol-oxidizing activity of the enzyme from horse liver[5] and of most preparations from human liver[46,51,53] is maximal between pH 10 and 11. However, the supernatant fluids after homogenization[60] and partially purified preparations[46] from some human liver specimens have been shown to exhibit a lower pH optimum, approximately 8.5. Based upon this difference in pH-activity behavior, von Wartburg *et al.*[60] designated this enzyme form or forms as "atypical." Electrophoretic examination of such preparations have revealed an activity band or bands migrating more cathodically than the $\beta_1\beta_1$ isoenzyme.[28,46,61] Smith *et al.*[28] suggested that "atypical" alcohol dehydrogenase is due to a variant allele at the ADH$_2$ gene locus coding for a β_2 subunit that is both electrophoretically and catalytically distinct from those isoenzymes containing β_1 subunits. The "atypical" isoenzyme is particularly prominent in Japanese populations, greater than 85%,[62] but appears in only 5 to 20% of Caucasian European populations.[28,63] The molecular basis for the difference between "atypical" or β_2 subunits and the "typical" or β_1 subunits has been preliminarily examined. In the β_2 subunit, there appears to be a substitution of proline for alanine-230 located in a sequence region that corresponds to the coenzyme binding site of horse liver enzyme.[52] The K_m for NAD$^+$ with the partially purified, "atypical" enzyme appeared somewhat elevated relative to enzyme preparations with an alkaline pH optimum.[60] However, the kinetic constants reported in that study did not agree well with more recent data.[64] Hence, the pH-dependent kinetic characteristics of the "atypical" forms must be examined in greater detail in order to establish their physiological significance.

C. Catalytic Mechanism

The reaction mechanism for two molecular forms of human liver alcohol dehydrogenase, the $\beta_1\beta_1$ isoenzyme[64] and Π-ADH[32] have been

TABLE II

Kinetic Constants for Human Liver $\beta_1\beta_1$ and Π-Alcohol Dehydrogenase Isoenzymes

	Constant[a]	Π-ADH (mM)	$\beta_1\beta_1$ (mM)
K_a	(NAD$^+$)	0.014	0.017
K_b	(ethanol)	34	1.8
K_{ia}	(NAD$^+$)	0.086	0.049
K_{ib}	(ethanol)	800	92–703
K_p	(acetaldehyde)	30	1.5
K_q	(NADH)	0.016	0.010
K_{ip}	(acetaldehyde)	9	1–160
K_{iq}	(NADH)	0.0019	0.0021

[a] Kinetic constants for Π-ADH were determined at pH 7.5, 0.1 M sodium phosphate (NaP$_i$)[32] and for $\beta_1\beta_1$ at pH 8.0, 0.1 M NaP$_i$.[64] Data were evaluated utilizing the statistical programs described by Cleland[67] assuming an ordered bi bi mechanism.

examined at pH 8.0 and 7.5, respectively. Alternate substrate and product inhibition steady-state kinetic studies indicate that these two forms follow an ordered bi bi mechanism with coenzyme adding first to form a binary enzyme·coenzyme complex, and alcohol or aldehyde adding second to form the catalytic ternary complex. Similar kinetic studies with the horse liver enzyme had previously indicated that it follows this mechanism.[65,66] The kinetic constants for the two human isoenzyme forms are shown in Table II.[32,64,67] The K_m for NAD^+ and NADH (K_a and K_q) and inhibition constant for NADH (K_{iq}) are similar in magnitude for $\beta_1\beta_1$ and Π-alcohol dehydrogenase. However, the K_m values for ethanol and acetaldehyde (K_b and K_p) are approximately 20-fold greater for Π-ADH than for the $\beta_1\beta_1$ isoenzyme. This suggests that Π-ADH should play an increasingly important role in alcohol oxidation *in vivo* at ethanol concentrations that are intoxicating and saturating for the $\beta_1\beta_1$ and presumably also for the other isoenzyme forms.

D. Substrate Specificity

Mammalian liver alcohol dehydrogenases are relatively specific for NAD^+ or NADH as coenzyme but exhibit broad substrate specificity with respect to alcohols and aldehydes. The horse liver enzyme readily oxidizes most primary and secondary aliphatic and aromatic alcohols[5]; tertiary and some bulky secondary alcohols are poor substrates. Alcohol dehydrogenase also readily reduces aldehydes and aromatic ketones. However, most long chain aliphatic ketones are not enzymatically reduced.[5] A comparison of saturated and unsaturated alcohols as substrates for the horse and human enzymes has been made.[68] In all instances, the 2-enoic alcohols exhibited lower K_m values than the saturated analogues. However, the saturated aldehydes were better substrates in the reductive direction than the unsaturated aldehydes.

The specificity of the enzyme from human liver toward primary alcohols has been examined with enzyme preparations containing mixtures of isoenzyme forms.[39,53,68,69] Increasing the primary alcohol chain length by 6-carbon units from methanol to 1-hexanol resulted in a 117-fold decrease in K_m from 7.0 to 0.06 mM (Table III). On the other hand, the V_m for primary alcohols with 2 to 6 carbons remained relatively constant. A similar dependence of K_m and V_m on primary alcohol chain-length was observed for Π-ADH (Table III). This suggests that the rate-limiting step does not involve the binding or release of alcohol or aldehyde intermediates. Previous studies of structure–activity relationships with the horse liver[70] and the yeast[71] enzymes have evaluated the role of electronic, apolar and steric effects on kinetic constants. Similar studies with

TABLE III

Substrate Specificity of Human Liver Alcohol Dehydrogenase

Alcohol	Π-ADH[a]		Isoenzyme mixture[b]	
	K_m (mM)	V_m	K_m (mM)	V_m
Methanol	N.D.[c]	N.D.	7.0	0.09
Ethanol	34	1.0	0.40	1.0
1-Propanol	3.0	0.8	0.10	0.9
1-Butanol	0.14	1.0	0.14	1.1
1-Pentanol	0.036	1.0	—	—
1-Hexanol	—	—	0.06	0.9
2-Hexene-1-ol	—	—	0.003	1.4
Ethylene glycol	N.D.	N.D.	30	0.4
1,3-Butanediol	—	—	6.8	0.6
1,6-Hexanediol	—	—	0.07	1.0

[a] Activity of Π-ADH determined in 0.1 M sodium phosphate at pH 7.5 with 2.4 mM NAD$^+$.[32] Values for V_m are expressed relative to ethanol which is defined as unity.

[b] Activity of an unidentified mixture of isoenzymes determined in 0.1 M sodium phosphate at pH 7.0 with 0.5 mM NAD$^+$.[53,68,69]

[c] No activity detected.

the human liver enzyme indicate a sharp dependence of K_m and k_{cat}/K_m for alcohol oxidation on the apolar character of the substrate.[53,72] Moreover, the K_m for ethanol determined over the range of 290° to 317°K decreased 3.3-fold, a response characteristic of apolar interactions.[53]

Human liver alcohol dehydrogenase also catalyzes the oxidation of polyalcohols (Table III): glycerol, ethylene glycol, and sugar alcohols such as D-ribose have been shown to be substrates.[27,69,73] Certain of these substrates have physiological importance. For example, reduction of glyceraldehyde to glycerol is an important step in fructose metabolism.[74] The specificity of oxidation of diols has also been examined with the human and horse liver enzymes.[69] As with primary alcohols, diols of long-chain length, e.g., 1,6-hexanediol (Table III), exhibit low K_m values. In animals and man, the diol, ethylene glycol is oxidized ultimately to oxalate, a product that is toxic in high amounts and crystallizes in kidneys.[7] By contrast, 1,3-butanediol is metabolized to the nontoxic ketone body, acetoacetate.[75]

There have been few detailed investigations of the specificity of human alcohol dehydrogenase isoenzymes with respect to substrates that are potentially of physiologic importance. NAD^+-dependent dehydrogenases capable of oxidizing retinol (Vitamin A_1) to retinal have been reported in the rat liver and eye.[76,77] The retinol dehydrogenase in rat retina, however, is electrophoretically distinct from the liver enzyme. Purified rat liver alcohol dehydrogenase exhibits a higher affinity for retinol, $K_m = 0.7$ μM, than for ethanol.[78] Human liver Π-ADH also reduces retinal but complete kinetic studies with this and other isoenzymes are not available (W. F. Bosron and T.-K. Li, unpublished observations). The substrate specificity of the horse, rat and human liver enzymes for hydroxy- and oxosteroids has been examined.[79,80] Only those molecular forms of the horse liver enzymes containing steroid-active or "S-type" subunits interconvert 3β-hydroxysteroids and 3β-ketosteroids.[80] However, no molecular form of human alcohol dehydrogenase has been found to exhibit steroid reactivity similar to that of the horse "SS" isoenzyme. Pietruszko et al.[48] reported that the activity of five molecular forms of the human liver enzyme for 5β-androstan-3β-ol-17-one was less than 7% of that for ethanol.

Recent reports indicate that differences exist in the substrate specificity of some of the molecular forms of human liver enzyme. Based upon the activity staining of isoenzymes on starch gel electrophoresis, Smith et al.[81] concluded that isoenzymes containing γ subunits are more reactive toward n-amyl alcohol than forms containing α or β-subunits. Furthermore, isoenzymes containing β_2-subunits appear to exhibit a lower ratio of activity with 3-pyridylcarbinol relative to ethanol than isoenzymes containing β_1 subunits.[46] Purified Π-ADH exhibits a more limited substrate specificity than the other molecular forms.[32] Neither methanol, glycerol, nor ethylene glycol are oxidized by Π-ADH at concentrations up to 100 mM.

E. Inhibitor Specificity

Human liver alcohol dehydrogenase activity is inhibited by several classes of compounds shown previously to be inhibitors of horse liver enzyme. Heterocyclic compounds, such as pyrazole, imidazole, and their derivatives, are potent inhibitors of the horse liver enzyme.[36] It has been demonstrated that pyrazole compounds form tight, deadend, ternary enzyme·NAD^+·inhibitor complexes. The K_i values of human liver alcohol dehydrogenase preparations containing mixtures of isoenzymes range from 0.12 to 2.6 μM for pyrazole and its 4-substituted analogs (Table IV). By contrast, the K_i value of Π-ADH for pyrazole is as much as 10,000-times greater than that for such mixtures (Table IV). However, Π-ADH exhibits much lower K_i values for pyrazole analogs containing apolar

TABLE IV

**Inhibition of Human Liver Alcohol
Dehydrogenase by Pyrazole Analogues**

Inhibition	K_i for Π-ADH[a] (μM)	K_i for isoenzyme mixture[b] (μM)
Pyrazole	30,000	2.6
4-Methoxypyrazole	6,300	—
4-Methylpyrazole	2,000	0.21
4-Nitropyrazole	27	—
4-Bromopyrazole	10	0.29
4-Iodopyrazole	—	0.12
4-Pentylpyrazole	4.3	—

[a] Activity of Π-ADH was determined with 2.4 mM NAD$^+$ in 0.1 M sodium phosphate at pH 7.5. Inhibition was competitive with respect to acetaldehyde for pyrazole, and its 4-methoxy and 4-methyl derivatives, and competitive with respect to ethanol for 4-bromo-, 4-nitro-, and 4-pentylpyrazole.[32]

[b] Activity of an unidentified mixture of isoenzymes was determined with 0.35 mM NAD$^+$ in 0.15 M sodium phosphate at pH 7.4. Inhibition was competitive with respect to ethanol for pyrazole and its 4-bromo and 4-methyl analogues, and noncompetitive with respect to ethanol for 4-iodopyrazole.[38]

(4-propyl- or 4-pentylpyrazole) or electron-withdrawing substituents (4-nitro- or 4-bromopyrazole). Hence, these compounds may be more suitable than pyrazole or 4-methylpyrazole for the differentiation of Π-ADH from dehydrogenase-independent, high K_m alcohol-oxidizing systems in man both *in vitro* and *in vivo*.[32]

Fatty acids and amides are also known to form dead end, ternary complexes with the horse liver enzyme in the presence of the nucleotide substrates[36,82] and, hence, are potent inhibitors. Halogenated substrates, such as trichloroethanol and chloral hydrate, show little catalytic activity but act as competitive inhibitors.[27,81] As with the horse liver enzyme, human alcohol dehydrogenase is a zinc metalloenzyme.[53] Chelating agents, such as *o*-phenanthroline, α,α'-bipyridine, and EDTA, are inhibitors. *o*-Phenanthroline forms an instantaneous and reversible enzyme complex at the active site metal atom.[53]

VII. REGULATION OF ETHANOL METABOLISM IN MAN

Ethanol in the form of alcoholic beverages is one of the most commonly used, as well as abused, drugs. While certain aspects of the short-term behavioral effects of ethanol consumption remain socially acceptable, chronic ethanol abuse is acknowledged as a major medical, social, and economic problem. The various pharmacologic, addictive, and pathologic consequences of alcohol consumption are directly related to the biochemical properties and rate of elimination of ethanol and the formation of its metabolic by-products. Hence, knowledge about regulation of the oxidation of ethanol as catalyzed by liver alcohol dehydrogenase in man is fundamental to our understanding of the etiology and underlying mechanisms of alcoholism.

There are several primary metabolic processes that together may regulate the rate of alcohol metabolism in normal individuals: (1) the absorption of ethanol in the small intestine, (2) the content or activity of liver alcohol dehydrogenase, (3) the content or activity of liver aldehyde dehydrogenase, and (4) the reoxidation of NADH in liver mitochondria. In both experimental animals and man it is generally accepted that regulation of dehydrogenase activity and regeneration of NAD^+ are the dominant rate-limiting factors for ethanol oxidation *in vivo*.[10,11,26] Administration of fructose has been shown to accelerate alcohol elimination in some human subjects, presumably by accelerating NADH oxidation and increasing the NAD^+/NADH ratio.[23] The extent to which changes in the NAD^+/NADH ratio can alter the oxidation rate of alcohol can be quantitatively described by the kinetic constants of the individual isoenzymes for substrates and the concentrations of the coenzyme and substrate pairs in liver. The overall ethanol oxidation rate should equal the sum of the activities of the individual isoenzymes under any given physiologic NAD^+/NADH concentration ratio.

In fed rats, the maximal activity of alcohol dehydrogenase *in vitro* is approximately equal to the alcohol elimination rate *in vivo*. With fasting, both enzyme activity and alcohol elimination rate have been shown to decrease in parallel.[83] These data indicate that the hepatic content of enzyme is a major rate-limiting factor of alcohol elimination under such conditions. A systematic correlation between total liver enzyme activity and ethanol elimination rate *in vivo* has not been performed in man. However, the identification of genetically determined isoenzymes of human liver alcohol dehydrogenase with widely divergent kinetic properties (*vide supra*) suggests that the regulation of alcohol metabolism is much more complex in man than in experimental animals. Indeed, in man, the pharmacokinetics of alcohol elimination *in vivo* do not fit a simple

Michaelis–Menten process with a single K_m, but are nonlinear.[84] This is consistent with the observed multiplicity of isoenzymes of human alcohol dehydrogenase, some of which exhibit different K_m values for ethanol. The relative importance and role of the individual isoenzymes in alcohol metabolism must await further examination of their individual kinetic properties as well as their frequency of occurrence in man.

There appears to be real and substantial differences among individuals in their rate of alcohol elimination.[10] As shown by comparison of the rates of metabolism of ethanol by identical and fraternal twins, variations in rate of ethanol elimination between nonalcoholic individuals can be attributed largely to genetic rather than environmental differences.[85] Moreover, ethanol elimination rate, blood ethanol or acetaldehyde concentration, and certain physiological responses to alcohol ingestion have been observed to vary among racial groups.[12-16] Variation in the frequency and distribution of alcohol dehydrogenase isoenzymes, some of which have unique catalytic properties, may underlie some of these differences.

ACKNOWLEDGMENT

Supported by United States Public Health Service Grants AA-02342 and AA-03243.

REFERENCES

1. Williams, R. T. (1959). "Detoxication Mechanisms." Wiley, New York.
2. Teschke, R., Hasumura, Y., and Lieber, C. S. (1975). Hepatic microsomal alcohol-oxidizing system affinity for methanol, ethanol, propanol and butanol. *J. Biol. Chem.* **250**, 7393–7404.
3. Teschke, R., Matsuzaki, S., Ohnishi, K., DeCarl, L., and Lieber, C. S. (1977). Microsomal ethanol oxidizing system (MEOS): Current status of its characterization and its role. *Alcohol.: Clin. Exp. Res.* **1**, 7–15.
4. Haggard, H. W., Miller, D. P., and Greenberg, L. A. (1945). The amyl alcohols and their ketones: Their metabolic fates and comparative toxicities. *J. Ind. Hyg.* **27**, 1–14.
5. Sund, H., and Theorell, H. (1963). Alcohol dehydrogenases. *In* "The Enzymes" (P. D. Boyer, H. Lardy, and K. Myrbäck, eds.), 2nd ed., Vol. 7, pp. 25–83. Academic Press, New York.
6. Mulder, G. J. (1979). Detoxification or toxification? Modification of the toxicity of foreign compounds by conjugation in the liver. *Trends Biol. Sci.* **4**, 86–90.
7. Wacker, W. E. C., Haynes, H., Druyan, R., Fisher, W., and Coleman, J. E. (1965). Treatment of ethylene glycol poisoning with ethyl alcohol. *J. Am. Med. Assoc.* **194**, 1231–1233.
8. Makar, A. B., and Tephyl, T. R. (1975). Inhibition of monkey liver alcohol dehydrogenase by 4-methylpyrazole. *Biochem. Med.* **13**, 334–342.

9. Thurman, R. G., and Brentzel, H. J. (1977). The role of alcohol dehydrogenase in microsomal ethanol oxidation and the adaptive increase in ethanol metabolism due to chronic treatment with ethanol. *Alcohol.: Clin. Exp. Res.* **1**, 33–38.

10. Li, T.-K. (1977). Enzymology of human alcohol metabolism. *Adv. Enzymol.* **46**, 427–483.

11. Lieber, C. S. (1977). Metabolism of ethanol. *In* "Metabolic Aspects of Alcoholism" (C. S. Lieber, ed.), pp. 1–29. Univ. Park Press, Baltimore, Maryland.

12. Partanen, J., Broon, K., and Markkanen, T. (1966). A study on intelligence, personality and use of alcohol of adult twins. Alcohol research in the northern countries. *Finn. Found. Alcohol Stud.* **14**, 1–159.

13. Reed, T. E. (1978). Racial comparisons of alcohol metabolism: Background, problems and results. *Alcohol.: Clin. Exp. Res.* **2**, 83–87.

14. Hanna, J. M. (1978). Metabolic responses of Chinese, Japanese and Europeans to alcohol. *Alcohol.: Clin. Exp. Res.* **2**, 89–92.

15. Schaefer, J. M. (1978). Alcohol metabolism and sensitivity reactions among the Reddis of South India. *Alcohol.: Clin. Exp. Res.* **2**, 61–69.

16. Seto, A., Tricomi, S., Goodwin, D. W., Kolodney, R., and Sullivan, T. (1978). Biochemical correlates of ethanol-induced flushing in orientals. *J. Stud. Alcohol* **39**, 1–11.

17. Moser, K., Papenberg, J., and von Wartburg, J.-P. (1968). Heterogenitat und organventeilung der alcohol-dehydrognease bei verschiedenen spezies. *Enzymol. Biol. Clin.* **9**, 447–458.

18. Hempel, J. D., and Pietruszko, R. (1979). Human stomach alcohol dehydrogenase: Isoenzyme composition and catalytic properties. *Alcohol.: Clin. Exp. Res.* **3**, 95–98.

19. Raskin, N. H., and Sokoloff, L. (1970). Alcohol dehydrogenase activity in rat brain and liver. *J. Neurochem.* **17**, 1677–1687.

20. Tabakoff, B., and von Wartburg, J.-P. (1975). Separation of aldehyde reductases and alcohol dehydrogenase from brain by affinity chromatography: Metabolism of succinic semialdehyde and ethanol. *Biochem. Biophys. Res. Comm.* **63**, 957–966.

21. Davis, V. E., Huff, J. A., and Brown, H. (1969). Alcohol and biogenic amines. *In* "Biochemical and Clinical Aspects of Alcohol Metabolism" (V. M. Sardesai, ed.), pp. 95–204. Thomas, Springfield, Illinois.

22. Van Thiel, D. H., Gavaler, J., and Lester, R. (1974). Ethanol inhibition of vitamin A metabolism in the testes: Possible mechanism for sterility in alcoholics. *Science* **186**, 941–942.

23. Tygstrup, N., Winkler, K., and Lundquist, F. (1965). The mechanism of the fructose effect of the ethanol metabolism of the human liver. *J. Clin. Invest.* **44**, 817–830.

24. Winkler, K., Lundquist, F., and Tygstrup, N. (1969). The hepatic metabolism of ethanol in patients with cirrhosis of the liver. *Scand. J. Clin. Lab. Invest.* **23**, 59–69.

25. Larsen, J. A. (1969). Determination of the hepatic blood flow by means of ethanol. *Scand. J. Clin. Lab. Invest.* **11**, 340–347.

26. Plapp, B. V. (1975). Rate-limiting steps in ethanol metabolism and approaches to changing these rates biochemically. *Adv. Exp. Med. Biol.* **56**, 77–109.

27. Blair, A. H., and Vallee, B. L. (1966). Some catalytic properties of human liver alcohol dehydrogenase. *Biochemistry* **5**, 2026–2034.

28. Smith, M., Hopkinson, D. A., and Harris, H. (1971). Developmental changes and polymorphism in human alcohol dehydrogenase. *Ann. Hum. Genet.* **34**, 251–271.

29. Smith, M., Hopkinson, D. A., and Harris, M. (1971). Alcohol dehydrogenase isoenzymes in adult human stomach and liver: Evidence for activity of the ADH_3 locus. *Ann. Hum. Genet.* **35**, 243–253.

30. Li, T.-K., and Magnes, L. J. (1975). Identification of a distinctive molecular form of alcohol dehydrogenase in human liver with high activity. *Biochem. Biophys. Res. Commun.* **63,** 202–208.
31. Bosron, W. F., Li, T.-K., Lange, L. G., Dafeldecker, W. P., and Vallee, B. L. (1977). Isolation and characterization of an anodic form of human liver alcohol dehydrogenase. *Biochem. Biophys. Res. Commun.* **74,** 85–91.
32. Bosron, W. F., Li, T.-K., Dafeldecker, W. P., and Vallee, B. L. (1979). Human liver Π-alcohol dehydrogenase: Kinetic and molecular properties. *Biochemistry* **18,** 1101–1105.
33. Li, T.-K., Bosron, W. F., Dafeldecker, W. P., Lange, L. G., and Vallee, B. L. (1977). Isolation of Π-alcohol dehydrogenase of human liver: Is it a determinant of alcoholism? *Proc. Natl. Acad. Sci. U.S.A.* **74,** 4378–4381.
34. Adinolfi, A., Adinolfi, M., Hopkinson, D. A., and Harris, H. (1978). Immunological properties of the human alcohol dehydrogenase (ADH) isozymes. *J. Immunogenet.* **5,** 283–296.
35. Lange, L. G., and Vallee, B. L. (1976). Double-ternary complex affinity chromatography: Preparation of alcohol dehydrogenases. *Biochemistry* **15,** 4681–4686.
36. Theorell, H., Yonetani, T., and Sjöberg, B. (1969). On the effects of some heterocyclic compounds on the enzymatic activity of liver alcohol dehydrogenase. *Acta Chem. Scand.* **23,** 255–260.
37. Dahlbom, R., Tolf, B. R., Akeson, A., Lundquist, G., and Theorell, H. (1974). On the inhibitory power of some further pyrazole derivatives of horse liver alcohol dehydrogenase. *Biochem. Biophys. Res. Commun.* **57,** 549–553.
38. Li, T.-K., and Theorell, H. (1969). Human liver alcohol dehydrogenase: Inhibition by pyrazole and pyrazole analogs. *Acta Chem. Scand.* **23,** 892–902.
39. Pietruszko, R. (1975). Human liver alcohol dehydrogenase inhibition of methanol activity by pyrazole, 4-methylpyrazole, 4-hydroxypyrazole and 4-carboxypyrazole. *Biochem. Pharmacol.* **24,** 1603–1607.
40. Theorell, H., and Tonetani, T. (1963). Liver alcohol dehydrogenase-DPN-pyrazole complex: A model of a ternary intermediate in the enzyme reaction. *Biochem. Z.* **338,** 537–553.
41. Lowe, C. R., and Dean, P. D. G. (1974). "Affinity Chromatography." Wiley, New York.
42. Kaplan, N. O., Everse, J., Dixon, J. E., Stolzenbach, F. E., Lee, C.-Y., Lee, C.-L. T., Taylor, S. S., and Mosbach, K. (1974). Purification and separation of pyridine nucleotide-linked dehydrogenases by affinity chromatography techniques. *Proc. Natl. Acad. Sci. U.S.A.* **71,** 3450–3454.
43. Andersson, L., Jornvall, H., Akeson, A., and Mosbach, K. (1974). Separation of isozymes of horse liver alcohol dehydrogenase and purification of the enzyme by affinity chromatography on an immobilized AMP-analogue. *Biochim. Biophys. Acta* **364,** 1–8.
44. Adinolfi, A., and Hopkinson, D. A. (1978). Blue Sepharose chromatography of human alcohol dehydrogenase: Evidence for interlocus and interallelic differences in affinity characteristics. *Ann. Hum. Genet.* **41,** 399–407.
45. Mourad, N., and Woronick, C. L. (1967). Crystallization of human liver alcohol dehydrogenase. *Arch. Biochem. Biophys.* **121,** 431–439.
46. Schenker, T. M., Teeple, L. J., and von Wartburg, J.-P. (1971). Heterogeneity and polymorphism of human liver alcohol dehydrogenase. *Eur. J. Biochem.* **24,** 271–279.
47. Mezey, E., and Holt, P. R. (1969). Loss of the characteristic features of atypical human liver alcohol dehydrogenase during purification. *Life Sci.* **8,** 245–251.

48. Pietruszko, R., Theorell, H., and DeZalinski, C. (1972). Heterogeneity of alcohol dehydrogenase from human liver. *Arch. Biochem. Biophys.* **153**, 279–293.
49. Smith, H., Hopkinson, D. A., and Harris, H. (1973). Studies on the subunit structure and molecular size of the human alcohol dehydrogenase isozymes determined by the different loci, ADH_1, ADH_2 and ADH_3. *Ann. Hum. Genet.* **36**, 401–414.
50. Branden, C.-I., Jornvall, H., Eklund, H., and Furugren, B. (1975). Alcohol dehydrogenases. *In* "The Enzymes" (P. D. Boyer, ed.), 3rd ed., Vol. 11, pp. 103–190. Academic Press, New York.
51. Von Wartburg, J.-P., Bethune, J. L., and Vallee, B. L. (1964). Human liver alcohol dehydrogenase: Kinetic and physicochemical properties. *Biochemistry* **3**, 1775–1782.
52. Berger, D., Berger, M., and von Wartburg, J.-P. (1974). Structural studied of human liver alcohol dehydrogenase isoenzymes. *Eur. J. Biochem.* **50**, 215–225.
53. Lange, L. G., Sytkowski, A. J., and Vallee, B. L. (1976). Human liver alcohol dehydrogenase: Purification, composition and catalytic features. *Biochemistry* **15**, 4687–4693.
54. Jornvall, H. (1970). Horse liver alcohol dehydrogenase. The primary structure of the protein chaim of the ethanol-active isoenzyme. *Eur. J. Biochem.* **16**, 25–40.
55. Jornvall, H., and Pietruszko, R. (1972). Structural studies of alcohol dehydrogenase from human liver. *Eur. J. Biochem.* **25**, 283–290.
56. Theorell, H., and Bonnichsen, R. (1951). Studies on liver alcohol dehydrogenase. I. Equilibria and initial reaction velocities. *Acta Chem. Scand.* **5**, 1105–1126.
57. Theorell, H., Nygaard, A. P., and Bonnichsen, R. (1955). Studies on liver alcohol dehydrogenase. III. The influence of pH and some anions on the reaction velocity constants. *Acta Chem. Scand.* **9**, 1148–1165.
58. Cantarow, W., and Stollar, B. D. (1976). The use of bacterial luciferase and a liquid scintillation spectrometer to assay the enzymatic synthesis of NAD^+. *Anal. Biochem.* **71**, 333–340.
59. Pietruszko, R., DeZalinski, C., and Theorell, H. (1976). Dissociation and rate constants of some human liver alcohol dehydrogenase isoenzymes. *Acta Chem. Scand., Ser. B.* **30**, 595–599.
60. Von Wartburg, J.-P., Papenberg, J., and Abei, H. (1965). Atypical human alcohol dehydrogenase. *Can. J. Biochem.* **43**, 889–898.
61. Harada, S., Agarwal, D. P., and Goedde, H. W. (1978). Human liver alcohol dehydrogenase isoenzyme variations. *Hum. Genet.* **40**, 215–220.
62. Stamatoyannopoulos, G., Chen, S.-H., and Fukui, M. (1975). Liver alcohol dehydrogenase in Japanese: High population frequency of atypical form and its possible role in alcohol sensitivity. *Am. J. Hum. Genet.* **27**, 789–796.
63. Von Wartburg, J.-P., and Schurch, P. M. (1968). Atypical human liver alcohol dehydrogenase. *Ann. N.Y. Acad. Sci.* **151**, 936–946.
64. Dubied, A., von Wartburg, J.-P., Bohlken, D. P., and Plapp, B. V. (1977). Characterization and kinetics of native and chemically activated human liver alcohol dehydrogenases. *J. Biol. Chem.* **252**, 1464–1470.
65. Wratten, C. C., and Cleland, W. W. (1963). Product inhibition studies on yeast and liver alcohol dehydrogenases. *Biochemistry* **2**, 935–941.
66. Dworschack, R. T., and Plapp, B. V. (1977). Kinetics of native and activated isoenzymes of horse liver alcohol dehydrogenase. *Biochemistry* **16**, 111–116.
67. Cleland, W. W. (1967). The statistical analysis of enzyme kinetic data. *Adv. Enzymol.* **29**, 1–32.
68. Pietruszko, R., Crawford, K., and Lester, D. (1973). Comparison of substrate specificity of alcohol dehydrogenases from human liver, horse liver and yeast towards saturated and z-enoil alcohols and aldehydes. *Arch. Biochem. Biophys.* **159**, 50–60.

69. Pietruszko, R., Voigtlander, K., and Lester, D. (1978). Alcohol dehydrogenase from human and horse liver—Substrate specificity with diols. *Biochem. Pharmacol.* **27,** 1296–1297.
70. Hansch, C., Schaeffer, J., and Kerley, R. (1972). Alcohol dehydrogenase structure-activity relationships. *J. Biol. Chem.* **247,** 4703–4710.
71. Klinman, J. P. (1976). Isotope effects and structure reactivity correlations in the yeast alcohol dehydrogenase reaction. A study of the enzyme-catalyzed oxidation of aromatic alcohols. *Biochemistry* **15,** 2018–2026.
72. Klyosov, A. A., Lange, L. G., Sytkowski, A. J., and Vallee, B. L. (1977). Unusual character of substrate specificity in alcohol dehydrogenases of different origin. *Bioorg. Khim.* **3,** 1141–1143.
73. Vallee, B. L., Frey, W. F., Dafeldecker, W. P., Bosron, W. F., and Li, T.-K. (1979). Substrate specificity and characteristics of π-alcohol dehydrogenase and other human liver ADH isoenzymes. *In* "Cyclotrons to Cytochromes" (N. O. Kaplan and W. E. MacElroy, eds.). Academic Press, New York (in press).
74. Holzer, H. S., Schneider, S., and Lange, K. (1955). Funktion der leber-alcohol dehydrase. *Angew. Chem.* **67,** 276–277.
75. Mehlman, M. A., Tobin, R. B., and Mackerer, C. R. (1975). 1,3-Butanediol catabolism in the rat. *Fed. Proc., Fed. Am. Soc. Exp. Biol.* **34,** 2182–2185.
76. Bliss, A. F. (1951). The equilibrium between vitamin A alcohol and aldehyde in the presence of alcohol dehydrogenase. *Arch. Biochem. Biophys.* **31,** 197–204.
77. Koen, A. L., and Shaw, C. R. (1966). Retinol and alcohol dehydrogenases in retina and liver. *Biochim. Biophys. Acta* **128,** 48–54.
78. Reynier, M. (1969). Pyrazole inhibition and kinetic studies of ethanol and retinol oxidation catalyzed by rat liver alcohol dehydrogenase. *Acta Chem. Scand.* **23,** 1119–1129.
79. Cronholm, T., Larsen, C., Sjovall, J., Theorell, H., and Akeson, A. (1975). Steroid oxidoreductase activity of alcohol dehydrogenases from horse, rat and human liver. *Acta Chem. Scand., Ser. B.* **29,** 571–576.
80. Ryzewski, C. N., and Pietruszko, R. (1977). Horse liver alcohol dehydrogenase SS: Purification and characterization of the homogeneous isoenzyme. *Arch. Biochem. Biophys.* **183,** 73–82.
81. Smith, M., Hopkinson, D. A., and Harris, H. (1973). Studies on the properties of the human alcohol dehydrogenase isozymes determined by the different loci ADH_1, ADH_2, ADH_3. *Ann. Hum. Genet.* **37,** 49–67.
82. Winer, A. D., and Theorell, H. (1960). Dissociation constants of ternary complexes of fatty acids and fatty acid amides with horse liver alcohol dehydrogenase-coenzyme complexes. *Acta Chem. Scand.* **14,** 1729–1742.
83. Lumeng, L., Bosron, W. F., and Li, T.-K. (1979). Quantitative correlation of ethanol elimination rates *in vivo* with liver alcohol dehydrogenase activities in fed, fasted and food-restricted rats. *Biochem. Pharmacol.* **28,** 1547–1551.
84. Wilkinson, P. K. (1979). Pharmacokinetics of ethanol: a review. *Alcoholism: Clin. Exptl. Res.* **4,** 6–21.
85. Vesell, E. S., Page, J. G., and Passananti, G. T. (1971). Genetic and environmental factors affecting ethanol metabolism in man. *Clin. Pharmacol. Ther.* **12,** 192–201.

Chapter 11

Aldehyde Reductase

*JEAN-PIERRE VON WARTBURG and BENDICHT
WERMUTH*

I. INTRODUCTION

Aldehydes and ketones are widely distributed in nature and have a surprisingly broad spectrum of biological functions. In addition to alcohol dehydrogenase (Chapter 10, this volume), a family of aldehyde reductases seems to contribute to the reductive biotransformation of such carbonyl compounds. These enzymes catalyze the following reaction:

$$R_1-\overset{\overset{\displaystyle O}{\|}}{C}-R_2 + NADPH + H^+ \rightarrow R_1-\underset{\underset{\displaystyle H}{|}}{\overset{\overset{\displaystyle OH}{|}}{C}}-R_2 + NADP^+$$

The main characteristics of aldehyde reductases include a monomeric structure of low molecular weight, cytoplasmic localization, broad substrate specificity, and the dependence on NADPH as coenzyme. En-

249

ENZYMATIC BASIS OF DETOXICATION, VOL. I

zymatic activities in reducing both xenobiotic and naturally occurring carbonyl compounds have been reported for many tissues and a large variety of animals. Only few have been purified to homogeneity and fully characterized in terms of possible overlapping substrate specificities. Nevertheless, it becomes clear that identical enzymes have been classified under several numbers of the enzyme nomenclature and that these aldehyde reductases can reduce both xenobiotics and normal metabolites.

A number of NADPH-linked aldehyde and ketone reductases with molecular weights between 30,000 and 40,000 have to be considered: aldehyde reductase (EC 1.1.1.2, alcohol:NADP oxidoreductase), L-gulonate dehydrogenase or L-hexonate dehydrogenase or glucuronate reductase (EC 1.1.1.19, L-gulonate:NADP oxidoreductase), aldose reductase (EC 1.1.1.21, alditol:NADP oxidoreductase), mevaldate reductase (EC 1.1.1.33, mevaldate:NADP oxidoreductase), lactaldehyde reductase (EC 1.1.1.55, 1,2-propanediol:NADP oxidoreductase), glycerol dehydrogenase, daunorubicin reductase, succinic semialdehyde reductase, or γ-hydroxybutyrate dehydrogenase. When the substrate specificity, the sensitivity toward inhibitors, the behavior during purification and the physicochemical properties of these enzymes are compared, it becomes evident that aldehyde reductase, glucuronate reductase, mevaldate reductase, lactaldehyde reductase, and daunorubicin reductase are in fact identical and represent different activities of the same enzyme protein. On the other hand, they are distinct from aldose reductase and glycerol dehydrogenase although they seem to have similar structural properties and overlapping substrate specificities and may be closely related in evolution.

II. DISTRIBUTION OF ALDEHYDE REDUCTASES

NADPH-dependent aldehyde reductase activity is widely distributed in nature and has been found in a considerable number of mammalian tissues (see Table I).[1-30] In addition, enzymes with the characteristics of aldehyde reductase have also been observed in all nonmammalian animal classes, as well as in yeast. In most species, kidney contains the highest level of activity, followed by liver and brain. In all other tissues the enzyme is present in only very small amounts.

III. PURIFICATION AND ASSAY

In kidney and liver the enzyme protein constitutes about 0.5 and 0.05%, respectively, of the total protein of the tissue homogenate. Accordingly

TABLE I

Occurrence of NADPH-Linked Aldehyde Reductase

Species	Tissue	Reference
Human	Kidney	1
	Liver	1–3
	Brain	4, 5
	Erythrocytes	6, 6a
	Platelets, Leucocytes	6a
Monkey	Brain	5, 8
Ox	Brain	1, 9–12
Pig	Kidney	1, 13–16
	Brain	17, 18
Sheep	Heart	19
Rat	Kidney	1, 5, 20, 21
	Liver	1, 3, 5, 7, 10, 20–25
	Brain	1, 4, 5, 10, 20, 26, 27
	Lung	5, 10
	Heart	5, 10, 20, 21
	Testes	5, 10, 20
	Seminal vesicle	20
	Pancreas	28
	Muscle	20, 21
	Spleen	20
Mouse	Liver	3, 29
Rabbit	Kidney	1, 30
	Liver	1, 3
	Brain	1
	Heart	1
Chicken	Kidney, liver, heart	21
	Brain	5
Reptiles	Kidney, liver, heart, muscle	21
	Brain	5
Amphibia	Kidney, liver, heart, muscle	21
Fish	Kidney, liver, muscle	21
	Brain	5
Drosophila		21
Yeast		21

aldehyde reductase was first purified to homogeneity from pig kidney.[13] It has also been obtained in a pure state from rat, chicken and frog kidney[21]; from human,[2] rat,[7,22] and mouse liver[29]; and from whole organisms of fruit fly and yeast.[21]

A. Purification Procedure

Principally, all purification procedures are based on a similar combination of standard protein isolation methods. For human liver aldehyde reductase,[2] the tissue is homogenized with 50 mM Na$_2$HPO$_4$, dialyzed against 5 mM sodium phosphate at pH 7.0 and applied to a column of DEAE-cellulose from which it is eluted with a sodium phosphate gradient at pH 7.0. The preparation is concentrated by ammonium sulfate precipitation and subjected to gel filtration (Sephadex G-100). The eluate is chromatographed on Cibacron 3 GA-Sepharose (Blue-Sepharose) and removed therefrom with a high concentration of ammonium sulfate at pH 7.0. The resulting preparation is used to charge a column of hydroxyapatite to which it does not bind and is eluted as a homogeneous protein. The purified enzyme is stable in the cold for several weeks, but loses its activity upon freezing and thawing concomitantly with precipitation of the protein.

B. Assay

Aldehyde reductase activity is measured by recording the change of NADPH absorbance as a function of time at 340 nm. The standard reaction mixture contains 0.1 M sodium phosphate at pH 7.0, 0.08 mM NADPH, 10 mM D-glucuronate (or any other aldehyde), and enzyme solution. With crude enzyme solutions, blank reactions may occur with NADPH in the absence of any aldehyde substrate; consequently the rate of this reaction must be recorded before addition of exogenous aldehyde. Depending on the substrate used and on the tissue analyzed, alcohol dehydrogenase and aldose reductase may interfere. Interference of alcohol dehydrogenase may be excluded by the addition of pyrazole. In the presence of aldose reductase, the ratio of the activities measured with glucuronate and xylose as a substrate may be used to distinguish between aldehyde reductase and aldose reductase. Furthermore, the activity contributed by aldehyde reductase can be inhibited by the addition of 1.0 mM phenobarbital or diphenylhydantoin. Oxidation of alcohols is assayed at pH 9.0 in 0.1 M sodium pyrophosphate containing 0.1 mM NADP$^+$ and 10 mM L-gulonate (or any other alcohol).

IV. STRUCTURAL PROPERTIES

The molecular weights of native aldehyde reductase from most of the species and tissues listed in Table I were determined by gel filtration and ultracentrifugation, yielding values between 29,000 and 44,000.[2,4,7,11-13,17,19,21,25,29] The only exception to these low molecular weights was reported for an aldehyde reductase from monkey brain, which, however, did not show the marked inhibition by phenobarbital typical for the other aldehyde reductases.[8] Determination of the molecular weight of homogeneous enzymes under denaturing and reducing conditions, also gave values between 33,000 and 40,000,[2,7,13,21,29] indicating a monomeric structure for the enzyme.

The amino acid composition has been reported for the enzymes from human,[2] rat,[7] and mouse[29] liver and from pig kidney.[14] Comparison by the difference index method of Metzger et al.[31] indicates considerable homology.[2,14,32] By this criterion, more distant relationships are suggested between mammalian aldehyde reductase and the enzymes from fruit fly and yeast.[32]

Circular dichroism spectra indicate a degree of helicity between 7 and 18%,[2,14] agreeing well with the high proline content of aldehyde reductase. A molar extinction coefficient of 54,300 M^{-1} cm^{-1} has been determined at 280 nm for the human liver enzyme.[2] Isoelectric focusing of purified aldehyde reductases yielded isoelectric points of 5.3, 5.7, 5.8, and 6.3 for the enzymes from human liver,[2] pig kidney,[13] pig brain,[17] and rat liver,[7] respectively.

Several functional groups have been found by chemical modifications. Incubation of aldehyde reductase with SH-blocking agents decreases the enzymatic activity. Detailed studies have been carried out for the reaction of human liver[2] and pig kidney[14] aldehyde reductase with p-mercuribenzoate. They suggested the presence of three differently reactive classes of cysteine residues. Modification of the most reactive residues occurred instantaneously, but did not affect the enzyme activity. Addition of further equivalents of p-mercuribenzoate caused a time-dependent inactivation of the enzyme which could be prevented by the presence of the coenzyme[2] or reversed by the addition of β-mercaptoethanol.[14] The least reactive cysteine residues appeared to be inaccessible to the modifying agent in the native enzyme and become modified only after protein denaturation.

The presence of amino groups in or close to the coenzyme binding domain of the active site of aldehyde reductase from human liver was indicated by modification studies with amino group specific imidoesters.[33] Amidination with monofunctional n-alkane methylimidates increased the activity by 10–30%, whereas analogous bifunctional imidoesters caused a loss of activity of about 80%. Both effects were prevented in the presence of the nucleotide, but not of the aldehyde substrate.

V. CATALYTIC PROPERTIES

A. Substrate Specificity

In view of the ubiquitous distribution of aldehyde reductases one might expect them to perform fundamental roles in cellular metabolism. In spite of the wide substrate specificity which includes aromatic, aliphatic, and sugar aldehydes, as well as some ketones, no metabolic role can be attributed with certainty as yet. Special attention has been paid to the enzymes from brain[4,5,8,11,17,26,34,35,35a] because of their preference for the aldehydes derived from the neurotransmitters containing a β-hydroxyl group on the side chain, e.g., octopamine and noradrenaline. Similar specificities for the aldehydes from β-hydroxylated amines were observed with the enzymes from heart and liver.[19,36]

This preference for glycolaldehydes is also reflected by a greater turnover of lactaldehyde and glyceraldehyde in comparison with propionaldehyde. The role of biogenic aldehydes is not yet understood, but it has been hypothesized that they may bind to macromolecules in brain and thereby exert physiological action.[37] They have also been shown to inhibit both, the Na^+/K^+- and Mg^{2+}-dependent ATPases.[38] Of further interest in brain function is the reduction of succinic semialdehyde to γ-hydroxybutyric acid. The semialdehyde, formed by transamination from γ-aminobutyric acid,[39] was shown to be reduced by two distinct enzymes,[27,39-41] one of which is identical with aldehyde reductase.[27,40-43] γ-Hydroxybutyrate occurs physiologically in brain and cerebrospinal fluids[44-46] and has been used as an anaesthetic adjuvant.[47] Its effects on the central nervous system lead to behavioral depression and an induction of an epileptic-like stupor.[48,49]

Whereas aldose reductase catalyzes the reduction of many sugars,[50] aldehyde reductase is essentially inactive toward aldohexoses and hardly active toward pentoses, but uses uronic acids and their lactones, such as glucuronate and glucuronolactone.[12,13,20,29] Based on this specificity for glucuronic acid, aldehyde reductase has been purified and characterized independently in the course of the elucidation of the biosynthetic pathway of L-ascorbic acid and the glucuronic acid pathway, and was described as glucuronate reductase or L-hexonate dehydrogenase (EC 1.1.1.19).[16,20,51]

Numerous important carbonyl-containing drugs undergo carbonyl reduction as a major biotransformation step in mammals.* In crude extracts NADPH-dependent reductions have been described for substituted aromatic aldehydes and ketones, aliphatic ketones (e.g., oxisuran and naloxone), methyl ketones (e.g., daunorubicin and warfarin), a tetralone (e.g., bulonol), and unsaturated ketones.[52] Purified aldehyde reductase from rat liver showed substantial activity toward aromatic aldehydes, camphoroquinone, daunorubicin, and adriamycin but exhibited negligible activity toward oxisuran, retinal, bunolol, cyclohexanone, and some steroid hormones.[7,23] Daunorubicin reductase activity was also observed in human, rabbit and mouse liver.[3] An aldehyde reductase from rabbit kid-

* Ketone reductase activity discussed in Chap. 13, this volume.

ney reducing p-chlorobenzaldehyde, propiophenone and dihydro-19-nortestosterone showed a different stereospecificity and therefore represents a different enzyme.[30] Reductases with A or B specificity were also observed with drugs, such as oxisuran, metapyrone, naloxone, naltrexone, and daunorubicin.[53]

3-Ketosteroids with A/B cis configuration are reduced by an NADPH-dependent enzyme to the corresponding 3α alcohol, thus differing from alcohol dehydrogenase which forms the 3β-OH-steroids.[24] Chloral hydrate is a substrate for both aldehyde reductase and alcohol dehydrogenase.[54] Since essentially no alcohol dehydrogenase activity occurs in brain,[27] an aldehyde reductase-mediated reduction of chloral hydrate to its active form, trichloroethanol, in brain may be of pharmacological significance. Hence, several enzymes reducing aldehydes and ketones may occur at the same localization. At this time, it is difficult to estimate their number before they have been isolated and fully characterized in terms of possible overlapping substrate specificity.

With human liver aldehyde reductase the highest catalytic constants (k_{cat}) for model substrates were observed for benzaldehydes containing an electron withdrawing group in the para position to the aldehyde function, e.g., p-nitrobenzaldehyde ($k_{cat} = 9.5$ sec^{-1}) and p-carboxybenzaldehyde ($k_{cat} = 7.3$ sec^{-1}).[2] Among the physiological substrates only the biogenic aldehydes derived from octopamine and noradrenaline were better substrates ($k_{cat} = \sim 12$ sec^{-1}).[36] Other possible physiological substrates, such as D-glucuronate, d,l-glyceraldehyde, and daunorubicin, had k_{cat} values of 6.5, 3, and 2.5 sec^{-1}, respectively. Similar values were observed for the corresponding compounds with aldehyde reductases from other species and tissues.

The lowest K_m values with human aldehyde reductase were observed for the aromatic model substrates p-carboxybenzaldehyde ($K_m = 0.025$ mM) and p-nitrobenzaldehyde ($K_m = 0.15$ mM)[2] as well as for the biogenic aldehydes derived from octopamine ($K_m = 0.05$ mM) and noradrenaline ($K_m = 0.04$ mM).[36] Biogenic aldehydes lacking the hydroxyl group on the side chain, i.e., the aldehydes from tyramine, dopamine and serotonin, exhibited K_m values of 1.1, 0.6, and 0.2 mM, respectively.[36] A similar preference for the biogenic aldehydes derived from β-hydroxylated amines was also observed for the aldehyde reductases from brain and heart of other species, although the differences in K_m values of the aldehydes derived from noradrenaline and dopamine were less pronounced.[8,11,17,19,26] K_m values in the millimolar range are observed for most sugar aldehydes. With the human liver enzyme, d,l-glyceraldehyde, D-glucuronate and D-ribose showed K_m values of 1.5, 3, and more than 100 mM, respectively. A K_m value of about 8 M has been reported for D-glucose with the rat liver enzyme.[23] The xenobiotic compounds, daunorubicin, phenylglyoxal and camphoroquinone, gave K_m values of 0.12, 0.23, and 1.3 mM, respectively, with the human liver enzyme.[2] Rat liver aldehyde reductase exhibited K_m values of 0.08 and 0.22 mM for the two anthracycline antibiotics, daunorubicin and adriamycin, respectively.[23]

All aldehyde reductases exhibit low K_m values in the range of 1–13 μM for the nucleotide substrate, NADPH.

B. Kinetic Properties

Studies of product inhibition of bovine brain aldehyde reductase showed that NADP+ or p-nitrobenzyl alcohol were competitive inhibitors with NADPH or p-nitrobenzaldehyde, respectively, as the variable substrate, suggesting a random order of addition of substrates to the enzyme.[55] In contrast, analysis of initial velocity and product inhibition, carried out with pig kidney aldehyde reductase, indicate an ordered bi bi reaction mechanism in which NADPH binds first.[15,56] Similar conclusions may be drawn from the product inhibition data obtained with human liver aldehyde reductase.[57] However, calculation of individual rate constants for a simple ordered bi bi reaction mechanism gave negative values. Similarly, when the oxidation of alcohol was investigated, product inhibition showed that NADPH and aldehyde (D-glucuronate) were competitive inhibitors with NADP+ or alcohol (L-gulonate). This inhibition pattern is not in accordance with an ordered bi bi mechanism. Human liver aldehyde reductase exhibited deuterium isotope effects (V_H/V_D) between 1.1 and 1.9 with several aldehydes (D-glucuronate, xylose, p-nitrobenzaldehyde, propanal) at different substrate and coenzyme concentrations and over the pH range of 6 to 8.[57]

Aldehyde reductase from pig kidney[14] and human liver[6a] catalyzes the hydrogen transfer from the pro-4R position on the dihydronicotinamide ring to the re face of the carbonyl carbon atom of the substrate. Transfer of the pro-4R hydrogen was also observed with aldehyde reductase from rat and rabbit liver;[53] rat, rabbit, and chicken kidney; fruit fly; and yeast.[21] This stereospecificity can be used to distinguish aldehyde reductase from other similar enzymes.[21,53]

The rate of reduction of most other carbonyl substrates in the presence of NADPH is maximal between pH 6.2 and 7.0, depending on the source of enzyme. Little influence appears to be exerted on the pH optimum by different buffers although certain salts, e.g., citrate or nitrate, greatly decrease enzymatic activity. A unique pH optimum of 9 was observed for the reduction of the ketone substrate daunorubicin by rat[7,23] and human liver aldehyde reductase. Oxidation of alcohols occurs at a maximum rate between pH 9 and 10. At pH 7 to 7.4, i.e., under physiological conditions, the rate of alcohol oxidation was less than 2% of the rate of aldehyde reduction.[2,4,11,13,17,19,23]

C. Inhibition

Anticonvulsant drugs, such as barbiturates, open-chained analogues of barbiturates, glutethimide, succinimides, hydantoins, and oxazolidinedione compounds, are all potent inhibitors of aldehyde reductase *in vitro;* they display noncompetitive patterns with either coenzyme or aldehydes, as variable substrates.[11,20,23,25,55,58-60] For inhibition, a minimal structure with the -CO-NH-CO- grouping and a lipophilic substitution are required[60]; inhibitor constants (K_i values) range from 1 to 3 $\times$ 10^{-4} M. Although it seems uncertain whether this *in vitro* inhibition is associated with the pharmacological action of these anticonvulsive drugs *in vivo,* inhibition may be used as a characteristic feature to distinguish aldehyde reductases from other similar enzymes. Aldehyde reductases are not inhibited by pyrazole and its derivatives (typical inhibitors of alcohol dehydrogenase), disulfiram, or diethyldithiocarbamate (typical for aldehyde dehydrogenase), or by metal chelating agents. On the other hand, tetramethyleneglutaric acid and flavonoids, e.g., quercetin, quercitrin, that have been used as strong inhibitors of aldose reductase, are also potent inhibitors of human brain aldehyde reductase.[41,46] Additional inhibitory compounds comprise phenothiazine derivatives, e.g., chlorpromazine[4,8,9,11,55,60] and diuretics.[19] Interestingly, biogenic acids such as 5-hydroxyindolacetic acid, 4-hydroxyphenylacetic acid or homovanillic acid, representing the products of the oxidative pathway for biogenic aldehydes, can inhibit the reduction of aldehydes by aldehyde reductase.[4,22,34,62]

REFERENCES

1. Bosron, W. F., and Prairie, R. L. (1973). Reduced triphosphopyridine-linked aldehyde reductase. II. Species and tissue distribution. *Arch. Biochem. Biophys.* **154,** 166–172.
2. Wermuth, B., Münch, J. D. B., and von Wartburg, J. P. (1977). Purification and properties of NADPH-dependent aldehyde reductase from human liver. *J. Biol. Chem.* **252,** 3821–3828.
3. Ahmed, N. K., Felsted, R. L., and Bachur, N. R. (1978). Heterogeneity of anthracycline antibiotic carbonyl reductases in mammalian livers. *Biochem. Pharmacol.* **27,** 2713–2719.
4. Ris, M. M., and von Wartburg, J. P. (1973). Heterogeneity of NADPH-dependent aldehyde reductase from human and rat brain. *Eur. J. Biochem.* **37,** 69–77.

5. Erwin, V. G. (1974). Oxidative-reductive pathways for metabolism of biogenic aldehydes. *Biochem. Pharmacol., Suppl.* **1**, 110–115.
6. Beutler, E., and Guinto, E. (1974). The reduction of glyceraldehyde by human erythrocytes. L-hexonate dehydrogenase activity. *J. Clin. Invest.* **53**, 1258–1264.
6a. Wermuth, B., Münch, J. D. B., and von Wartburg, J. P. (1979). *Experientia* **35**, 1288–1289.
7. Felsted, R. L., Gee, M., and Bachur, N. R. (1974). Rat liver daunorubicin reductase. An aldo-keto reductase. *J. Biol. Chem.* **249**, 3672–3679.
8. Bronaugh, R. L., and Erwin, V. G. (1973). Partial purification and characterization of NADPH-linked aldehyde reductase from monkey brain. *J. Neurochem.* **21**, 809–815.
9. Erwin, V. G., Heston, W. D. W., and Tabakoff, B. (1972). Purification and characterization of an NADH-linked aldehyde reductase from bovine brain. *J. Neurochem.* **19**, 2269–2278.
10. Erwin, V. G., and Deitrich, R. A. (1972). Heterogeneity of alcohol-dehydrogenase enzymes in various tissues. *Biochem. Pharmacol.* **21**, 2915–2924.
11. Tabakoff, B., and Erwin, V. G. (1970). Purification and characterization of a reduced nicotinamide adenine dinucleotide phosphate-linked aldehyde reductase from brain. *J. Biol. Chem.* **245**, 3263–3268.
12. Moonsammy, G. I., and Stewart, M. A. (1967). Purification and properties of brain aldose reductase and L-hexonate dehydrogenase. *J. Neurochem.* **14**, 1187–1193.
13. Bosron, W. F., and Prairie, R. L. (1972). Triphosphopyridine nucleotide-linked aldehyde reductase. *J. Biol. Chem.* **247**, 4480–4485.
14. Flynn, T. G., Shires, J., and Walton, D. J. (1975). Properties of the nicotinamide adenine dinucleotide phosphate-dependent aldehyde reductase from pig kidney. *J. Biol. Chem.* **250**, 2933–2940.
15. Davidson, W. S., and Flynn, T. G. (1979). Kinetics and mechanism of action of aldehyde reductase from pig kidney. *Biochem. J.* **177**, 595–601.
16. York, J. L., Grollman, A. P., and Bublitz, C. (1961). TPN-L-gulonate dehydrogenase. *Biochim. Biophys. Acta* **47**, 298–306.
17. Turner, A. J., and Tipton, K. F. (1972). The purification and properties of an NADPH-linked aldehyde reductase from pig brain. *Eur. J. Biochem.* **30**, 361–368.
18. Turner, A. J., and Tipton, K. F. (1972). The characterization of two reduced nicotinamide-adenine dinucleotide phosphate-linked aldehyde reductase from pig brain. *Biochem. J.* **130**, 765–772.
19. Smolen, A., and Anderson, A. D. (1976). Partial purification and characterization of a reduced nicotinamide adenine dinucleotide phosphate-linked aldehyde reductase from heart. *Biochem. Pharmacol.* **25**, 317–323.
20. Mano, Y., Suzuki, K., Yamada, K., and Shimazono, N. (1961). Enzymic studies of TPN L-hexonate dehydrogenase from rat liver. *J. Biochem. (Tokyo)* **49**, 618–634.
21. Davidson, W. S., Walton, D. J., and Flynn, T. G. (1978). A comparative study of the tissue and species distribution of NADPH-dependent aldehyde reductase. *Comp. Biochem. Physiol. B* **60**, 309–315.
22. Turner, A. J., and Hick, P. E. (1976). Metabolism of daunorubicin by a barbiturate-sensitive aldehyde reductase from rat liver. *Biochem. J.* **159**, 819–822.
23. Felsted, R. L., Richter, D. R., and Bachur, N. R. (1977). Rat liver aldehyde reductase. *Biochem. Pharmacol.* **26**, 1117–1124.
24. Pietruszko, R., and Chen, F. F. (1976). Aldehyde reductase from rat liver is a 3α-hydroxysteroid dehydrogenase. *Biochem. Pharmacol.* **25**, 2721–2725.
25. Beedle, A. S., Rees, H. R., and Goodwin, T. W. (1974). Some properties and a suggested reclassification of mevaldate reductase. *Biochem. J.* **139**, 205–209.

26. Tabakoff, B., Anderson, R., and Alivisatos, G. A. (1973). Enzymatic reduction of biogenic aldehydes in brain. *Mol. Pharmacol.* **9**, 428–437.
27. Tabakoff, B., and von Wartburg, J. P. (1975). Separation of aldehyde reductases and alcohol dehydrogenase from brain by affinity chromatography: Metabolism of succinic semialdehyde and ethanol. *Biochem. Biophys. Res. Commun.* **63**, 957–966.
28. Gabbay, K. H., and Cathcart, E. S. (1974). Purification and immunologic identification of aldose reductases. *Diabetes* **23**, 460–468.
29. Tulsiani, D. R. P., and Touster, O. (1977). Resolution and partial characterization of two aldehyde reductases of mammalian liver. *J. Biol. Chem.* **252**, 2545–2550.
30. Culp, H. W., and McMahon, R. E. (1968). Reductase for aromatic aldehydes and ketones. *J. Biol. Chem.* **243**, 848–852.
31. Metzger, H., Shapiro, M. B., Mosimann, J. E., and Vinton, J. E. (1968). Assessment of compositional relatedness between proteins. *Nature (London)* **219**, 1166–1168.
32. Davidson, W. S., and Flynn, T. G. (1979). Compositional relatedness of aldehyde reductase from several species. *Fed. Proc., Fed. Am. Soc. Exp. Biol.* **38**, 563 (1770).
33. Wermuth, B., Münch, J. D. B., Hajdu, J., and von Wartburg, J. P. (1979). Amidination of amino groups of aldehyde reductase from human liver. *Biochim. Biophys. Acta* **566**, 237–244.
34. Anderson, R. A., Meyerson, L. R., and Tabakoff, B. (1976). Characteristics of enzymes forming 3-methoxy-4-hydroxy-phenylethyleneglycol (MOPEG) in brain. *Neurochem. Res.* **1**, 525–540.
35. Reyes, E., and Erwin, V. G. (1977). Distribution and properties of NADPH-linked aldehyde reductases from rat brain synaptosomes. *Neurochem. Res.* **2**, 87–97.
35a. Huffman, D. H., and Bachur, N. R. (1972). *Cancer Res.* **32**, 600–605.
36. Wermuth, B., and Münch, J. D. B. (1979). Reduction of biogenic aldehydes by aldehyde reductase and alcohol dehydrogenase from human liver. *Biochem. Pharmacol.* **28**, 1431–1433.
37. Tabakoff, B., Ungar, F., and Alivisatos, S. G. A. (1972). Aldehyde derivatives of indoleamines and the enhancement of their binding onto brain macromolecules by pentobarbital and acetaldehyde. *Nature (London), New Biol.* **238**, 126–128.
38. Tabakoff, B. (1974). Inhibition of sodium–potassium- and magnesium-activated ATP-ases by acetaldehyde and biogenic aldehydes. *Res. Commun. Clin. Pathol. Pharmacol.* **7**, 621–624.
39. Gold, B. I., and Roth, R. H. (1977). Kinetics of *in vivo* conversion of γ-(³H)aminobutyric acid to γ-(³H)hydroxybutyric acid by rat brain. *J. Neurochem.* **28**, 1069–1073.
40. Cash, C., Maítre, M., and Mandel, P. (1978). Purification de deux semi-aldéhydes succinique réductases de cerveau humain. *C. R. Hebd. Seances Acad. Sci. Ser. D* **286**, 1829–1832.
41. Hoffman, P. L., Wermuth, B., and von Wartburg, J. P. (1980). Human brain aldehyde reductases: Relationship to SSA reductase and aldose reductase. *J. Neurochem.*, in press.
42. Anderson, R. A., Ritzmann, R. F., and Tabakoff, B. (1977). Formation of γ-hydroxybutyrate in brain. *J. Neurochem.* **28**, 633–639.
43. Kaufman, E. E., Nelson, T., Goochee, C., and Sokoloff, L. (1979). Purification and characterization of an NADP⁺-linked alcohol oxido-reductase which catalyzes the interconversion of γ-hydroxybutyrate and succinic semialdehyde. *J. Neurochem.* **32**, 699–712.
44. Roth, R. H., and Giarman, N. J. (1970). Natural occurrence of γ-hydroxybutyrate in mammalian brain. *Biochem. Pharmacol.* **19**, 1087–1093.
45. Doherty, J. D., Hattox, S. E., Snead, O. C., and Roth, R. H. (1978). Identification of

endogenous γ-hydroxybutyrate in human and bovine brain in its regional distribution in human, guinea pig and rhesus monkey brain. *J. Pharmacol. Exp. Ther.* **207**, 130–139.

46. Tabakoff, B., and Radulovacki, M. (1976). γ-Hydroxybutyrate in CSF during sleep and wakefulness. *Res. Commun. Chem. Pathol. Pharmacol.* **14**, No. 3, 587–590.
47. Helrich, M., McAslan, R. C., and Skolnik, S. (1964). Correlation of blood levels of 4-hydroxybutyrate with states of consciousness. *Anesthesiology* **23**, 771–775.
48. Snead, O. C., Yu, R. K., and Huttenlocher, P. R. (1976). Gamma-hydroxybutyrate, correlation of serum and cerebrospinal fluid levels with electroencephalographic and behavioral effects. *Neurology* **26**, 51–56.
49. Snead, O. C. (1978). Gamma-hydroxybutyrate in the monkey. I. Electroencephalographic, behavioral and pharmacokinetic studies. *Neurology* **28**, 636–648.
50. Gabbay, K. H., and Kinoshita, J. H. (1975). Aldose reductases from mammalian tissues. *In* "Methods in Enzymology" (W. A. Wood, ed.), Vol. 41, Part B, pp. 159–165. Academic Press, New York.
51. Burns, J. J. (1967). Ascorbic acid. *In* "Metabolic Pathways" (D. M. Greenberg, ed.), 3rd ed., Vol. 1, pp. 394–411. Academic Press, New York.
52. Bachur, N. R. (1976). Cytoplasmic aldo-keto-reductases: A class of drug metabolizing enzymes. *Science* **193**, 595–597.
53. Felsted, R. L., Richter, D. R., and Bachur, N. R. (1978). Stereospecificity of hydride transfer by aldehyde and ketone reductases. *Fed. Proc., Fed. Am. Soc. Exp. Biol.* **37**, 1692 (2332).
54. Tabakoff, B., Vugrincic, C., Anderson, R., and Alivisatos, S. G. A. (1973). The reduction of chloral hydrate to trichloroethanol in brain extracts. *Biochem. Pharmacol.* **23**, 455–460.
55. Bronaugh, R. L., and Erwin G. (1972). Further characterization of a reduced nicotinamide-adenine dinucleotide phosphate-dependent aldehyde reductase from bovine brain. *Biochem. Pharmacol.* **21**, 1457–1464.
56. Prairie, R. L., and Lai, D. K. (1973). Kinetic properties of TPNH-linked aldehyde reductase. *Fed. Proc., Fed. Am. Soc. Exp. Biol.* **32**, 606 (2203).
57. Wermuth, B., and von Wartburg, J. P. (1979). Kinetic studies on NADPH-linked aldehyde reductase from human liver. *Proceedings of Int. Symp. Alcohol Aldehyde Metab. Syst. 3rd, 1979*, Academic Press, in press.
58. Erwin, V. G., and Deitrich, R. A. (1973). Inhibition of bovine brain aldehyde reductase by anticonvulsant compounds *in vitro*. *Biochem. Pharmacol.* **22**, 2615–2624.
59. Erwin, V. G., Tabakoff, B., and Bronaugh, R. L. (1971). Inhibition of a reduced nicotinamide adenine dinucleotide phosphate-linked aldehyde reductase from bovine brain by barbiturates. *Mol. Pharmacol.* **7**, 169–176.
60. Ris, M. M., Deitrich, R. A., and von Wartburg, J. P. (1975). Inhibition of aldehyde reductase isoenzymes in human rat brain. *Biochem. Pharmacol.* **24**, 1865–1869.
61. Tabakoff, B. (1977). Brain aldehyde dehydrogenases and reductases. *In* "Structure and Function of Monoamine Enzymes" (E. Usdin, N. Weiner, and M. B. H. Youdim, eds.), pp. 629–649. Dekker, New York.
62. Turner, A. J., and Hick, P. E. (1975). Inhibition of aldehyde reductase by acidic metabolites of the biogenic amines. *Biochem. Pharmacol.* **24**, 1731–1733.

Chapter 12

Aldehyde Oxidizing Enzymes

HENRY WEINER

I. INTRODUCTION

Aldehydes are produced as intermediates in many biological reactions. For example, the monoamine oxidase catalyzed oxidation of biogenic amines, the metabolism of plasmalogen in brain, and the peroxidation of polyunsaturated lipids, all produce aldehydes. These and other aldehydes are reactive intermediates which are not normally found in any appre-

ENZYMATIC BASIS OF DETOXICATION, VOL. I

ciable concentration in the body. Although one can be exposed to or actually ingest aldehydes directly, the bulk of the aldehydes encountered physiologically are derived from catabolism of other compounds. The most commonly found aldehyde, at least in humans, is acetaldehyde derived from the metabolism of ethanol which is ingested voluntarily by a large portion of the population. Formaldehyde and glycolaldehyde are two others encountered by humans after ingestion of methanol and ethylene glycol, respectively. One can also be exposed to formaldehyde directly as it is used in insulating materials.[1]

In general, aldehydes are toxic substances. In addition to being oxidized to acids, the subject of this chapter, or being reduced to alcohols, the subject of Chapter 10, this volume, they combine with cellular material.[2,3] The physiological significance of their binding to proteins and amino acids is not fully understood. Thus, we have another example of detoxication of a foreign compound producing an intermediate that may be more toxic to the host than the parent compound.

Although about 100 organic acids have been detected in the urine of humans,[4] many are not produced directly from the oxidation of an aldehyde; lactic and hydroxybutyric acids are not derived from aldehyde precursors. Aldehydes that are oxidized to acids are in general formed from a detoxication reaction or an inactivation reaction. An example of the latter is the catabolism of biogenic amines to acids.

There are two major aldehyde oxidizing systems in the mammalian cell: the NAD-dependent aldehyde dehydrogenases and the flavin requiring aldehyde oxidase (see also Chapter 14). Both systems are widely distributed in the tissues. In fact, aldehyde dehydrogenase is not only found in every organ[5] but is found in different subcellular organelles.[6] A separate type of aldehyde dehydrogenase exists which appears to have the sole function of oxidizing formaldehyde.[7-9]

The endogenous substrates of aldehyde dehydrogenase and aldehyde oxidase are not known. Conceivably, some of the toxic effects produced by the metabolism of foreign compounds whose metabolic degradation results in aldehydes could be that their presence interferes with endogenous metabolism. For example, biogenic amines are oxidized to acids after being converted to aldehydes; foreign aldehydes could inhibit this pathway causing an accumulation of the biogenic aldehyde.

By simple examination of structure it is not possible to determine if a compound will be converted to an aldehyde. A large number of pesticides are metabolized to yield an acid as the ultimate excretion product (Table I).[10] (It is possible that some of these compounds are metabolized only by a plant or bacteria enzyme and not in a mammalian system.) Simple

TABLE I

Pesticides Which Are Metabolized to Acids[a]

Aldicarb (Temik)
Benthiocarb
BPMSMC (Chevron-11775)[b]
Chloral hydrate
Dinoseb
Ethylene dibromide
Fluenethyl
Furamethrin
Phenothrin
Picloram
Prolan
Pronamide
Resmethrin
Robenidine hydrochloride

[a] From Menzie.[10]

[b] The aldehyde intermediate is reduced to an alcohol.

aldehydes, such as acetaldehyde and benzaldehyde, can contribute to one's discomfort; others can be lethal. Some, e.g., glicidaldehyde,[11] macrosamine,[12] 3,4,5-trimethoxycinnamaldehyde,[13] and an aldehyde metabolite of safrole,[13] have been identified as mutagens or carcinogens. That the organism has an effective way of ridding itself of these potentially toxic and chemically reactive compounds is literally vital.

Exposure to drugs and herbicides can cause the increased synthesis of aldehyde dehydrogenase, although the mechanism or physiological role for induction is not known. In fact, ethanol is reported to induce aldehyde dehydrogenase synthesis in rat liver;[14-19] with this compound one can rationalize a physiological significance for the induction of an enzyme that would ultimately contribute to the metabolism of the inducer. However, phenobarbital[20-27] and 2,3,7,8-tetrachlorodibenzo-p-dioxin[28-30] are potent inducers of different isozymes of rat liver aldehyde dehydrogenase; the enzyme is not involved in the metabolism of either compound. Tumor cells have an altered aldehyde dehydrogenase isozyme pattern,[30-34] but the cause of this phenomena also is not understood.

In 1963, a comprehensive review of the enzymology of aldehyde dehydrogenase was published.[35] A more recent review[6] primarily discusses the enzyme with respect to its role in acetaldehyde metabolism. Aldehyde oxidase was reviewed in 1973;[36] a detailed discussion of this enzyme is presented in Chapter 14, this volume.

II. ALDEHYDE DEHYDROGENASES

A. General

There are two major types of aldehyde dehydrogenase in mammals: one oxidizes formaldehyde that is complexed with glutathione[7-9] and is referred to as formaldehyde dehydrogenase; the other uses the uncomplexed aldehyde as a substrate with a very broad substrate specificity and is referred to as aldehyde dehydrogenase. The glutathione-bound substrate appears to be specifically involved in the oxidation of formaldehyde in mammals;[7-9] no evidence could be found to show that the adduct with other aldehydes is important for their *in vivo* oxidation. Each dehydrogenase utilizes NAD as a cofactor, although some isozymes of aldehyde dehydrogenase have been reported to use NADP.[37] Therefore, by simply adding aldehyde to an homogenate and following the appearance of NAD(P)H or the formation of acid, it is not possible to determine which enzyme is operative. Since so many isozymes of aldehyde dehydrogenase exist in any one cell, it is impossible to identify unequivocally which isozyme or within which subcellular organelle, the metabolism of a particular aldehyde is taking place. Although the topic will not be discussed, it is possible that intestinal bacteria also contribute to the oxidation of ingested aldehyde.

Aldehyde dehydrogenase is found in every organ, but the best characterized enzymes have been isolated from liver and brain. In 1972, the horse liver enzyme was purified to homogeneity.[38] Prior to that, the enzyme was investigated from the liver of many species including man,[39,40] rabbit,[41-43] cow,[44,45] mouse,[46] and rat[5,47,48] and from the brain of the cow[49] and pig.[50]

Isozymes of rat liver aldehyde dehydrogenase are found in cytosol, mitochondria, and microsomes.[51] It is not known if the isozymes are products of different genes, but the levels of some are differentially altered by the administration of drugs.[20,28] The mitochondrial enzymes are in all probability products of separate genes, since they are differentially inhibited by disulfiram[52] and have different specificities toward NAD and NADP.[37] Although two forms of aldehyde dehydrogenase have been described from bovine brain[49] and at least two from rat heart have been shown to exist,[53] the isozyme pattern in nonhepatic cells has not been well described.

No evidence could be found to suggest that there are isozymes of formaldehyde dehydrogenase. This enzyme seems to be localized in the cytosol[9] as is the esterase that hydrolyzes formylglutathione, the product of the dehydrogenase reaction.[54]

The generalization to be made is that acetaldehyde is primarily oxidized in the mitochondria.[55-58] A similar conclusion has been reached for the oxidation of 3,4-dihydroxyphenylacetaldehyde, derived from dopamine,[59] as well as for many other aldehydes.[60] Inasmuch as aldehydes derived from drugs, pesticides, and other foreign compounds are difficult to obtain, the subcellular location of their oxidation has not been investigated in depth. The use of inhibitors could help delineate this question. The mitochondrial enzyme is linked to the electron transport system (ETS). Thus, an inhibitor of the ETS such as rotenone will drastically inhibit the oxidation of aldehyde in mitochondria.[51,61] Disulfiram,[62-64] cyanamide,[64-66] and coprine[67-69] are *in vivo* inhibitors of aldehyde dehydrogenase although disulfiram (Antabuse), a drug given to deter alcoholics from drinking, does not inhibit all isozymes of the dehydrogenase to the same degree.[52] It is conceivable, therefore, that an investigator could selectively inhibit isozymes of aldehyde dehydrogenase and then measure the metabolism of exogenously added or endogenously formed aldehydes in order to localize the site of oxidation.

B. General Mechanism of Action of Aldehyde Dehydrogenase

The oxidation of an aldehyde to an acid requires the removal of two electrons and a hydrogen from carbon followed by the addition of water. In most biological reactions in which two electrons and a proton are transferred, hydrogen is in the form of a hydride ion $(H:)^-$. The acceptor (A_{ox}) of the hydride is either flavine or NAD. A simple formulation describing this process is shown in Eq. (1). The acylonium ion, however, is not a typical intermediate in a biological or chemical reaction.

$$R-\overset{\overset{O}{\parallel}}{C}-H \; + \; A_{ox} \longrightarrow R\overset{\overset{O}{\parallel}}{C}{}^+ \; + \; AH_{red} \qquad (1)$$

It was originally demonstrated with glyceraldehyde-3-phosphate dehydrogenase that a thiohemiacetal intermediate was formed between substrate and enzyme which was subsequently oxidized to the level of an acid.[70] The ester could be hydrolyzed by the enzyme itself to produce a

$$ESH \; + \; R\overset{\overset{O}{\parallel}}{C}H \longrightarrow ES-\underset{\underset{H}{\mid}}{\overset{\overset{OH}{\mid}}{C}}-R \xrightarrow{NAD} ES-\overset{\overset{O}{\parallel}}{C}-R \; + \; NADH \qquad (2)$$

free acid. In the case of glyceraldehyde-3-phosphate dehydrogenase, phosphate could attack the thioester to form an acid phosphate product, 1,3-diphosphoglycerate.

The same general mechanism was proposed for aldehyde dehydrogenase,[35] the major difference being that phosphate is not involved. Thus the substrate, for the actual oxidation step is the enzyme-bound hemiacetal.

$$\begin{array}{c} OH \\ | \\ (E-N-C-R) \\ | \\ H \end{array}$$

With the nonspecific aldehyde dehydrogenase, the nucleophile (N) is an integral part of the enzyme (E). Although never proved, it has been suggested that the nucleophile is the SH group of a protein cysteine moiety.[35,71,72]

Formaldehyde dehydrogenase *requires* glutathione for activity, since the substrate for oxidation is not an enzyme-bound hemiacetal but the formaldehyde glutathione adduct[7-9] where G is glutathione. Thus, the

$$\begin{array}{c} OH \\ | \\ H-C-SG \\ | \\ H \end{array}$$

product of the enzyme reaction is not formic acid but the formylthioester.

$$\begin{array}{c} O \\ \| \\ H-C-SG \end{array}$$

Unlike the nonspecific aldehyde dehydrogenase which possesses esterase activity,[73-77] i.e., can hydrolyze the acyl–enzyme complex, a separate enzyme must hydrolyze the formyl ester to formic acid.[54] An animal could have a deficiency in the level of either formaldehyde dehydrogenase or in the esterase; obviously either deficiency would in turn affect formaldehyde oxidation by the animal.

III. NONSPECIFIC ALDEHYDE DEHYDROGENASE

A. Isolation and Purification of Aldehyde Dehydrogenase

In 1972, the purification to homogeneity of horse liver aldehyde dehydrogenase was reported[38] and since then a number of other aldehyde dehydrogenases have been purified to homogeneity. These include human,[74,78,79] sheep,[80] horse,[38,71,81] cow,[82] rat,[32,37,83] and rabbit.[76] The purification of the non-membrane-bound enzyme appears to be relatively straightforward, and essentially similar techniques have been employed.

Most investigators using livers start with an ammonium sulfate precipitation step which is followed by chromatography on carboxymethyl cellulose (CM-cellulose) and DEAE-cellulose. Chromatography on DEAE-cellulose appears to be the major step in the purification of the enzyme and the separation of isozymes. Isoelectric focusing has been employed for the final purification step by some.[32,37,38,82] Affinity chromatography on either NAD- or AMP-Sepharose has also been used[74,76,79,83]; the enzyme can be eluted from the affinity column with NAD or by changing pH. The rabbit liver enzyme was purified from an AMP-agarose column with NAD as the eluent.[76] The human enzyme, however, was eluted from AMP-Sepharose by simply raising the pH from 5 to 8.[79] Alternatively, the human enzyme may be eluted with NAD.[74]

A mitochondrial enzyme was purified from the liver of both sheep[80] and rat.[37] In each case the mitochondria from fresh liver were isolated by differential centrifugation. To purify the sheep enzyme, mitochondria were first lysed by freeze-thaw techniques, followed by salt precipitation, DEAE-cellulose and CM-cellulose chromatography. This process yielded a homogeneous enzyme.

Two enzyme activities were purified from rat liver mitochondria.[37] One enzyme form was membrane bound, while the other was located in the matrix space. The mitochondria were isolated, lysed, and fractionated by conventional methods with isoelectric focusing as the final step.

Enzyme from rat liver microsomes has also been purified to homogeneity.[83] The isolated microsomes were suspended in a buffer containing 20% glycerol and 0.5% sodium cholate, and the solubilized proteins subjected to chromatography on aminohexyl-Sepharose, followed by affinity chromatography on AMP-Sepharose. For the latter, 0.2% Triton X-100 and 20% glycerol were added to the buffer; active enzyme was eluted with NAD.

Both isoelectric focusing[38] and affinity chromatography have been used (unpublished observations) as the final step in purifying the horse liver enzyme. Homogeneity has not been obtained without one or the other of these steps, although other investigators appear to be able to purify liver enzymes without them.[71,80] When homogeneity is required, an affinity chromatography step should be included. However, if partially purified enzymes are adequate, simple ammonium sulfate fractionation followed by chromatography on DEAE-cellulose should yield a reasonably pure product. Because the molecular weight is greater than 200,000, a gel filtration step on Sephadex G200 or its equivalent is recommended to provide additional purification.

B. Assay for Aldehyde Dehydrogenase Activity

Aldehyde dehydrogenase can be assayed either spectrophotometrically or fluorometrically by following the formation of NADH since the reaction produces 1 mole of NADH for every mole of aldehyde oxidized.

Insoluble aldehydes may be added with Triton X-100 or in an organic solvent such as methanol.[84]

An alternative method for assaying enzyme activity is to measure acid formation. If a radioactive aldehyde or aldehyde precursor is employed in an incubation, then radioactivity in the isolated acid product would be a measure of aldehyde dehydrogenase activity; this technique has been used to measure the *in vivo* and *in vitro* metabolism of 3,4-dihydroxyphenylacetaldehyde derived from dopamine.[84]

C. Properties of Purified Aldehyde Dehydrogenase

Aldehyde dehydrogenases are generally stable enzymes. Some investigators store them in 20% glycerol, while others have reported that the enzyme may be stored in aqueous solution at −20° for many weeks. It is also possible to simply keep the enzyme in the refrigerator in a suitable buffer. In general, thiols, e.g., dithiothreitol or mercaptoethanol, have to be added to maintain full activity. Most investigators add a reducing agent to the isolation buffers as well as to all chromatography buffers.

In all cases, the activity of the enzyme increases with increasing pH. Activity at pH 9 is 5 to 10 times greater than that at pH 7[38] and is not substantially affected by buffer composition. Phosphate at low pH and pyrophosphate at high pH are the buffers most frequently used, whereas Tris and other nitrogen-containing buffers should not be employed since they may complex with the substrate and could inhibit the enzyme. In a similar manner, compounds such as cyanide which might be included to inhibit other enzyme systems in a crude assay mixture should be avoided because they also will complex with the substrate.

Aldehyde dehydrogenase appears to be a tetrameric enzyme with a molecular weight of 200,000 to 260,000 (Table II). A notable exception to this generalization is the enzyme isolated from rat mitochondria.[37] One form is reported to have a molecular weight of 320,000, while another form is 67,000. The enzyme isolated from rat liver microsome was shown to have a subunit molecular weight of 51,000 but in aqueous solution appeared to aggregate to a very high molecular weight species.[83] An enzyme isolated from a liver tumor appears to be a dimer.[32] The subunit molecular weight is, therefore, between 50,000 and 60,000, independent of species or subcellular organelle.

The isoelectric point for the enzyme has been determined since isoelectric focusing is often used to purify the enzyme. No generalization can be made in that the p*I* varies from acidic, 5.0, for the horse enzyme[38] to basic, 7.5, for an isozyme from rat liver cytosol.[59]

TABLE II

Properties of Pure Liver Aldehyde Dehydrogenase from Different Species

Species[a]	pI	Molecular weight (10^{-3}) Subunit	Enzyme	Specific activity[b]	pH	Reference
Human						
—[c]		55	245	0.58	9	79
—[c]		54	224	1.0	9	79
Horse						
—	5	57	250	0.72	9	38
—	6			0.3	9	Unpublished results
C[d]		52	230	0.35	9	71
M[d]		53	240	0.80	9	71
Sheep						
M		48	198	0.2	8	80
Rabbit						
—		48	194	0.5	7	76
Cow						
—	5	55	220	1.14	8	82
Rat						
M[e]	6.06		320	0.23	9	37
M[f]	6.64		67	0.8	9	37
P				0.02	9	83
H	6.9	53	106			32

[a] C, cytosolic; M, mitochondrial; P, microsomal; H, hepatomas.
[b] Specific activity reported as in μmoles NADH formed per minute per milligram protein at the approximate pH indicated.
[c] Represents two separate isozymes.
[d] These authors claim the pI5 form corresponds to the cytosolic enzyme.
[e] Matrix space.
[f] Membrane bound.

D. Substrate Specificity of Aldehyde Dehydrogenase

Aldehyde dehydrogenase has very broad substrate specificity. The K_m for most aldehydes fall in the micromolar range. The bulk of the studies with this enzyme have employed acetaldehyde or propionaldehyde as substrates, although the limited number of commercially available aldehydes have been used to investigate the specificity of the individual enzymes.

With the enzyme from horse liver[73] and bovine brain[49] the velocity of oxidation for different substrates could be correlated with the Taft electron withdrawing constant. That is, the velocity of oxidation is faster for an aldehyde of general structure $X\text{-}CH_2\text{-}CHO$ if X contains electron-withdrawing groups. Similarly, in the aromatic series the velocity is faster if a substitution in the para position has an electron withdrawing group such as nitro rather than methoxy.

In Table III are listed some V_{max} and K_m values for different substrates for horse liver aldehyde dehydrogenase. These are presented only to illustrate the broad specificity of the enzyme; similar results could be expected with the enzyme from different species.

Acetaldehyde is primarily oxidized in mitochondria although the cytosol enzymes also have this ability. Most investigators report that in the rat both high K_m and low K_m acetaldehyde-oxidizing system exist. The enzyme with low K_m is mitochondrial, while the high K_m forms are localized in the cytosol or microsomes. The evidence supporting these generalization have been reviewed.[9,85] Both cytosol and mitochondrial enzymes from horse have a low K_m (μM) for acetaldehyde[81] and only the human mitochondrial enzyme has a low K_m.[79]

The microsomal enzyme from rat has a high K_m (mM) for acetaldehyde. The enzyme form that is isolated from rat liver shows greater activity with high molecular weight aldehydes (C_{10} or C_{12}) than with low molecular weight aldehydes.[83] The rat liver microsomal enzyme also has a high K_m

TABLE III

Maximum Velocities and Michaelis Constants for Aldehydes with Horse Liver Aldehyde Dehydrogenase[a]

Aldehyde	K_m (μM)[b]	Velocity[c]
Acetaldehyde	0.3	80[d]
Propionaldehyde	0.4	100
Isobutyraldehyde	0.4	100
Phenylacetaldehyde	0.4	110
Chloroacetaldehyde	5.9	260
Glycolaldehyde		250[d]
Glyceraldehyde	200	130
Furfural	0.5	12
Benzaldehyde	0.3	20
o-Nitrobenzaldehyde	0.5	3
p-Nitrobenzaldehyde		55[d]

[a] Feldman and Weiner.[38]

[b] K_m apparent; data uncorrected for hemiacetal formation.

[c] Propanaldehyde, 100 assay at pH 9.

[d] At pH 7.4 (R. I. Feldman and H. Weiner, unpublished values).

for 3,4-dihydroxyphenylacetaldehyde.[59] The microsomal enzyme may be physiologically involved in the metabolism of aldehydes derived from steroids and fatty alcohols. It was recently shown that fatty alcohols were converted to acids in the microsomes.[86] Although a specific formaldehyde oxidizing dehydrogenase is present, the nonspecific enzymes can oxidize this aldehyde[87] and one enzyme form, isolated form the matrix space of mitochondria, is capable of oxidizing formaldehyde but not acetaldehyde.[88] Another enzyme form was shown to have a K_m of 31 μM for formaldehyde.[37]

E. Distribution and Regulation of Aldehyde Dehydrogenase

Aldehyde dehydrogenase is found in every organ, although the liver contains more activity per weight than other organs; kidney contains approximately 20% of the enzyme activity present in liver.[51] Age and sex do not seem to affect greatly the level or isozyme distribution, but the isozyme pattern does change during the development of the rat fetus to the adult.[33] There is some indication that female rats have more enzyme activity than do males.[60] Pregnant mice have twice the cytosol activity of nonpregnant animals.[89]

Drugs, such as phenobarbital, and herbicides, such as tetrachlorodibenzo-p-dioxin, greatly enhance the activity of the cytosol enzyme in rats (see Section I) and aminofluorine will induce tumors which contain different isozymes of the enzyme.[30] It has been observed that rats with different preferences for voluntary consumption of alcohol also possess different patterns of liver cytosol enzyme[90]; the physiological significance or consequences of these differences in isozyme patterns is unknown.

Recently it has been shown that enzyme activity is stimulated twofold by the presence of divalent ions.[91] A form of the rabbit liver enzyme is inhibited by steroids.[41,43,76] Insufficient work has been done with metal activators and steroid inhibitors to know if all forms of the enzyme are affected in a similar manner.

IV. FORMALDEHYDE DEHYDROGENASE

A. General

Formaldehyde has been implicated as the toxic agent in methanol poisoning.[92] The aldehyde is, however, very rapidly removed from the body ($t_{1/2}$ = 90 sec), while the irreversible damage due to methanol poisoning seems to occur, at least in the monkey, 12 to 24 h after the animal received the alcohol[93]; formaldehyde cannot be detected in the animal at that time. Thus, formate may be the toxic agent rather than the aldehyde.

The oxidation of formaldehyde can occur in all tissues, although liver appears to possess the largest capacity.[9] Formaldehyde can be formed not only by oxidation of methanol but also from N-demethylation reactions catalyzed by cytochrome P-450[88,94-96]; the formaldehyde derived from the microsomal N-demethylation reactions is oxidized in mitochondria,[88] whereas formaldehyde derived from methanol appears to be oxidized to formate in the cytosol.

The cytosol contains the glutathione-dependent formaldehyde dehydrogenase[9]; the mitochondria contains a non-glutathione-requiring enzyme.[37,88] In man there are equal activities of each enzyme, but the formaldehyde dehydrogenase activity in the rat is approximately threefold higher than aldehyde dehydrogenase activity.[8] Human liver has twice the combined activity per weight when compared to rat liver.

B.   Purification of Formaldehyde Dehydrogenase

The enzyme has been purified by salt fractionation chromatography on DEAE-cellulose, isoelectric focusing, gel filtration, and chromatography on hydroxyapatite and QAE-Sephadex to yield a homogeneous protein.[9] Goodman et al.[8] used an ethanol–chloroform precipitation step after salt fractionation.

C.   Assay for Formaldehyde Dehydrogenase Activity

The enzyme is assayed by following the increase in NADH formation. It is necessary to include a blank reaction in the absence of glutathione since the nonspecific aldehyde dehydrogenase, if present, will oxidize formaldehyde while producing NADH. If alcohol dehydrogenase contamination is suspected, it will be necessary to include an inhibitor of the enzyme (pyrazole or isobutyramide, see Chapter 10, this volume) to prevent formaldehyde and NADH from producing methanol and thereby removing NADH.

It is possible to assay the enzyme by following the disappearance of labeled formaldehyde. Some investigators have utilized this technique when studying demethylation reactions.[96]

D. Characterization of Formaldehyde Dehydrogenase

The human enzyme contains two subunits with a total molecular weight of 81,400[9] and an isoelectric point of 6.35. The enzyme is remarkably stable in the presence of added NADH with no loss in activity at zero degrees for up to 2.5 months; in the absence of added NADH, activity is lost within a week.

Unlike the nonspecific aldehyde dehydrogenase whose velocity continually increases as the pH is raised, formaldehyde dehydrogenase appears to have a maximum velocity between pH 8 and 9 with half the maximal velocity at about pH 7.[8]

E. Substrate Specificity of Formaldehyde Dehydrogenase

In vitro the enzyme will oxidize aldehydes other than formaldehyde, although it does not act on a simple aldehyde such as acetaldehyde. However, methylglyoxal has been used in some studies.[9] The concentration of substrates necessary to produce 50% maximum velocity is listed in Table IV. It can be seen that formaldehyde is the best of the limited substrates tested. Both rat and human liver enzymes were shown in a different study to have a K_m for formaldehyde of 7 to 8 μM.[8]

F. Regulation of Formaldehyde Dehydrogenase

The regulation of the enzyme has not been investigated in depth, although all organs are noted as containing the enzyme.[9] Any understand-

TABLE IV

Kinetic Properties of Different Substrates with
Human Liver Formaldehyde
Dehydrogenase[a]

Aldehyde	$[S]_{0.5}$ $(\mu M)^b$	V^c
Formaldehyde	9	100
Methylglyoxal	80	85
Glyoxal	960	10
Hydroxypyruvaldehyde	2000	61

[a] Votila and Koivusalo.[9]

[b] Substrate concentration reported to produce 50% V_{max}.

[c] Formaldehyde, 100 corresponding to 3.2 I.U./mg at pH 8.

ing of the metabolism of formaldehyde requires reference to the specific esterase that hydrolyzes the formylglutathione ester. This esterase has been purified to homogeneity[54] and is a dimeric enzyme of molecular weight 52,500 which can be isolated from the cytosol of human liver. The esterase will hydrolyze the acyl ester of glutathione at 0.5% of the rate of the formyl ester.

V. ALDEHYDE OXIDASE

As early as 1946, evidence was presented for a flavin-dependent aldehyde dehydrogenase.[97] However, it was later shown that this enzyme, though capable of oxidizing aldehydes,[98-101] preferentially oxidized purines and other heterocyclic compounds.[102] Thus, aldehyde oxidase is similar to xanthine oxidase in both function and properties. Xanthine and aldehyde oxidase are discussed in depth in Chapter 14, this volume.

Although the enzyme will oxidize a variety of aldehydes, no data could be found which leads one to believe that the oxidase catalyzes this reaction *in vivo*. In general, the K_m for corresponding aldehydes are much higher than with aldehyde dehydrogenase. For example, acetaldehyde and salicylaldehyde have K_m of 3.5 and 0.15 mM, respectively,[98] whereas palmitaldehyde and stearaldehyde have a K_m of 0.35 mM.[101] The fact that the K_m for many purines is very much lower[102] suggests that their presence in the cell would prevent the enzyme from oxidizing any aldehydes.

VI. COMMENTS

The nonspecific aldehyde dehydrogenases located in microsome, cytosol, and mitochondria of all tissue appear to be capable of oxidizing any aldehyde which they encounter. The absolute substrate specificity of the various forms of each of the enzymes is not known. Presumably the different isozymes have unique specificity such that they are selectively involved in the detoxication of specific classes of compounds. Conceivably, an individual missing a specific isozyme, would be more susceptible to toxic reactions potentially produced by the presence of a particular aldehyde. The formaldehyde-specific enzyme seems to be present with sufficient activity to remove all formaldehyde formed in the animal. It would appear that evolution has resulted in a sufficient battery of enzymes for the removal of these toxic and reactive intermediates.

ACKNOWLEDGMENTS

The author is a recipient of a Research Scientist Development Award from the National Institutes of Mental Health (NIAAA K02 AA00028). The author thanks Belinda Wyss for editorial assistance.

REFERENCES

1. Morin, N. C., and Kubinski, H. (1978). Potential toxicity of materials used for home insulation. *Ecotoxicol. Environ. Saf.* **2**, 133–141.
2. Alivisatos, S. G. A., and Arora, R. C. A. (1975). Formation of aberrant neurotransmitters and its implications for alcohol addiction and intoxication. *In* "Biochemical Pharmacology of Ethanol" (E. Majchrowicz, ed.), pp. 255–263. Plenum, New York.
3. Weiner, H., Tank, A. W., von Wartburg, J. P., and Weber, S. (1979). Interactions of aldehydes and proteins. *Abstr., Int. Symp. Alcohol Aldehyde Metab. Syst., 3rd, 1979* p. 264.
4. Gates, S. C., Dendramis, N., and Swelley, C. C. (1978). Automated metabolic profiling of organic acids in human urine. 1. Description of methods. *Clin. Chem.* **24**, 1674–1679.
5. Deitrich, R. A. (1966). Tissue and subcellular localization of mammalian aldehyde oxidizing capacity. *Biochem. Pharmacol.* **15**, 1911–1922.
6. Weiner, H. (1979). Aldehyde dehydrogenase: Mechanism of action and possible physiological roles. *In* "Biochemistry and Pharmacology of Ethanol, Vol. 1" (E. Majchrowicz and E. I. Noble, eds.). Plenum, New York (in press).
7. Strittmater, P., and Ball, E. G. (1955). Formaldehyde dehydrogenase, a glutathione-dependent enzyme system. *J. Biol. Chem.* **213**, 445–461.
8. Goodman, J. I., and Tephly, T. R. (1971). A comparison of rat and human liver formaldehyde dehydrogenase. *Biochim. Biophys. Acta* **252**, 489–505.
9. Uotila, L., and Koivusalo, M. (1974). Formaldehyde dehydrogenase from human liver. Purification, properties, and evidence for the formation of glutathione thiol esters by the enzyme. *J. Biol. Chem.* **249**, 7653–7663.
10. Menzie, C. M. (1978). Metabolism of pesticides—update, II. *U. S., Fish Wildl. Serv., Spec. Sci. Rep.: Wildl.* **212**.
11. Corbett, T. H., Heidelberger, C., and Dove, W. F. (1970). Determination of the mutagenic activity to bacteriophage T4 of carcinogenic and noncarcinogenic compounds. *Mol. Pharmacol.* **6**, 667–679.
12. Schoental, R. (1976). Carcinogens in plants and microorganisms. *In* "Chemical Carcinogens" (C. E. Searle, ed.), Publ. No. 173, pp. 626–689. Am. Chem. Soc., Washington, D.C.
13. Schoental, R., and Gibbard, S. (1972). Nasal and other tumors in rats given 3,4,5-trimethyoxy-cinnamaldehyde, a derivative of sinapaldehyde and of other α,β-unsaturated aldehyde wood lignin constituents. *Br. J. Cancer* **26**, 504–505.
14. Horton, A. A. (1971). Induction of aldehyde dehydrogenase in a mitochondrial fraction. *Biochim. Biophys. Acta* **253**, 514–517.
15. Koivula, T., and Lindros, K. O. (1975). Effects of long-term ethanol treatment on aldehyde and alcohol dehydrogenase activities in rat liver. *Biochem. Pharmacol.* **24**, 1937–1942.

16. Hasumura, Y., Teschke, R., and Lieber, C. S. (1976). Characteristics of acetaldehyde oxidation in rat liver mitochondria. *J. Biol. Chem.* **251,** 4908–4913.
17. Horton, A. A., and Barrett, M. C. (1976). Rates of induction of mitochondrial aldehyde dehydrogenase in rat liver. *Biochem. J.* **156,** 177–179.
18. Greenfield, N. J., Pietruszko, R., Lin, G., and Lester, D. (1976). The effect of ethanol ingestion on the aldehyde dehydrogenases of rat liver. *Biochim. Biophys. Acta* **428,** 627–632.
19. Amir, S. (1978). Brain aldehyde dehydrogenase: Adaptive increase following prolonged ethanol administration in rats. *Neuropharmacology* **17,** 463–467.
20. Deitrich, R. A. (1971). Genetic aspects of increase in rat liver aldehyde dehydrogenase induced by phenobarbital. *Science* **173,** 334–336.
21. Deitrich, R. A., Collins, A. C., and Erwin, V. G. (1972). Genetic influence upon phenobarbital in rat liver supernatant aldehyde dehydrogenase activity. *J. Biol. Chem.* **247,** 7232–7236.
22. Deitrich, R. A., Troxell, P. A., and Erwin, V. G. (1975). Characteristics of the induction of aldehyde dehydrogenase in rat liver. *Arch. Biochem. Biophys.* **166,** 543–548.
23. Koivula, T., and Koivusalo, M. (1975). Partial purification and properties of a phenobarbital-induced aldehyde dehydrogenase of rat liver. *Biochim. Biophys. Acta* **410,** 1–11.
24. Eriksson, C. J., Marselos, M., and Koivula, T. (1975). Role of cytosolic rat liver aldehyde dehydrogenase in the oxidation of acetaldehyde during ethanol metabolism *in vivo*. *Biochem. J.* **152,** 709–712.
25. Petersen, D. R., Collins, A. C., and Deitrich, R. A. (1977). Role of liver cytosolic aldehyde dehydrogenase isozymes in control of blood acetaldehyde concentrations. *J. Pharmacol. Exp. Ther.* **201,** 471–481.
26. Torronen, R., Nousiainen, U., and Marselos, M. (1977). Inducible aldehyde dehydrogenases in the hepatic cytosol of the rat. *Acta Pharmacol. Toxicol.* **41,** 263–272.
27. Nakanishi, S., Shiohara, E., and Tsukada, M. (1978). Rat liver aldehyde dehydrogenases: Strain differences in the response of the enzymes to phenobarbital treatment. *Jpn. J. Pharmacol.* **28,** 653–659.
28. Deitrich, R. A., Bludeau, P., Stock, T., and Roper, M. (1977). Induction of different rat liver supernatant aldehyde dehydrogenases by phenobarbital and tetrachlorodibenzo-*p*-dioxin. *J. Biol. Chem.* **252,** 6169–6176.
29. Marselos, M., and Torronen, R. (1978). Increase of hepatic and serum aldehyde dehydrogenase activity after TCDD treatment. *Arch. Toxicol. 1,* Suppl., 271–273.
30. Lindahl, R., Roper, M., and Deitrich, R. A. (1978). Rat liver aldehyde dehydrogenase—immunochemical identity of 2,3,7,8-tetrachlorodibenzo-*p*-dioxin inducible normal liver and 2-acetylaminofluorene inducible hepatoma isozymes. *Biochem. Pharmacol.* **27,** 2463–2465.
31. Feinstein, R. N., and Cameron, E. C. (1972). Aldehyde dehydrogenase activity in a rat hepatoma. *Biochem. Biophys. Res. Commun.* **48,** 1140–1146.
32. Lindahl, R., and Feinstein, R. N. (1976). Purification and immunochemical characterization of aldehyde dehydrogenase from 2-acetylaminofluorene-induced rat hepatomas. *Biochim. Biophys. Acta* **452,** 345–355.
33. Lindahl, R. (1977). Aldehyde dehydrogenase in 2-acetamidofluorene-induced rat hepatomas. Ontogeny and evidence that the new isoenzymes are not due to normal gene de-repression. *Biochem. J.* **164,** 119–123.
34. Lindahl, R. (1978). Aldehyde dehydrogenase in 2-acetylaminofluorene-induced rat hepatomas. Characterization of antigens recognized by anti-hepatoma aldehyde dehydrogenase sera. *Biochim. Biophys. Acta* **525,** 9–17.

35. Jakoby, W. B. (1963). Aldehyde dehydrogenase. *In* "The Enzymes" (P. D. Boyer, H. Lardy, and K. Myrbäck, eds.), 2nd ed., Vol. 7, pp. 203–221. Academic Press, New York.
36. Massey, V. (1973). Iron-sulfur flavoprotein hydroxylases. *In* "Iron-Sulfur Proteins" (W. Lovenberg, ed.), Vol. 1, pp. 301–360. Academic Press, New York.
37. Siew, C., Deitrich, R. A., and Erwin, V. G. (1976). Localization and characteristics of rat liver mitochondrial aldehyde dehydrogenases. *Arch. Biochem. Biophys.* **176**, 638–649.
38. Feldman, R. I., and Weiner, H. (1972). Horse liver aldehyde dehydrogenase. I. Purification and characterization. *J. Biol. Chem.* **247**, 260–266.
39. Kraemer, R. J., and Deitrich, R. A. (1968). Isolation and characterization of human liver aldehyde dehydrogenase. *J. Biol. Chem.* **243**, 6402–6408.
40. Blair, A. H., and Bodley, F. H. (1969). Human liver aldehyde dehydrogenase: Partial purification and properties. *Can. J. Biochem.* **47**, 265–272.
41. Maxwell, E. S., and Topper, Y. (1961). Steroid-sensitive aldehyde dehydrogenase from rabbit liver. *J. Biol. Chem.* **236**, 1032–1037.
42. Raison, S. K., Henson, G., and Rienitz, K. C. (1966). The oxidation of gentisaldehyde by NAD specific aromatic aldehyde dehydrogenase from rabbit liver. *Biochim. Biophys. Acta* **118**, 285–293.
43. Douville, A. W., and Warren, C. (1968). Steroid–protein interaction at sites which influence catalytic activity. *Biochemistry* **7**, 4056–4059.
44. Freda, C. E., and Stoppani, A. O. (1970). Kinetics of bovine liver aldehyde dehydrogenase. Effect of coenzyme and aldehyde structure. *Enzymologia* **38**, 225–242.
45. Robbins, J. H. (1966). Electrophoresis of mammalian aldehyde dehydrogenase. *Arch. Biochem. Biophys.* **114**, 585–592.
46. Sheppard, J. R., Albersheim, P., and McClearn, G. (1970). Aldehyde dehydrogenase and ethanol preference in mice. *J. Biol. Chem.* **245**, 2876–2882.
47. Smith, L., and Packer, L. (1972). Aldehyde oxidation in rat liver mitochondria. *Arch. Biochem. Biophys.* **148**, 270–276.
48. Marjanen, L. (1972). Intracellular localization of aldehyde dehydrogenase in rat liver. *Biochem. J.* **127**, 633–639.
49. Erwin, V. G., and Deitrich, R. A. (1966). Brain aldehyde dehydrogenase. Localization, purification, and properties. *J. Biol. Chem.* **241**, 3533–3539.
50. Duncan, R. J., and Tipton, K. F. (1971). The purification and properties of the NAD-linked aldehyde dehydrogenase from pig brain. *Eur. J. Biochem.* **22**, 257–262.
51. Tank, A. W. (1976). The effects of ethanol on dopamine metabolism in the rat liver and brain. Ph.D. Thesis, Purdue University, West Lafayette, Indiana.
52. Berger, D., and Weiner, H. (1977). *In vivo* interactions of chloralhydrate and disulfiram with the metabolism of catecholamines. *In* "Currents in Alcoholism," **1**, 231–241.
53. Krug, E. L. (1977). The preferential reduction of norepinephrine in rat heart, liver and brain: Effects of aldehyde reductase inhibition *in vivo*. M.S. Thesis, Purdue University, West Lafayette, Indiana.
54. Uotila, L., and Koivusalo, M. (1974). Purification and properties of S-formylglutathione hydrolase from human liver. *J. Biol. Chem.* **249**, 7664–7672.
55. Parrilla, R., Okawa, K., Lindros, K. O., Zimmerman, U. J., Kobayashi, K., and Williamson, J. R. (1974). Functional compartmentation of acetaldehyde oxidation in rat liver. *J. Biol. Chem.* **249**, 4926–4933.
56. Grunnet, N., Thieden, H. I. D., and Quistorff, B. (1976). Metabolism of 1-^{3}H-ethanol by isolated liver cells. Time-course of the transfer of tritium from R,S-1-^{3}H-ethanol to lactate and S-hydroxybutyrate. *Acta Chem. Scand.* **30**, 345–352.

57. Corrall, R. J., Havre, P., Margolis, J. M., Kong, M., and Landau, B. R. (1976). Subcellular site of acetaldehyde oxidation in rat liver. *Biochem. Pharmacol.* **25**, 17–20.
58. Havre, P., Margolis, J. M., and Abrams, M. A. (1976). Subcellular site of acetaldehyde oxidation in monkey liver. *Biochem. Pharmacol.* **25**, 2757–2758.
59. Tank, A. W., and Weiner, H. (1977). Intercellular location of 3,4-dihydroxy-phenylacetaldehyde and acetaldehyde oxidation in rat liver. *In* "Alcohol and Aldehyde Metabolizing Systems" (R. G. Thurman, J. R. Williamson, H. R. Drott, and B. Chance, eds.), Vol. 2, pp. 175–183. Academic Press, New York.
60. Horton, A. A., and Packer, L. (1976). Mitochondrial metabolism of aldehydes. *Biochem. J.* **116**, 19P–20P.
61. Hasumura, Y., Teschke, R., and Lieber, C. S. (1976). Characteristics of acetaldehyde oxidation in rat liver mitochondria. *J. Biol. Chem.* **251**, 4908–4913.
62. Kitson, T. M. (1977). The disulfiram-ethanol reaction. A review. *J. Stud. Alcohol* **38**, 96–113.
63. Tottmar, O., and Marchner, H. (1976). Disulfiram as a tool in the studies on the metabolism of acetaldehyde in rats. *Acta Pharmacol. Toxicol.* **38**, 366–375.
64. Marchner, H., and Tottmar, O. (1978). A comparative study on the effects of disulfiram, cyanamide and 1-aminocyclopropanol on the acetaldehyde metabolism in rats. *Acta Pharmacol. Toxicol.* **43**, 219–232.
65. Deitrich, R. A., Troxell, P. A., Worth, W. S., and Erwin, V. G. (1976). Inhibition of aldehyde dehydrogenase in brain and liver by cyanamide. *Biochem. Pharmacol.* **25**, 2733–2737.
66. Tottmar, O., Marchner, H., and Karlsson, N. (1978). The presence of an aldehyde dehydrogenase inhibitor in animal diets and its effects on the experimental results in alcohol studies. *Br. J. Nutr.* **39**, 317–324.
67. Tottmar, O., and Lindberg, P. (1977). Effects on rat liver acetaldehyde dehydrogenases *in vitro* and *in vivo* by coprine, the disulfiram-like constituent of *Coprinus atramentarius*. *Acta Pharmacol. Toxicol.* **40**, 476–481.
68. Carlsson, A., Henning, M., Lindberg, P., Martinson, P., Trolin, G., Waldeck, B., and Wickberg, B. (1978). On the disulfiram-like effect of coprine, the pharamacologically active principle of *Coprinus atramentarius*. *Acta Pharmacol. Toxicol.* **42**, 292–297.
69. Wiseman, J. S., and Abeles, R. H. (1979). Mechanism of inhibition of aldehyde dehydrogenase by cyclopropanone hydrate and the mushroom toxin coprine. *Biochemistry* **18**, 427–435.
70. Harris, J. I., and Waters, M. (1976). Glyceraldehyde-3-phosphate dehydrogenase. *In* "The Enzymes" (P. D. Boyer, ed.), 3rd ed., Vol. 13, pp. 1–49. Academic Press, New York.
71. Eckfeldt, J., Mope, L., Takio, K., and Yonetani, T. (1976). Horse liver aldehyde dehydrogenase. Purification and characterization of two isozymes. *J. Biol. Chem.* **251**, 236–240.
72. Deitrich, R. A. (1967). Diphosphopyridine nucleotide-linked aldehyde dehydrogenase. 3. Sulfhydryl characteristics of the enzyme. *Arch. Biochem. Biophys.* **119**, 253–263.
73. Feldman, R. I., and Weiner, H. (1972). Horse liver aldehyde dehydrogenase. II. Kinetics and mechanistic implications of the dehydrogenase and esterase activity. *J. Biol. Chem.* **247**, 267–272.
74. Sidhu, R. S., and Blair, A. H. (1975). Human liver aldehyde dehydrogenase. Esterase activity. *J. Biol. Chem.* **250**, 7894–7898.
75. Weiner, H., Hu, J. H. J., and Sanny, C. G. (1976). Rate-limiting steps for the esterase and dehydrogenase reaction catalyzed by horse liver aldehyde dehydrogenase. *J. Biol. Chem.* **251**, 3853–3855.

76. Duncan, R. J. S. (1977). The action of progesterone and diethylstilboestrol on the dehydrogenase and esterase activities of a purified aldehyde dehydrogenase from rabbit liver. *Biochem. J.* **161**, 123–130.
77. MacGibbon, A. K., Haylock, S. J., Buckley, P. D., and Blackwell, L. F. (1978). Kinetic studies on the esterase activity of cytoplasmic sheep liver aldehyde dehydrogenase. *Biochem. J.* **171**, 533–538.
78. Sidhu, R. S., and Blair, A. H. (1975). Human liver aldehyde dehydrogenase. Kinetics of aldehyde oxidation. *J. Biol. Chem.* **250**, 7899–7904.
79. Greenfield, N. J., and Pietruszko, R. (1977). Two aldehyde dehydrogenases from human liver. *Biochim. Biophys. Acta* **483**, 35–45.
80. Hart, G. J., and Dickinson, F. M. (1977). Some properties of aldehyde dehydrogenase from sheep liver mitochondria. *Biochem. J.* **163**, 261–267.
81. Eckfeldt, J. H., and Yonetani, T. (1976). Subcellular localization of the F1 and F2 isozymes of horse liver aldehyde dehydrogenase. *Arch. Biochem. Biophys.* **175**, 717–722.
82. Leicht, W., Heinz, F., and Freimuller, B. (1978). Purification and characterization of aldehyde dehydrogenase from bovine liver. *Eur. J. Biochem.* **83**, 189–196.
83. Nakayasu, H., Mihara, K., and Sato, R. (1978). Purification and properties of a membrane-bound aldehyde dehydrogenase from rat liver microsomes. *Biochem. Biophys. Res. Commun.* **83**, 697–703.
84. Tank, A. W., Weiner, H., and Thurman, J. A. (1976). Ethanol-induced alterations of dopamine metabolism in rat liver. *Ann. N.Y. Acad. Sci.* **273**, 219–226.
85. Lindros, K. O. (1978). Acetaldehyde—Its metabolism and role in the action of alcohol. *In* "Research Advances in Alcohol and Drug Problems" (Y. Israel, F. B. Glaser, H. Kalant, and R. G. Smart, eds.), Vol. 4, pp. 111–176. Plenum, New York.
86. Lee, T-C. (1979). Characterization of fatty alcohol: NAD oxidoreductase from rat liver. *J. Biol. Chem.* **254**, 2892–2896.
87. Koivula, T., and Koivusalo, M. (1978). Characteristics of the phenobarbital induced aldehyde dehydrogenase activity in rat liver. *In* "The Role of Acetaldehyde in the Action of Ethanol" (K. O. Lindros and C. J. P. Eriksson, eds.), pp. 37–46. Finn. Found. Alcohol Stud., Helsinki.
88. Cinti, D. L., Keyes, S. R., Lemelin, M. A., Denk, H., and Schenkman, J. B. (1976). Biochemical properties of rat liver mitochondrial aldehyde dehydrogenase with respect to oxidation of formaldehyde. *J. Biol. Chem.* **251**, 1571–1577.
89. Petersen, D. R., Panter, S. S., and Collins, A. C. (1977). Ethanol and acetaldehyde metabolism in the pregnant mouse. *Drug Alcohol Depend.* **2**, 409–420.
90. Berger, D., and Weiner, H. (1977). Relationship between alcohol preference and biogenic aldehyde metabolizing enzymes in rats. *Biochem. Pharmacol.* **26**, 841–846.
91. Takahashi, K., Weiner, H., and Brown, C. (1979). Metal activation of p*I* 5 isozyme of horse liver aldehyde dehydrogenase. *Abstr., Int. Symp. Alcohol Aldehyde Metab. Syst., 3rd, 1979* p. 264.
92. Goodman, L. S., and Gilman, A., eds. (1971). "The Pharmacological Basis of Therapeutics," 4th ed., pp. 145–146. Macmillan, New York.
93. McMartin, K. R., Martin-Amat, G., Noker, P. E., and Tephly, T. R. (1979). Lack of a role for formaldehyde in methanol poisoning in the monkey. *Biochem. Pharmacol.* **28**, 645–649.
94. Tephly, T. R., Watkins, W. D., and Goodman, S. F. (1974). The biochemical toxicology of methanol. *Essays Toxicol.* **5**, 149–177.
95. Jones, D. P., Thor, H., Andersson, B., and Orrenius, S. (1978). Detoxification reactions in isolated hepatocytes. *J. Biol. Chem.* **253**, 6031–6037.

96. Waydhas, C., Weigl, K., and Sies, H. (1978). The disposition of formaldehyde and formate arising from drug N-demethylations dependent on cytochrome P-450 in hepatocytes and in perfused rat liver. *Eur. J. Biochem.* **89,** 143–150.
97. Knox, W. E. (1946). The quinine-oxidizing enzyme and liver aldehyde oxidase. *J. Biol. Chem.* **163,** 699–711.
98. Rajagopalan, K. V., and Handler, P. (1964). Hepatic aldehyde oxidase. *J. Biol. Chem.* **239,** 2027–2038.
99. Lakshmanan, M. R., Vaidyanathan, C. S., and Cama, H. R. (1964). Oxidation of vitamin A_1 aldehyde and vitamin A_2 aldehyde to the corresponding acids by aldehyde oxidase from different species. *Biochem. J.* **90,** 569–573.
100. Johns, D. G. (1967). Human liver aldehyde oxidase: Differential inhibition of oxidation of charged and uncharged substrates. *J. Clin. Invest.* **46,** 1492–1505.
101. Ho, C. Y., and Clifford, A. J. (1976). Digestion and absorption of bovine milk xanthine oxidase and its role as an aldehyde oxidase. *J. Nutr.* **106,** 1600–1609.
102. Krenitsky, T. A., Neil, S. M., Elion, G. B., and Hitchings, G. H. (1972). A comparison of the specificities of xanthine oxidase and aldehyde oxidase. *Arch. Biochem. Biophys.* **150,** 585–599.

Chapter 13

Ketone Reductases

RONALD L. FELSTED and NICHOLAS R. BACHUR

I. INTRODUCTION

Carbonyl containing substances which are not biologically oxidized are reduced in the detoxication process.[1] Since carbonyl compounds frequently are hydrophobic and may be retained in tissues, their reduction to hydrophilic alcohols and subsequent conjugation are critical to their elimination. The potential for reductive metabolism of the carbonyl function in normal cell physiology is suggested by numerous naturally occurring carbonyl compounds and the excretion of naturally occurring and xenobiotic carbonyl compounds as alcohols and/or conjugates *in vivo*.[2] A portion of this reduction is mediated by well-known oxidoreductases (e.g., alcohol dehydrogenase, EC 1.1.1.1) but in recent years enzymes have been described which apparently function *in vivo* primarily as carbonyl reductases. Although probably related to the better known dehydrogenases, the carbonyl reductases have properties which distinguish them as a separate class. Based on the predictable differences in chemical reactivities of aldehydes and ketones, consideration of this class of reductases naturally resolves into a discussion of aldehyde reductases and

281

ENZYMATIC BASIS OF DETOXICATION, VOL. I
ISBN 0-12-380001-3

ketone reductases. A description of the relatively well-characterized aldehyde reductases (EC 1.1.1.2) and their role in carbonyl detoxication is presented elsewhere (Chapter 12 in this volume). We will complete the consideration of carbonyl detoxication enzymes with our discussion about the generally uncharacterized ketone reductases.

The *in vivo* metabolism of xenobiotic ketones to free alcohols or conjugated alcohols has been demonstrated for aromatic,[3-5] aliphatic,[3] alicyclic,[6] and unsaturated ketones.[7] Until recently, however, little of the enzymatic characteristics of xenobiotic ketone reduction *in vitro* has been reported. A number of ketone reductases which have been shown to produce alcohols *in vitro* are summarized in Table I. Although few have been extensively purified, specific properties which characterize this unique group are included in Table I.[5,7-14] These properties include ubiquitous tissue distribution, cytoplasmic localization, preference for NADPH as cofactor, and low molecular weight.[2] Although these properties characterize the carbonyl reductases collectively, the ketone reductases can be distinguished from aldehyde reductases on the basis of substrate specificity, inhibitor specificity, and cofactor hydrogen stereospecificity.[5,13,15]

II. ASSAY

Ketone reductase activities are most conveniently assayed spectrophotometrically by monitoring the oxidation of the reduced pyridine nucleotide cofactor, NADPH, at 340 nm. However, due to the strong absorbance of NADPH, the high natural absorbance of many aromatic ketones, the high concentrations of ketone usually required for enzyme saturation, and low enzyme activities necessitating an expanded spectrophotometer absorption scale, the spectrophotometric assay of ketone reductases usually requires an instrument with low stray light (less than 0.001%). Although rapid and convenient, the spectrophotometric assay suffers from a lack of sensitivity and follows the ketone reduction indirectly by observing the oxidation of pyridine nucleotide and assuming a one to one stoichiometry of ketone reduced to cofactor oxidized. In addition, the presence of nonspecific NADPH oxidases and pyridine nucleotide transhydrogenases in crude cell homogenates necessitates rigid reliance on background controls.

Greater sensitivity and specificity are achieved for individual reactions by analyzing for alcohol product formation directly. In these cases, the reaction mixtures are usually extracted and the isolated alcohol products separated from ketone substrates by gas–liquid, high pressure liquid, or

TABLE I

Mammalian Ketone Reductases

Trivial name	Source			Enzyme Properties			
	Animals	Tissues	Molecular weight	Cofactor	Location	References
Dihydromorphinone ketone reductase	Chicken, rabbit	Liver, kidney	38,000	NADPH	Cytosol	5,8
Warfarin reductase	Rabbit, rat	Liver, kidney	—	NADPH	Cytosol	9
Oxisuran reductase	Rabbit, human	Liver, kidney	32,000	NADPH	Cytosol	10
Bunolol reductase	Human	Liver	—	NADPH	Cytosol	11,12
Daunorubicin (pH 6.0)[a] reductase	Rabbit, human	Liver, kidney	33,000	NADPH	Cytosol	5,13
α,β-Unsaturated ketone reductase	Dog, human, rat	Red cells, liver	—	NADPH	Cytosol	7
4-Nitroacetophenone reductase	Dog, sheep, rat, human	Red cells	—	NADPH	Cytosol	14
Metyrapone reductase	Rabbit, human	Liver, kidney	33,000	NADPH	Cytosol	5
3,7-Dimethyl-1-(5-oxohexyl)xanthine reductase	Rabbit, human	Liver, kidney	38,000	NADPH	Cytosol	5
Aflatoxin reductase	Rabbit	Liver	—	NADPH	Cytosol	—[b]

[a] The reduction of the daunorubicin carbonyl with an optimum of pH 6.0.
[b] R. L. Felsted, and N. R. Bachur, unpublished observations.

thin layer chromatography followed by quantification of radioactivity, native absorbance or fluorescence or colorimetric reaction. Although more time consuming, these procedures have the advantage of greater sensitivity and are a direct measure of reductase activity. For example, although warfarin reduction was not detected in rat liver by the spectrophotometric assay,[5] its *in vitro* reduction has been described in rat liver by a more sensitive fluorometric thin layer plate assay. Daunorubicin and adriamycin reduction are similarly monitored with thin-layer systems[13] or by high pressure liquid chromatography.[16] In crude cell extracts, all assay procedures usually require a cofactor regenerating system for maximal activity.

III. PURIFICATION

Most ketone reductases have been studied in crude cell homogenates or partially purified tissue extracts. Only the human placental 9-ketoprostaglandin reductase has been purified to apparent homogeneity.[17] Daunorubicin reductase was purified to homogeneity from rat liver,[18] but subsequent characterization led to the conclusion that the purified enzyme was an aldehyde reductase rather than a ketone reductase.[19] A number of aromatic ketone reductase activities have been copurified 200- to 300-fold from rabbit kidney cortex.[4] Human erythrocyte 4-nitroacetophenone[14] and dog erythrocyte α,β-unsaturated ketone[7] reductases also have been extensively purified. Other partially purified reductases with specificity for oxisuran, 3,7-dimethyl-l-(5-oxohexyl)xanthine, naloxone, naltrexone, and metyrapone have been reported[5,15] (Table II).

The instability and heterogeneity of ketone reductases present serious difficulties when applying conventional procedures for the purification of these enzymes. Low recoveries of activities on purification and storage increases the difficulty. The inclusion of 0.5 mM dithiothreitol has been found to be an effective and practical means of stabilizing rabbit liver ketone reductases. Obviously, a rapid and specific affinity absorption approach offers the ultimate solution. Present attempts include the combination of gel filtration, specific affinity adsorption to the group-specific NADPH covalently bound to agarose, followed by separation of the individual species by ion exchange chromatography or by high resolution isoelectric focusing (R. L. Felsted and N. R. Bachur, unpublished observations). It is hoped that such an approach will facilitate the rapid resolution and purification of a number of ketone reductases to homogeneity.

An interesting but perplexing characteristic of ketone reductases is the

TABLE II

Ketone Containing Compounds[a]

Substrate structure	Name	Clinical applications
	Daunorubicin	Anticancer antibiotic
	Oxisuran	Immuno-suppressant
Naloxone Naltrexone	Narcotic antagonist
	Metyrapone	Inhibitor of hydroxysteroid synthesis
	3,7-Dimethyl-1-(5-oxohexyl)xanthine	Vasodilator

[a] The reducible carbonyls are indicated in boldface.

large number of proteins from a single tissue that are capable of performing this catalytic function. This complexity has been demonstrated by gel filtration, ion exchange chromatography, and by isoelectric focusing.[5,13,15] An illustration of the isoelectric focusing of several ketone reductase activities from rabbit liver is present in Fig. 1. The majority of ketone reductases [daunorubicin (pH 6.0), oxisuran, metyrapone, and 3,7-dimethyl-1-(5-oxohexyl)xanthine] are focused as a major peak of pI 4.8–5.0 and are essentially resolved from the major aldehyde reductase [as

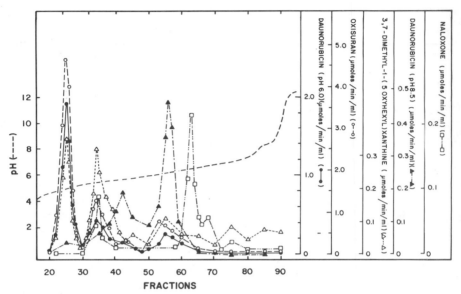

Fig. 1. Isoelectric focusing of rabbit liver carbonyl reductases. Rabbit liver homogenate was fractionated with ammonium sulfate (0.30–0.60), dialyzed against 1% glycine containing 0.5 mM dithiothreitol and isoelectrofocused for 40 h on a pH 4–7 gradient (---). The carbonyl reductases tested: daunorubicin (pH 6.0) reductase, (●——●); oxisuran reductase, (○- -○); 3,7-dimethyl-1-(5-oxyhexyl)xanthine, (△---△); daunorubicin (pH 8.5) reductase, (▲-··-▲); and naloxone reductase, (□-··-□).

defined by daunorubicin (pH 8.5) activity] (p*I* 6.0–6.3). The naloxone reductase is separate and distinct from both groups (p*I* 6.4–7.2). In addition, numerous minor forms of most activities are evident. The general patterns are not effected when proteolytic enzyme inhibitors are included in the homogenization buffers (R. L. Felsted and N. R. Bachur, unpublished observations).

IV. DISTRIBUTION

Despite the lack of extensive characterization, enough is known of the ketone reductases to establish their close relationship to the more extensively studied aldehyde reductases and to distinguish both groups from other common dehydrogenases. One characteristic of carbonyl reductases is their apparent ubiquitous distribution within the animal kingdom. For example, aldehyde reductases have been demonstrated in microorganisms, invertebrates, and vertebrates.[20] Although aromatic ketone reductases have been described in microorganisms,[21] most work has been restricted to higher animals and include trout,[22] chicken,[8] dog,[7] sheep,[14] rat,[9] human,[10] rabbit,[9] guinea pig,[23] and mouse.[13] Fourteen ketone-containing drugs, known to yield alcohol metabolites *in vivo* were assayed spectrophotometrically as potential ketone reductase substrates with rabbit, human, mouse, and rat liver preparations[13]; nearly all were reduced by rabbit liver with highest rates observed for oxisuran, 3,7-dimethyl-1-(5-oxohexyl)xanthine, naloxone, naltrexone, and metyrapone. In contrast, intermediate activities were seen with human liver and lowest activities with rat and mouse livers. These observations are consistent with the very active *in vivo* carbonyl reduction activity demonstrated in animal drug disposition studies[1,6,24] and suggest that the rabbit represents a good model for the prediction or study of carbonyl metabolism in humans. The widespread distribution of these reductases among so many species also suggest that the enzymes are universal components of cells.

Ketone reductases have been detected in virtually every mammalian tissue examined. With *p*-nitroacetophenone as a model ketone substrate and the spectrophotometric assay, ketone reductase activity was detected in mouse liver, kidney, heart, spleen, and brain with liver and kidney having the highest activity (R. L. Felsted and N. R. Bachur, unpublished observations). The widespread distribution of daunorubicin and adriamycin reductase activities among mammalian tissues are examples[25] as are the reductases of the immunosuppressant, oxisuran.[10] Reductases have been reported from human placenta[17] and human, rat,[12] and dog erythrocytes.[14] Thus the general reductase potential appears to be widely distributed among mammalian tissues. It is likely that the absence of a specific reductase activity in a particular tissue merely reflects the lack of sufficient assay sensitivity. While it appears that ketone reduction occurs throughout the animal, the high specific activity in the liver and kidney plus the large quantity of, albeit less active, blood reductases undoubtedly account for the bulk of drug ketone metabolism *in vivo*.

Another characteristic of carbonyl reductases is their subcellular location. Although the study of drug metabolism and detoxification has traditionally centered on the inducible enzyme systems of the cell microsomal fraction, most ketone reductases are localized to the cytosol. Exceptions to this generalization may include a microsomal NADPH-dependent D-3-oxosphinganine oxidoreductase[26] (EC 1.1.1.102) and a minor oxisuran reductase of mitochondrial origin.[5,10] Lysosomal or nuclear compartmentalized carbonyl reductases do not appear to have been described. Despite these apparent exceptions, most ketone reductases have been reported as soluble enzymes (Table I).

V. SPECIFICITY AND CHARACTERIZATION

These enzymes prefer NADPH rather than NADH as cosubstrate. This is true of the aromatic ketone (Table I) as well as aldehyde reductases. Aliphatic ketone reduction is less understood but appears to be mediated by both NADPH and NADH-dependent enzymes.[3] Other exceptions include a cytoplasmic aromatic and keto acid reductase from beef and dog heart[27] and a cytoplasmic 4-ketoproline reductase from rabbit kidney.[28] For the most part, however, the preference for NADPH is a highly distinctive feature of the ketone reductases. For example, rabbit liver daunorubicin (pH 6.0), metyrapone, and naltrexone ketone reduction require NADPH rather than NADH. Rabbit liver oxisuran and 3,7-dimethyl-1-(5-oxohexyl)xanthine reductases use NADH at only 10–25% of the activity seen with NADPH.[5]

Molecular weight is another property used to distinguish the carbonyl reductases from other mammalian dehydrogenases. Most mammalian dehydrogenases exist as oligomeric enzymes composed of multiple subunit polypeptides of 30,000–40,000 daltons each. Aldehyde reductases have been recognized as a group of monomeric oxidoreductases with native molecular weights of 30,000–40,000. We recently found that the major ketone reductases of rabbit and human liver also have low molecular weights.[5] Since the enzymes were only partially purified, the molecular weight determinations were made by gel filtration chromatography with comparison to standard globular proteins of known molecular weight. By this method, naloxone, naltrexone, metyrapone, and 3,7-dimethyl-1-(5-oxohexyl)xanthine reductases eluted as symmetrical peaks suggesting molecular weights about 38,000 and oxisuran and daunorubicin (pH 6.0) reductases both eluted as distinct activity peaks corresponding to a molecular weight of 32,000. Preliminary data suggests that, although two distinct molecular components account for the aromatic ketone reduction in rabbit liver, both are composed of a single catalytic subunit.

Since homogeneous enzymes are not available, the question of substrate specificity remains undefined. Nevertheless, recent studies indicate that the carbonyl reductases can be divided into subgroups of aldehyde reductases and ketone reductases on the basis of their natural preference for aldehydes or ketones, respectively. Because of the greater chemical reactivity of aldehydes relative to ketones, and assuming otherwise similar structures, it might be expected that aldehyde reductases would not reduce the relatively less reactive ketones. For example, although aldehyde reductases reduce an extensive variety of aldehyde substrates, these enzymes have very limited ketone reductase activity. This is confirmed for rat liver aldehyde reductase which is very active with p-nitrobenzaldehyde but has negligible activity with p-nitro-acetophenone.[19] An exception is the ketone anthracycline antibiotic, daunorubicin, which is reduced by rat liver aldehyde reductase, but only at pH 8.5, i.e., 2.5 pH units above the typical pH optima of most aldehyde reductases.[19] Furthermore, in rabbit and human liver, a separate daunorubicin reductase with an optimum of pH 6.0, typical of most carbonyl reductases, has been resolved from the aldehyde reductases and designated as a true ketone reductase.[13] In other words, rabbit and human have two daunorubicin reductases; one with an optimum of pH 8.5 [daunorubicin (pH 8.5)] designated as an aldehyde reductase and one with an optimum of pH 6.0 [daunorubicin (pH 6.0)] designated as a ketone reductase.

By the same reasoning, it might be expected that ketone reductases also catalyze reactions with the highly reactive aldehyde substrates. This overlapping pattern of substrates was suggested for the aromatic aldehyde reductase activity that copurified with the aromatic ketone reductase activity from rabbit kidney cortex.[4] p-Nitrobenzaldehyde reduction was also coincidently associated with each of the distinctive rabbit liver ketone reductase fractions resolved by isoelectric focusing.[15] It, therefore, appears that, although aldehydes are reduced by aldehyde and ketone reductases, ketones are reduced only by ketone reductases.

Aldehyde and ketone reductases are further distinguished by inhibitor specificity. Inhibition of aldehyde reductases by barbiturates has been used to differentiate this carbonyl reducing enzyme from similar activities mediated by classic alcohol dehydrogenase. Pyrazole, a specific inhibitor of alcohol dehydrogenase, similarly has no significant effect on aldehyde reductase. Both compounds have been tested with rabbit liver carbonyl reductases; while barbiturate inhibits daunorubicin (pH 8.5) reductase (aldehyde reductase) as expected (N. K. Ahmed, R. L. Felsted, and N. R. Bachur, unpublished observations), it was without effect on daunorubicin (pH 6.0), oxisuran, 3,7-dimethyl-1-(5-oxohexyl)xanthine, metyrapone, or naltrexone reductases.[5] Insensitivity to barbiturates thus distinguishes the ketone reductases from the aldehyde reductases. Pyrazole was without significant effect on any of the ketone reductases.[5] A similar distinction was observed with the plant flavonoid, quercitrin, which is a potent apparent inhibitor of bovine lens aldose reductase[29] and rabbit liver ketone reductases[5] but has little effect on rabbit liver aldehyde reductase (N. R. Ahmed, R. L. Felsted, and N. R. Bachur, unpublished observations).

The carbonyl reductases are distinguished also by their stereospecificity for the A-(pro-4R) or B-(pro-4S) hydrogen of triphosphopyridine nucleotides. Like alcohol dehydrogenase, the aldehyde reductases from several species exhibit A-hydrogen stereospecificity.[21] In contrast, an aromatic aldehyde ketone reductase from rabbit kidney cortex exhibits B-hydrogen stereospecificity.[4] We have examined the hydrogen stereospecificity of rat and rabbit liver carbonyl reductases with enzymatically prepared A- and B-labeled [4-^{3}H]NADPH by examining the retention of radioactive label in the resulting oxidized cofactors and the transfer of label to alcohol products.[15] The major and minor forms of oxisuran (including α- and β-diastereomers), metyrapone, and daunorubicin (pH 6.0) reductases display B-hydrogen stereospecificity. Some enzymes reducing 3,7-dimethyl-1-(5-oxohexyl)xanthine, p-nitroacetophenone, and p-nitrobenzaldehyde are also B-stereospecific while others are A-stereospecific. Only daunorubicin (pH 8.5), naloxone, and naltrexone reductases (both α- and β-disastereomers) are exclusively A-stereospecific. Several of these enzymes represent exceptions to the generalization that enzymes which catalyze the same reaction have the same stereospecificity.[15,30] Despite the exceptions noted, these results confirm the A-hydrogen stereospecificity of aldehyde reductases and the B-hydrogen stereospecificity of the majority of the rabbit liver ketone reductases. Thus, a general distinction between aldehyde and ketone reductases can be made on the basis of their hydrogen stereospecificity. Considering the otherwise very similar properties of the carbonyl reductases, this difference may be an important criterion.

Stereospecificity for enantiomer ketones as well as stereocatalytic reduction of ketone substrates have been described for several ketone reductases. The reduction of the anticoagulant drug warfarin by rat and human tissue cytosol is selective for the $R(+)$ enantiomer. Reduction of the $S(-)$-warfarin occurs only at elevated concentrations of substrate. The actual catalytic carbonyl reduction also appears to be stereocatalytic. The R-warfarin is reduced mainly to the R,S-alcohol, whereas the S-warfarin is reduced to a mixture with slightly more S,S-alcohol than S,R-alcohol.[9] In rabbit liver, D-camphor is used preferentially over L-camphor with the production of borneal; L-camphor was reduced mainly to isborneal. In addition, rat and rabbit liver cytosol reduces 2,5-bornaneodione preferentially to 5-endohydroxycamphor rather than 5-exohydroxycamphor.[6] The aromatic ketone reductases of rabbit kidney cortex reduces acetophenone to a mixture of 76% $S(-)$- and 24% $R(+)$- methylphenyl carbinol.[4] Rabbit liver also reduces oxisuran into two alcohol diastereomers in an approximately 70:30 ratio.[10] The narcotic antagonists, naloxone and naltrexone, are reduced at the 6-keto group to 6 α-hydroxy products in chicken and to 6β-hydroxy products in rabbit *in vivo*.[31] Several forms of the dihydromörphinone ketone reductases have been extracted from rabbit liver which yield ratios of α-OH/β-OH naloxol and naltrexol diastereomers varying from 0.1 to 0.01.[15] Most of the above stereospecific properties were established with crude extracts or, at best, only partially purified enzymes. As a result, the lack of stereospecificity or the apparent partial stereospecificities are possibly the result of the activity of multiple enzymes with different absolute stereospecific or stereocatalytic properties.

The precise relationship among the carbonyl reductases requires primary structural information which is not yet available.

In addition to their relationship with other oxidoreductases, structural and immunological analysis will also clarify the question of heterogeneity. For example, is the apparent heterogeneity of rabbit liver aromatic ketone reductase activity a result of multiple gene products or posttranscriptional modifications?

VI. COMMENTS

The need to convert toxic and lipid-soluble carbonyls into less reactive alcohols and/or soluble conjugates as well as the need for the synthesis of biologically important alcohol metabolites apparently require a widespread presence of reductases. Although cells maintain an overall potential for oxidation in the form of a high ratio of NAD^+ and $NADP^+$ to NADH and NADPH, cells retain a selective reduction potential by designing reductases specific for NADPH which is present in excess over $NADP^+$.[32]

The restriction of any enzyme to a particular species, tissue, or subcellular fraction is presumptive evidence of a very specialized enzymatic role in the normal physiology of that animal or organ. Examples are (1) cellulases in plants and microorganisms, (2) cholineesterase in the central nervous system, (3) amylase in salivary glands, (4) collogenases in connective tissue, (5) respiratory enzymes in mitochondria, and (6) permeases in membranes. Even ubiquitous enzyme systems may take on specialized cellular function through specific induction, e.g., in xenobiotic metabolism. On the other hand, if the ubiquitous enzyme system is also constitutive, the principle of biological conservation requires that the system fill an important role in normal cell physiology. Reductases are ubiquitous and apparently constitutive[3] enzymes. Although undoubtedly involved in the metabolism of xenobiotics, the availability of this omnipresent reductive potential also must play an important part in normal cellular physiology. Because of the intracellular multiplicity of reductases, it is possible that numerous, specific as well as nonspecific, enzymes exist simultaneously and, whatever their normal function, their collective presence is necessary for cellular viability.

REFERENCES

1. Williams, R. T. (1959). "Detoxication Mechanisms," pp. 88 and 322. Wiley, New York.
2. Bachur, N. R. (1976). Cytoplasmic aldo-keto reductases: A class of drug metabolizing enzymes. *Science* **193**, 595–597.
3. Leibman, K. C. (1971). Reduction of ketones in liver cytosol. *Xenobiotica* **1**, 97–104.
4. Culp, H. W., and McMahon, R. E. (1968). Reductase for aromatic aldehydes and ketones. The partial purification and properties of a reduced triphosphopyridine nucleotide-dependent reductase from rabbit kidney cortex. *J. Biol. Chem.* **243**, 848–852.
5. Ahmed, N. K., Felsted, R. L., and Bachur, N. R. (1979). Comparison and characterization of mammalian xenobiotic ketone reductases. *J. Pharmacol. Exp. Ther.* **209**, 12–19.
6. Leibman, K. C., and Ortiz, E. (1973). Mammalian metabolism of terpenoids. I. Reduction and hydroxylation of camphor and related compounds. *Drug Metab. Dispos.* **1**, 543–551.
7. Fraser, I. M., Peters, M. A., and Hardinge, M. G. (1967). Purification and some characteristics of an α,β-unsaturated ketone reductase from dog erythrocytes and human liver. *Mol. Pharmacol.* **3**, 233–247.
8. Roerig, S., Fujimoto, J. M., Wang, R. I. H., Pollack, S. H., and Lange, D. (1976). Preliminary characterization of enzymes for reduction of naloxone and naltrexone in rabbit and chicken liver. *Drug Metab. Dispos.* **4**, 53–58.
9. Moreland, T. A., and Hewick, D. S. (1975). Studies on a ketone reductase in human and rat liver and kidney soluble fraction using warfarin as a substrate. *Biochem. Pharmacol.* **24**, 1953–1957.
10. Bachur, N. R., and Felsted, R. L. (1976). Oxisuran reduction by rabbit tissue preparations. *Drug Metab. Dispos.* **4**, 239–243.
11. Leinweber, F. J., Greenough, R. C., Schwender, C. F., Kaplan, H. R., and DiCarlo, F. J. (1972). Bunolol metabolism by cell-free preparations of human liver: Biosynthesis of dihydrobunolol. *Xenobiotica* **2**, 191–202.
12. Leinweber, F. J., and DiCarlo, F. J. (1974). Bunolol metabolism by human and rat red blood cells and extrahepatic tissues. *J. Pharmacol. Exp. Ther.* **189**, 271–277.
13. Ahmed, N. K., Felsted, R. L., and Bachur, N. R. (1978). Heterogeneity of anthracycline antibiotic carbonyl reductases in mammalian livers. *Biochem. Pharmacol.* **27**, 2713–2719.
14. Cohen, G. M., and Flockhart, I. R. (1975). The partial purification and properties of a human erythrocyte 4-nitroacetophenone reductase. *Xenobiotica* **5**, 213–222.
15. Felsted, R. L., Richter, D. R., Jones, D., and Bachur, N. R. Isolation and characterization of rabbit liver xenobiotic carbonyl reductases. *Biochem. Pharmacol.* (in press).
16. Andrews, P. A., Brenner, D. E., Chou, F. E., Kubo, H., and Bachur, N. R. (1980). Facile and definitive determination of human adriamycin and daunorubicin metabolites by high pressure liquid chromatography. *Drug Metab. Dispos.* (in press).
17. Lin, Y. M., and Jarabak, J. (1978). Isolation of two proteins with 9-ketoprostaglandin reductase and NADP-linked 15-hydroxyprostaglandin dehydrogenase activities and studies on their inhibition. *Biochem. Biophys. Res. Commun.* **81**, 1227–1234.

18. Felsted, R. L., Gee, M., and Bachur, N. R. (1974). Rat liver daunorubicin reductase. An aldo-keto reductase. *J. Biol. Chem.* **249**, 3672–3679.
19. Felsted, R. L., Richter, D. R., and Bachur, N. R. (1977). Rat liver aldehyde reductase. *Biochem. Pharmacol.* **26**, 1117–1124.
20. Davidson, W. S., Walton, D. J., and Flynn, T. G. (1978). A comparative study of the tissue and species distribution of NADPH-dependent aldehyde reductase. *Comp. Biochem. Physiol. B* **60**, 309–315.
21. Prelog, V. (1964). Specification of the stereospecificity of some oxido-reductases by diamond lattice sections. *Pure Appl. Chem.* **9**, 119–130.
22. Schoenhard, G. L., Lee, D. J., Howell, S. E., Pawlowski, N. E., Libbey, L. M., and Sinnhuber, R. O. (1976). Aflatoxin B₁ metabolism to aflatoxicol and derivatives lethal to *Bacillus subtilis* GSY 1057 by rainbow trout (*Salmo gairdneri*) liver. *Cancer Res.* **36**, 2040–2045.
23. Roerig, S. C., Fujimoto, J. M., and Wang, R. I. H. (1977). The stimulatory effect of morphine on metabolism of naloxone to 6α-naloxol in the guinea pig. *Drug Metab. Dispos.* **5**, 454–463.
24. Elliott, T. H., Tao, R. C. C., and Williams, R. T. (1965). Stereochemical aspects of the metabolism of the iosmeric methylcyclohexanols and methylcycloheanones. *Biochem. J.* **95**, 59–69.
25. Loveless, H., Arena, E., Felsted, R. L., and Bachur, N. R. (1978). Comparative mammalian metabolism of adriamycin and daunorubicin. *Cancer Res.* **38**, 593–598.
26. Stoffel, W., Lekim, D., and Sticht, G. (1968). Stereospecificity of the NADPH-dependent reduction reaction of 3-oxodihydrosphingosine (2-amino-1-hydroxy-octadecane-3-one). *Hoppe-Seyler's Z. Physiol. Chem.* **349**, 1637–1644.
27. Zannoni, V. G., and Weber, W. W. (1966). Isolation and properties of aromatic α-keto acid reductase. *J. Biol. Chem.* **241**, 1340-1344.
28. Smith, T. E., and Mitoma, C. (1962). Partial purification and some properties of 4-ketoproline reductase. *J. Biol. Chem.* **237**, 1177–1180.
29. Varma, S. D., and Kinoshita, J. H. (1976). Inhibition of lens aldose reductase by flavonoids—their possible role in the prevention of diabetic cataracts. *Biochem. Pharmacol.* **25**, 2505–2513.
30. Bently, R. (1970). "Molecular Asymmetry in Biology," Vol. 2, p. 5. Academic Press, New York.
31. Fujimoto, J. M., Roerig, S., Wang, R. I. H., Chatterjie, N., and Inturrisi, C. E. (1975). Narcotic antagonist activity of several metabolites of naloxone and naltrexone tested in morphine dependent mice (38558). *Proc. Soc. Exp. Biol. Med.* **148**, 443–446.
32. Jacobson, K. B., and Kaplan, N. O. (1957). Pyridine coenzymes of subcellular tissue fractions. *J. Biol. Chem.* **226**, 603–613.

Chapter 14

Xanthine Oxidase and Aldehyde Oxidase

K. V. RAJAGOPALAN

I. INTRODUCTION

Xanthine oxidase (EC 1.2.3.2) and aldehyde oxidase (EC 1.2.3.1) are metalloflavoproteins which are strikingly similar to each other and distinctly different from other metalloflavoproteins. Both enzymes are large proteins with molecular weights of about 300,000; are composed of two subunits of identical size; and contain molybdenum, FAD, and iron (as Fe/S clusters) as prosthetic groups in a ratio of $1:1:4$ per subunit[1-3]. The two enzymes coexist in several species including man. They catalyze a unique reaction, involving oxidative hydroxylation of the substrate, in which the oxygen atom incorporated into the product is derived from

295

ENZYMATIC BASIS OF DETOXICATION, VOL. I

water rather than 0_2. The electrons derived from the substrate are transferred to any of a number of electron acceptors

$$SH + H_2O \rightarrow SOH + 2e^- + 2H^+$$
$$A + 2e^- + 2H^+ \rightarrow AH_2$$

The above mechanism differentiates these ezymes from the group of enzymes known as mixed-function oxidases which effect hydroxylation of their substrates using O_2 as the source of the hydroxyl group. The uniqueness of xanthine oxidase and aldehyde oxidase in this respect stems from the presence of molybdenum in these enzymes, since it is now known that substrate hydroxylation occurs at the molybdenum center of these enzymes.[1-3] The FAD and Fe/S centers are segments of internal electron transfer chains in both enzymes, and electron egress is considered to occur from any one of these centers, depending on the nature of the electron acceptor.[1-3] Both enzymes can oxidize a variety of nitrogen-containing heterocyclic compounds as well as aliphatic and aromatic aldehydes. While xanthine oxidase derives its name from its well-established role in the conversion of xanthine to uric acid, aldehyde oxidase must be considered to be misnamed. While the true physiological function of aldehyde oxidase remains to be determined, detailed information on the substrate specificity of the enzyme is available. Since xanthine oxidase and aldehyde oxidase have overlapping substrate specificities and since many mammalian tissues contain both enzymes, a detailed knowledge of the specificities of each enzyme is important for identifying the enzyme responsible for an observed *in vivo* metabolic reaction involving potential substrates for the two enzymes.

Lack of specificity also characterizes the electron transferring activity of these enzymes, though xanthine oxidizing enzymes from various sources do exhibit some specificity for physiological electron acceptors. In order to place the role of xanthine oxidase and aldehyde oxidase in detoxication in proper perspective it is necessary to consider in some detail the specificities of these enzymes for reducing and oxidizing substrates.

II. ELECTRON DONOR SPECIFICITY

The enzymes for which detailed information is available are bovine milk xanthine oxidase and rabbit liver aldehyde oxidase. The substrate specificity of milk xanthine oxidase has been studied by several groups over many years,[4-11] while a comparative study of xanthine oxidase and aldehyde oxidase has been carried out by Krenitsky et al.[11] Some of the data available to date are summarized in Table I.[2,8,11,12] Since the role of these enzymes in the oxidation of aldehydes is noted in Chapter 12 in this volume, only the heterocyclic substrates are included in the list.

The data presented in Table I show that, despite the fact that both xanthine oxidase and aldehyde oxidase use similar groups of compounds as reducing substrates, they show significant differences in their ability to oxidize specific compounds. Some notable similarities and dissimilarities are discussed below.

A. Purines

Xanthine, the classic substrate for xanthine oxidase, is not a substrate for aldehyde oxidase. Hypoxanthine, another good substrate for xanthine oxidase, is oxidized to xanthine at a much lower rate by aldehyde oxidase. Purine itself is a good substrate for both enzymes, being converted to hypoxanthine, xanthine, and uric acid in succession by xanthine oxidase and to 8-hydroxypurine by aldehyde oxidase. Thus, aldehyde oxidase presents a paradox because of its inability to hydroxylate the C-8 position of hypoxanthine and xanthine, while hydroxylating purine at the C-8 position only.

N-Methylation of hypoxanthine or xanthine has variable effects on oxidizability by xanthine oxidase and aldehyde oxidase. 1-Methyl-hypoxanthine is as efficient a substrate as hypoxanthine for xanthine oxidase but more efficient than hypoxanthine as substrate for aldehyde oxidase. Even more strikingly, methylation at N-3 results in a dramatic increase in reactivity with aldehyde oxidase while having the opposite effect on reactivity with xanthine oxidase. It is apparent that the whimsicalities of these enzymes are beyond the reach of predictability.

Among other purines, adenine and guanine are poor substrates for the two enzymes. Krenitsky et al.[11] have observed that certain nucleosides such as inosine, 2'-deoxyinosine, and purine ribonucleoside serve as low efficiency substrates for aldehyde oxidase and, to a lesser extent, for xanthine oxidase. Theophylline (1,3-dimethylxanthine) is efficiently converted to 1,3-dimethyluric acid by xanthine oxidase but not by aldehyde oxidase.

B. Other Heterocyclic Systems

Quaternary pyridinium compounds, typified by N^1-methyl-nicotinamide, are good substrates for aldehyde oxidase but not for xanthine oxidase. However, as the pH is increased in the region 9 to 11 xanthine oxidase acquires increasing ability to oxidize N^1-

TABLE I

Comparative Substrate Specificities of Bovine Milk Xanthine Oxidase and Rabbit Liver Aldehyde Oxidase[a]

Substrates	Xanthine oxidase		Aldehyde oxidase
Pyridines			
Pyridine	—[b]	[c]	22[a]
Nicotinamide	<14		<22
N^1-Methylnicotinamide	<14		755
1-(4'-Pyridyl)pyridinium ion	<14		111
Nicotine	<14		<22
Pyrimidines			
Pyrimidine	<14		22
2-hydroxy	134		6220
2-amino	<14		<20
4-hydroxy	623		2980
4-hydroxy-6-amino	436		510
5-methyl	<14		400
2-hydroxy-5-fluoro			+++[43]
Purines			
Purine	480	163	2220
2-hydroxy	255	130	3110
6-hydroxy	625	570	67
8-hydroxy	115	12	<22
2,6-dihydroxy	815	815	<22
2,8-dihydroxy	14	2	22
6,8-dihydroxy	575	815	<22
Hypoxanthine			
1-methyl	600	0	755
3-methyl	<14	1	15,800
7-methyl	19	0	600
9-methyl	<14	0	45
Xanthine			
1-methyl	815	367	<22
3-methyl	<14	0	<22
7-methyl	19	0	22
Adenine	29	5[1]	44
Guanine	<14	0[9]	<22
6-Mercaptopurine	82	31[7]	355
6-Thioxanthine	200	380[7]	<22
Azathioprine	<14		178
Theophylline	574		<22
Theobromine	<14		<22
Caffeine	<14		<22
Pyrazolo[3,4-d]pyrimidine			
Pyrazolo[3,4-d]pyrimidine	187		800
4-hydroxy (Allopurinol)	380		355
4,6-dihydroxy (oxipurinol)	<14		<22
Allopurinol ribonucleoside	<14		488
Allopurinol ribonucleotide	<14		<22

(Cont.)

TABLE I (Cont.)

Substrates	Xanthine oxidase		Aldehyde oxidase
8-Azapurine	29	114[6]	710
6-hydroxy	235	335[6]	44
2-hydroxy	19	53[6]	
6-amino	10	25[6]	
2-amino	58	170[6]	
2-methylamino	19	61[6]	
Pteridines			
Pteridine	270		5330
4-hydroxy	410	430[5]	3640
2,4-dihydroxy	385		<22
2-amino-4-hydroxy	480	367[5]	<22
Methotrexate			+ +[43]
3′,5′-dichloro			+ +[43]
Quinolines			
Quinoline	<14		200
N-Methylquinolinium salt	<14		1780

[a] The values shown are turnover numbers (mole substrate oxidized per minute per mole active site) calculated from the data available in the literature and are based on a turnover number of 815 for xanthine oxidase at pH 8.5 at 25° with 0.1 mM xanthine as substrate[2] and 755 for aldehyde oxidase at pH 7.8 at 25° with 2.5 mM N^1-methylnicotinamide as substrate.[12] Under standard assay conditions, AFR (activity/flavin ratio, given as $\Delta A_{295\text{ nm}}$ per minute for the conversion of xanthine to urate/$A_{450\text{ nm}}$ of the enzyme sample for xanthine oxidase and $\Delta A_{300\text{ nm}}$ per minute for the conversion of N-methylnicotinamide to the pyridone/$A_{450\text{ nm}}$ of the enzyme sample for aldehyde oxidase) values of 210 for xanthine oxidase[2] and 100 for aldehyde oxidase[12] should be observable for fully functional enzymes. The percent functionality of any purified enzyme sample from the above sources can be determined from the above values. These values are not directly applicable to enzymes from other sources. K_m values for selected substrates are given in Krenitsky et al.

[b] Data in column from Krenitsky et al., unless otherwise indicated by superscript reference number.

[c] Data in column from Bergmann et al., unless otherwise indicated by superscript reference number.

methylnicotinamide.[13] In accord with this, aldehyde oxidase is the sole enzyme catalyzing the *in vivo* conversion of N^1-methylnicotinamide to the 2- and 4-pyridones in most mammals.[14] In man and rat, however, xanthine oxidase appears to contribute to the *in vivo* formation of additional 2-pyridone.[14]

Suitably substituted pyrimidines are excellent substrates for aldehyde oxidase. Quinoline and N-substituted quinolines are also substrates for aldehyde oxidase, but not for xanthine oxidase. Neither enzyme is capable of N-hydroxylation, as determined from studies on pyrazolo[3,4-*d*] pyrimidines and 8-azapurines. Pyrazolo[3,4-*d*]pyrimidine is converted to allopurinol and then to oxipurinol by both enzymes. This reaction is of use in the unequivocal differentiation between xanthine oxidase and aldehyde oxidase, since oxipurinol is a tight-binding, active-site directed inhibitor of xanthine oxidase[15-17] but has no inhibitory effect on aldehyde oxidase,[18] both *in vitro* and *in vivo*. Pteridines are good substrates for both xanthine oxidase and aldehyde oxidase.

Substrates such a purines and pteridines present multiple potential sites for hydroxylation. For example, purine contains three possible such sites, and xanthine oxidase hydroxylates all three producing uric acid as the end product. Interestingly, the conversion of purine to uric acid by xanthine oxidase follows the preferred sequential pathway of C-6 → C-2 → C-8. The accumulation of detectable amounts of hypoxanthine and xanthine during the reaction shows that after each hydroxylation step the product dissociates from the enzyme before another hydroxylation is effected. With either 2-hydroxypurine or 8-hydroxypurine the C-6 position is the final site of hydroxylation during conversion to uric acid by xanthine oxidase.

The pteridine ring contains four potential sites for hydroxylation: C-2, C-4, C-6, and C-7. Hydroxylation by xanthine oxidase can occur at C-2, C-4, and C-7, whereas aldehyde oxidase hydroxylates only at C-2 and C-4. The pathway followed during multiple hydroxylations depends on the substituents already present.[2-7]

III. ELECTRON ACCEPTOR SPECIFICITY

Xanthine-oxidizing enzymes from various sources are termed either as oxidases or as dehydrogenases in relation to their physiological electron acceptors.[1] Purified preparations from mammalian sources invariably use O_2 as an efficient electron acceptor and are described as oxidases. Enzymes from avian sources utilize NAD as the physiological electron acceptor[19,20] and are called dehydrogenases. The enzyme from *Micrococcus lactilyticus* (now called *Veillonella alcalescens*), which uses ferredoxin as the physiological electron acceptor,[21] is also called a dehydrogenase. In general, enzymes for which O_2 is not an efficient electron acceptor are called dehydrogenases.

The distinction between oxidases and dehydrogenases is now known to be more mechanistic than functional, since Stirpe and co-workers have shown that mammalian xanthine-oxidizing enzymes are NAD -dependent dehydrogenases *in vivo*, but are modified to become O_2-utilizing oxidases during purification.[22,23] Waud and Rajagopalan[24] have purified the NAD-dependent form of the enzyme from rat liver and have studied the factors involved in the conversion of the dehydrogenase form to the oxidase form.[25] The human liver enzyme also undergoes similar conversion[26] as does indeed the enzyme from bovine milk,[27] the classic xanthine oxidase. Whether the oxidase activity of these enzymes observed *in vitro* has physiological relevance is a moot question. There is no report of a dehydrogenase form of aldehyde oxidase.

In addition to the physiological electron acceptor, xanthine oxidase and aldehyde oxidase are capable of transferring electrons to any of a battery of oxidizing agents. The electron acceptor specificity of milk xanthine oxidase was first investigated by Dixon[28] who reported that the enzyme was capable of reducing O_2, methylene blue, nitrate, quinones, H_2O_2, dinitrobenzene, picric acid, permanganate, iodine, and alloxan. Dichlorophenol indophenol,[29] ferricyanide,[29] cytochrome c,[30] phenazine methosulfate,[31] and tetrazolium salts[32] have also been found to be electron acceptors for the enzyme. Xanthine oxidase has also been shown to be capable of transferring electrons to nicotinamide N-oxide[33] and purine N-oxides.[34] Where tested, rabbit liver aldehyde oxidase has been found to have electron acceptor specificity similar to that of xanthine oxidase, with the notable exception that quinones such as menadione, which are excellent electron acceptors for xanthine oxidase, are potent inhibitors of aldehyde oxidase.[35]

Reduction of O_2 by xanthine oxidase and aldehyde oxidase results in the stoichiometric formation of H_2O_2. The extensive studies of Fridovich and coworkers (see Chapter 15, this volume) have shown that both enzymes generate O_2^-, the superoxide anion, in the course of reduction of O_2 to H_2O_2. Wolpert *et al.*[18] have shown that reduction of organic nitro compounds by these enzymes generates the corresponding hydroxyamino derivatives. Nicotinamide N-oxide is reduced by xanthine oxidase to nicotinamide and H_2O.[33] Surprisingly, xanthine oxidase appears to catalyze the direct transfer of oxygen from nicotinamide N-oxide to xanthine to yield uric acid.[33] This is the only instance in which a donor other than H_2O is used as the hydroxylating species by xanthine oxidase.

The generation of O_2^- by these enzymes has some interesting consequences. Thus, aerobically the reduction of cytochrome c by xanthine oxidase or aldehyde oxidase is by an indirect mechanism involving the mediation of the superoxide ion.[36] The aerobic reduction of nitroblue tetrazolium is also mediated by superoxide.[37] Neither enzyme is capable of transferring electrons to cytochrome c anaerobically, but nitroblue tetrazolium does serve as an electron acceptor under anaerobic conditions.[37]

IV. INHIBITORS

Active Site Directed Inhibitors

Since the hydroxylation of substrates occurs at the molybdenum center, this group of inhibitors includes substrate analogues which produce competitive inhibition and reagents which undergo some type of interaction with the molybdenum center and act either as reversible competitive inhibitors or as tight-binding, stoichiometric inhibitors. Substrate analogues producing competitive inhibition include a wide variety of purines, pteridines, and other heterocyclic molecules.[2] Notable among these is 2-amino-4-hydroxy-6-formylpteridine which may be considered to be a stoichiometric inhibitor of xanthine oxidase.[38] Quinacrine is a powerful competitive inhibitor of aldehyde oxidase but a weak noncompetitive inhibitor of xanthine oxidase.[39] Cyanide, arsenite, methanol, and sulfhydryl reagents act as active site directed inhibitors by reason of their ability to interact with the molybdenum center.[3,40] Inactivation by cyanide occurs by a novel reaction in which a unique sulfur moiety associated with the molybdenum center is removed as thiocyanate.[41] The resultant desulfo enzyme is catalytically incompetent but can be reactivated by treatment with inorganic sulfide.[41] Most purified preparations of xanthine oxidase and aldehyde oxidase contain variable amounts of desulfo enzyme and are, therefore, only partially functional (see Table I). Arsenite is a reversible competitive inhibitor of aldehyde oxidase but a progressive, tight-binding inactivator of xanthine oxidase.[40] Inactivation by mercurials and by methanol is dependent on turnover of the enzyme and hence is syncatalytic.[39]

While the types of inhibition mentioned above are common to both xanthine oxidase and aldehyde oxidase, the latter enzyme is unique in its susceptibility to a group of inhibitors including nonionic detergents, amytal, antimycin A, oligomycin, estradiol, and progesterone.[42] These reagents have no effect on the interaction of the reducing substrate with the enzyme, but have differential inhibitory effects on electron transfer to various acceptors and appear to act as inhibitors of the internal electron transport chain. Any activity susceptible to inhibition by these compounds, when crude extracts are used as the source of activity, may be attributed to aldehyde oxidase rather than to xanthine oxidase.

V. HUMAN ENZYMES

As mentioned earlier, the xanthine-oxidizing enzyme of human liver exists as the NAD$^+$-dependent dehydrogenase *in vivo*. The enzyme has been partially purified from autopsy samples of liver (W. R. Waud and K. V. Rajagopalan, unpublished data) and has properties quite similar to those of the enzyme from rat liver.[24,25] The dehydrogenase form is converted to the oxidase form during purification, and the process is reversed on treatment with dithiothreitol. The purified protein cross-reacts efficiently with rabbit antibody directed against rat liver xanthine oxidase.

Human liver aldehyde oxidase has also been partially purified.[43] It would appear that the enzyme is largely inactivated in frozen human liver samples, necessitating the use of fresh tissue samples.[43] The partially purified enzyme is capable of oxidizing aldehydes as well as aromatic heterocyclic compounds, although it is somewhat less efficient toward the latter group of substrates than is the rabbit liver enzyme.[43] Oxidation of substrates containing quaternary nitrogen is unaffected by the electron transport inhibitors mentioned earlier, whereas oxidation of other heterocyclic and aldehydic substrates is inhibited by those reagents. This duality in the mode of action appears to be unique to the human enzyme.

VI. RELEVANCE TO DETOXICATION

While the role of xanthine oxidase in the generation of uric acid, a urinary end product, is well established, the question arises as to whether the enzyme performs some other vital detoxication function in the normal state. Lack of such a role for the enzyme is strongly suggested by two lines of evidence. The genetic deficiency of the enzyme in humans, termed xanthinuria, is apparently benign, since the deficient individuals are physically and mentally healthy and do not suffer from any debilitating pathological conditions.[44] Along the same lines, rats which are rendered deficient in xanthine oxidase by administration of high levels of tungsten, a molybdenum antagonist, grow and reproduce normally.[45] While the benign nature of xanthine oxidase deficiency argues against a role for the enzyme in the detoxication of endogenous toxicants, it does not negate the possibility of detoxication of xenobiotic compounds by the enzyme. While human genetic deficiency of aldehyde oxidase has not been reported, the above conclusions also can be reached regarding that enzyme on the basis of the findings with the molybdenum-deficient rats.

The fact that purified xanthine oxidase and aldehyde oxidase produce H_2O_2 when using O_2 as electron acceptor has led to the suggestion that these enzymes might serve *in vivo* to generate the H_2O_2 required for detoxication reactions using H_2O_2 as the oxidizing agent.[46] However, the results obtained with molybdenum-deficient rats, and the observation that the native form of xanthine oxidase is in fact a dehydrogenase with low reactivity with O_2, weaken this argument considerably. The possible role of these enzymes as generators of O_2^- *in vivo* is also of questionable validity for the same reasons.

Between them, xanthine oxidase and aldehyde oxidase encompass a wide spectrum of substrate specificity for oxidizable substrates, including aldehydes and substituted pyridines, pyrimidines, quinolines, purines, azapurines, pyrazalopyrimidines, pteridines, and even more heterocyclic systems. Since hydroxylation in general is a detoxication reaction in itself or a preliminary step before detoxication, it is conceivable that these enzymes are designed to cover a wide range of specificity for any non-physiological compounds in these classes for detoxication purposes. To quote Krenitsky *et al.*,[11] "Although it is not known what compounds are the primary substrates of aldehyde oxidase *in vivo*, the extremely high levels of this enzyme in some mammalian herbivores, compared with the low levels in carnivores, suggests that plants may be rich sources of these compounds (substituted pyrimidines)." Until toxicological studies are carried out with substituted pyrimidines such a role for these enzymes remains in the realms of speculation.

It is known that the rabbit has unusually high resistance to toxicity from methotrexate.[47] This resistance correlates well with the ability of rabbit liver aldehyde oxidase to hydroxylate methotrexate and the decreased toxicity of the hydroxylated product.[48] Methotrexate is not a good substrate for human liver aldehyde oxidase.

Among the electron acceptors, organic nitro compounds are converted to the hydroxyamino derivatives by both enzymes. This reaction could conceivably minimize the formation of nitroso derivatives, and could fall into the class of detoxication reactions. More definitively, purine *N*-oxides are known to be oncogenic,[49] and the ability of xanthine oxidase, and presumably of aldehyde oxidase, to reduce these compounds to the free bases is certainly a detoxication process. It would be of interest to see whether the purine *N*-oxides are better carcinogens for xanthine oxidase-deficient animals than for control animals.

The experimental model of xanthine oxidase deficiency produced in laboratory animals by tungsten treatment[45] should prove to be highly useful in assessing the detoxication roles of xanthine oxidase and aldehyde oxidase. This technique cannot by itself differentiate between aldehyde oxidase and xanthine oxidase, since both activities would be abolished by tungsten treatment. Such differentiation can, however, be achieved by further experimentation by allopurinol treatment of the animals, since such a treatment will inhibit xanthine oxidase activity but not aldehyde oxidase activity.

VII. PURIFICATION AND ASSAY

Since human liver xanthine oxidase, like the rat liver enzyme, functions as a NAD-utilizing dehydrogenase but is converted into an O_2-utilizing oxidase during purification, it is of interest to review the procedures used for the purification of the rat liver enzyme. The use of fractionation techniques such as heat treatment and ammonium sulfate precipitation leads to rapid conversion of the dehydrogenase to the oxidase,[23] and procedures which have included such steps have permitted purification of the oxidase form of the enzyme.[24, 50-52] Under certain circumstances the purified oxidase can be reconverted to the dehydrogenase form by treatment with dithiothreitol.[52] The dehydrogenase form of the rat liver enzyme has also been purified through the selective use of fractionation procedures.[24] Essentially similar procedures have been used for purification of human liver xanthine oxidase in both the oxidase and the dehydrogenase forms (W. R. Waud and K. V. Rajagopalan, unpublished data).

The standard assay procedure for xanthine oxidase involves the spectrophotometric assay of the aerobic conversion of xanthine to uric acid, by measuring the increase in $A_{295\ nm}$. Since the enzyme is a poor oxidase *in vivo,* this procedure yields erroneous values in extracts which contain the dehydrogenase form of the enzyme and has to be modified to include NAD as the electron acceptor.[24]

Aldehyde oxidase does not appear to exist as a dehydrogenase. Purification procedures have been described for the rabbit liver enzyme,[42,53] and partial purification of human liver aldehyde oxidase has been reported.[43] Since the substrate specificity of aldehyde oxidase appears to vary markedly depending on the species, it is important to test a variety of substrates for detecting the enzyme in a new source. Inhibition by electron transport chain inhibitors[42,43] is a good criterion for specifying a reaction carried out by aldehyde oxidase.

VIII. REGULATION

The complex nature of xanthine oxidase and aldehyde oxidase could, in theory, permit regulation of their activities through several mechanisms. These would include processes such as altered rates of synthesis and degradation with or without hormonal control, activation and deactivation through control of essential active site components such as the molybdenum moiety or the cyanolysable sulfur, modulation of the dehydrogenase and oxidase forms, and effect of small molecules such as product inhibitors. In rats, hepatic xanthine oxidase levels are affected by the protein content of the diet, and the increase in activity observed when the animals are transferred from a low protein diet to a high protein diet is controlled at the transcriptional level.[50] The levels of aldehyde oxidase in the mouse are under both genetic and hormonal control.[54,55] An interesting aspect of the control of xanthine oxidase is the observation that while the enzyme activity is expressed by cultured cells from many mammalian sources, human cells in culture are totally devoid of xanthine oxidase activity.[56] The activity of the enzyme is also greatly decreased in hepatomas.[57]

IX. COMMENTS

The extremely wide substrate specificity of xanthine oxidase and aldehyde oxidase for reducing substrates as well as for oxidizing substrates makes them potential candidates for detoxication systems. While their role as detoxifying agents are established in some instances, further investigation along these lines is rendered possible by the availability of experimental models for enzyme deficiency.

REFERENCES

1. Rajagopalan, K. V., and Handler, P. (1965). Metalloflavoproteins. *In* "Biological Oxidations" (T. P. Singer, ed.), pp. 301–337. Wiley (Interscience), New York.
2. Massey, V. (1973). Iron-sulfur flavoprotein hydroxylases. *In* "Iron-Sulfur Proteins" (W. Lovenberg, ed.), Vol. 1, pp. 301–360. Academic Press, New York.
3. Bray, R. C. (1975). Molybdenum iron-sulfur flavin hydroxylases and related enzymes. *Enzymes* 12, 299–419.
4. Wyngaarden, J. B. (1957). 2,6-Diaminopurine as substrate and inhibitor of xanthine oxidase. *J. Biol. Chem.* 224, 453–462.
5. Bergmann, F., and Kwietny, H. (1959). Pteridines as substrates for xanthine oxidase. II. Pathways and rates of oxidation. *Biochim. Biophys. Acta* 33, 29–46.
6. Bergmann, F., Levin, G., and Kwietny, H. (1959). 8-Azapurines as substrates of mammalian xanthine oxidase. *Arch. Biochem. Biophys.* 80, 318–325.
7. Bergmann, F., and Ungar, H. (1960). The enzymatic oxidation of 6-mercaptopurine to 6-thiouric acid. *J. Am. Chem. Soc.* 82, 3957–3960.
8. Bergmann, F., Kwietny, H., Levin, G., and Brown, D. J. (1960). The action of mammalian xanthine oxidase on N-methylated purines. *J. Am. Chem. Soc.* 82, 598–605.
9. Bergmann, F., Levin, G., Kwietny-Gorvin, H., and Ungar, H. (1961). The influence of mono- and dimethylamino substituents on the enzymic oxidation of purines. *Biochim. Biophys. Acta* 47, 1–17.
10. Valerino, D. M., and McCormack, J. J. (1969). Studies of the oxidation of some aminopteridines by xanthine oxidase. *Biochim. Biophys. Acta* 184, 154–163.
11. Krenitsky, T. A., Shannon, M. N., Elion, G. B., and Hitchings, G. H. (1972). A

comparison of the specificities of xanthine oxidase and aldehyde oxidase. *Arch. Biochem. Biophys.* **150**, 585–599.

12. Branzoli, U., and Massey, V. (1974). Evidence for an active site persulfide residue in rabbit liver aldehyde oxidase. *J. Biol. Chem.* **249**, 4346–4349.
13. Greenlee, L., and Handler, P. (1964). Xanthine oxidase. VI. Influence of pH on substrate specificity. *J. Biol. Chem.* **239**, 1090–1095.
14. Stanulović, M., and Chaykin, S. (1971). Aldehyde oxidase: Catalysis of the oxidation of N^1-methylnicotinamide and pyridoxal. *Arch. Biochem. Biophys.* **145**, 27–34.
15. Elion, G. B. (1966). Enzymatic and metabolic studies with allopurinol. *Ann. Rheum. Dis.* **25**, 608–614.
16. Spector, T., and Johns, D. G. (1970). Stoichiometric inhibition of reduced xanthine oxidase by hydroxypyrazolo(3,4-*d*)pyrimidines. *J. Biol. Chem.* **245**, 5079–5085.
17. Massey, V., Komai, H., Palmer, G., and Elion, G. B. (1970). On the mechanism of inactivation of xanthine oxidase by allopurinol and other pyrazolo(3,4-*d*)-pyrimidines. *J. Biol. Chem.* **245**, 2837–2844.
18. Wolpert, M. K., Althaus, J. R., and Johns, D. G. (1973). Nitroreductase activity of mammalian liver aldehyde oxidase. *J. Pharmacol. Exp. Ther.* **185**, 202–213.
19. Rajagopalan, K. V., and Handler, P. (1967). Purification and properties of chicken liver xanthine dehydrogenase. *J. Biol. Chem.* **242**, 4097–4107.
20. Cleere, W. F., and Coughlan, M. P. (1975). Avian xanthine dehydrogenases. I. Isolation and characterization of the turkey liver enzyme. *Comp. Biochem. Physiol. B* **50**, 311–322.
21. Smith, S. T., Rajagopalan, K. V., and Handler, P. (1967). Purification and properties of xanthine dehydrogenase from *Micrococcus lactilyticus*. *J. Biol. Chem.* **242**, 4108–4117.
22. Della Corte, E., and Stirpe, F. (1968). Regulation of xanthine oxidase in rat liver: Modifications of the enzyme activity of rat liver supernatant on storage at $-20°$. *Biochem. J.* **108**, 349–351.
23. Stirpe, F., and Della Corte, E. (1969). The regulation of rat liver xanthine oxidase: Conversion *in vitro* of the enzyme activity from dehydrogenase (Type D) to oxidase (Type O). *J. Biol. Chem.* **244**, 3855–3863.
24. Waud, W. R., and Rajagopalan, K. V. (1976). Purification and properties of NAD$^+$-dependent (Type D) and O$_2$-dependent (Type O) forms of rat liver xanthine dehydrogenase. *Arch. Biochem. Biophys.* **172**, 354–364.
25. Waud, W. R., and Rajagopalan, K. V. (1976). The mechanism of conversion of rat liver xanthine dehydrogenase from an NAD$^+$-dependent form (Type D) to an O$_2$-dependent form. *Arch. Biochem. Biophys.* **172**, 365–379.
26. Della Corte, E., Gozzetti, G., Novello, F., and Stirpe, F. (1969). Properties of the xanthine oxidase from human liver. *Biochim. Biophys. Acta* **191**, 164–166.
27. Battelli, M. G., Lorenzoni, E., and Stirpe, F. (1973). Milk xanthine oxidase Type D (dehydrogenase) and Type O (oxidase): Purification, interconversion and some properties. *Biochem. J.* **131**, 191–198.
28. Dixon, M. (1926). Studies on xanthine oxidase. VII. The specificity of the system. *Biochem. J.* **20**, 703–718.
29. Mackler, B., Mahler, H. R., and Green, D. E. (1954). Studies on metalloflavoproteins. I. Xanthine oxidase, a molybdoflavoprotein. *J. Biol. Chem.* **210**, 149–164.
30. Horecker, B. L., and Heppell, L. A. (1949). The reduction of cytochrome *c* by xanthine oxidase. *J. Biol. Chem.* **178**, 683–690.
31. Fridovich, I., and Handler, P. (1958). Xanthine oxidase. IV. The participation of iron in internal electron transport. *J. Biol. Chem.* **233**, 1581–1585.
32. Anderson, A. D., and Patton, R. L. (1954). Determination of xanthine oxidase in insects with tetrazolium salts. *Science* **120**, 956.

33. Murray, K. N., Watson, J. G., and Chaykin, S. (1966). Catalysis of the direct transfer of oxygen from nicotinamide N-oxide to xanthine by xanthine oxidase. *J. Biol. Chem.* **241**, 4798–4801.
34. Stohrer, G., and Brown, G. B. (1969). The reduction of purine N-oxides by xanthine oxidase. *J. Biol. Chem.* **244**, 2498–2502.
35. Mahler, H. R., Fairhurst, A. S., and Mackler, B. (1955). Studies on metalloflavoproteins. IV. The role of the metal. *J. Am. Chem. Soc.* **77**, 1514–1521.
36. Fridovich, I., and Handler, P. (1962). Xanthine oxidase. V. Differential inhibition of the reduction of various electron acceptors. *J. Biol. Chem.* **237**, 916–921.
37. Rajagopalan, K. V., and Handler, P. (1964). Hepatic aldehyde oxidase. II. Differential inhibition of electron transfer to various electron acceptors. *J. Biol. Chem.* **239**, 2022–2026.
38. Lowry, O. H., Bessey, O. A., and Crawford, E. J. (1949). Pterine oxidase. *J. Biol. Chem.* **180**, 399–410.
39. Rajagopalan, K. V., and Handler, P. (1964). Hepatic aldehyde oxidase. III. The substrate-binding site. *J. Biol. Chem.* **239**, 2027–2035.
40. Coughlan, M. P., Rajagopalan, K. V., and Handler, P. (1969). The role of molybdenum in xanthine oxidase and related enzymes: Reactivity with cyanide, arsenite and methanol. *J. Biol. Chem.* **244**, 2658–2663.
41. Massey, V., and Edmondson, D. (1970). On the mechanism of inactivation of xanthine oxidase by cyanide. *J. Biol. Chem.* **245**, 6595–6598.
42. Rajagopalan, K. V., Fridovich, I., and Handler, P. (1962). Hepatic aldehyde oxidase. I. Purification and properties. *J. Biol. Chem.* **237**, 922–928.
43. Johns, D. G. (1967). Human liver aldehyde oxidase: Differential inhibition of oxidation of charged and uncharged substrates. *J. Clin. Invest.* **46**, 1492–1505.
44. Wyngaarden, J. B. (1978). Hereditary xanthinuria. *In* "The Metabolic Basis of Inherited Disease" (J. B. Stanbury, J. B. Wyngaarden, and D. S. Frederickson, eds.), 4th ed., pp. 1037–1044. McGraw-Hill, New York.
45. Johnson, J. L., Rajagopalan, K. V., and Cohen, H. J. (1974). Molecular basis of the biological function of molybdenum. Effect of tungsten on xanthine oxidase and sulfite oxidase in the rat. *J. Biol. Chem.* **249**, 859–866.
46. Fried, R., Fried, L. W., and Babin, D. R. (1973). Biological role of xanthine oxidase and tetrazolium reductase inhibitor. *Eur. J. Biochem.* **33**, 439–445.
47. Redetzki, H. M., Redetzki, J. E., and Elias, A. L. (1966). Resistance of the rabbit to methotrexate: Isolation of a drug metabolite with decreased toxicity. *Biochem. Pharmacol.* **15**, 425–433.
48. Johns, D. G., Iannotti, A. T., Sartorelli, A. C., Booth, B. A., and Bertino, J. R. (1965). The identity of rabbit liver methotrexate oxidase. *Biochim. Biophys. Acta* **105**, 380–382.
49. Sugiura, K., and Brown, G. B. (1967). Purine N-oxides. XIX. On the oncogenic N-oxide derivatives of guanine and xanthine and a non-oncogenic isomer of xanthine N-oxide. *Cancer Res.* **27**, 925–931.
50. Rowe, P. B., and Wyngaarden, J. B. (1966). The mechanism of dietary alterations in rat hepatic xanthine oxidase levels. *J. Biol. Chem.* **241**, 5571–5576.
51. Johnson, J. L., Waud, W. R., Cohen, H. J., and Rajagopalan, K. V. (1974). Molecular basis of the biological function of molybdenum: Molybdenum-free xanthine oxidase from livers of tungsten-fed rats. *J. Biol. Chem.* **249**, 5056–5061.
52. Della Corte, E., and Stirpe, F. (1972). The regulation of rat liver xanthine oxidase. Involvement of thiol groups in the conversion of the enzyme activity from dehydrogenase (Type D) into oxidase (Type O) and purification of the enzyme. *Biochem. J.* **126**, 739–745.

53. Felsted, R. L., En-Yuen Chu, A., and Chaykin, S. (1973). Purification and properties of the aldehyde oxidases from hog and rabbit livers. *J. Biol. Chem.* **248,** 2580–2587.
54. Huff, S. D., and Chaykin, S. (1967). Genetic and androgenic control of N^1-methylnicotinamide oxidate activity in mice. *J. Biol. Chem.* **242,** 1265–1270.
55. Gluecksohn-Waelsch, S., Greengard, P., Quinn, G. P., and Teicher, L. S. (1967). Genetic variation of an oxidase in mammals. *J. Biol. Chem.* **242,** 1271–1273.
56. Brunschede, H., and Krooth, R. S. (1973). Studies on the xanthine oxidase activity of mammalian cells. *Biochem. Genet.* **8,** 341–350.
57. Prajda, N., and Weber, G. (1975). Malignant transformation-linked imbalance: Decreased xanthine oxidase activity in hepatomas. *FEBS Lett.* **59,** 245–249.

Chapter 15

Superoxide Dismutases: Detoxication of a Free Radical

H. MOUSTAFA HASSAN and IRWIN FRIDOVICH

I. INTRODUCTION

Most of the life forms we see are absolutely dependent on molecular oxygen, yet all living cells are prone to oxygen toxicity. The toxicity of oxygen has been related to the intermediates of oxygen reduction including the superoxide anion radical (O_2^-), hydrogen peroxide (H_2O_2), and the hydroxyl radical ($OH\cdot$). These are very reactive substances which can directly or indirectly cause substantial damage to living cells. It is obvious, therefore, that the first and the best defense against oxygen toxicity

311

ENZYMATIC BASIS OF DETOXICATION, VOL. I

would be avoidance of the generation of these reactive intermediates. Indeed, this is the case and most of the oxygen consumed by respiring cells is used by the cytochrome oxidase system which accomplishes the tetravalent reduction of oxygen to water, without the release of any free reactive intermediates. Nevertheless, both O_2^- and H_2O_2 are normal, if minor, products of the biological reduction of oxygen, and aerobic organisms are able to survive by virtue of a unique set of defensive enzymes that scavenge these reactive intermediates of oxygen reduction. Obligate anaerobes and mutants lacking these enzymes are killed upon exposure to atmospheric oxygen. The superoxide radical is eliminated by the superoxide dismutases which catalyze Eq. (1) while the hydrogen peroxide is removed by catalases and/or peroxidases. This chapter will attempt to summarize our current knowledge on the nature of superoxide dismutases, and on their physiological role in providing protection against oxygen toxicity.

$$O_2^- + O_2^- + 2H^+ \rightarrow H_2O_2 + O_2 \tag{1}$$

II. INTERMEDIATES OF OXYGEN REDUCTION

Molecular oxygen, in the ground state, is paramagnetic because it contains two unpaired parallel electronic spins. It is thus a diradical. These parallel electronic spins hinder the direct divalent reduction of oxygen which, without inversion of one electron spin, would result in the forbidden situation of two electrons with parallel spins occupying the same orbital. This spin restriction makes molecular oxygen much less reactive as an oxidant than we would otherwise expect, permitting the coexistence of organic matter and free oxygen. One way to avoid the spin restriction is to electronically excite molecular oxygen. This elevates an electron to a higher orbital and at the same time inverts its spin state. The resultant singlet oxygen is vastly more reactive than ground state dioxygen. There are two readily accessible singlet states of dioxygen. One of these ($^1\Delta_g$) lies 23 kcal above the ground state, while the second ($^1\epsilon_g$) is at 37 kcal above ground. Energies of this magnitude are available in visible light, and the phenomenon of photosensitized oxidation often involves the production of singlet oxygen. Common ''dark'' chemistry very seldom involves singlet oxygen because energies of activation of 23 kcal or more cannot be provided thermally at ordinary temperatures. A second way to circumvent the spin restriction is to ligate the dioxygen to a transition metal which bears unpaired spins of its own. Delocalization of electrons within the complex allows effective pairing of spins. This is an essential feature of catalysis by metal-containing oxidases, such as the cytochrome

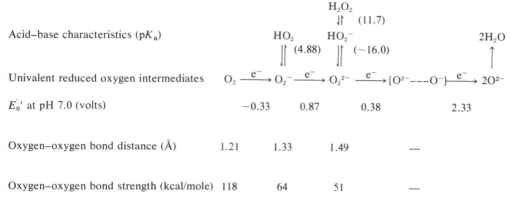

Fig. 1. Chemical and physical properties of reduced oxygen intermediates.

oxidase. There is a third way, and it involves adding electrons to dioxygen one at a time. Electronic spins can be inverted, but in a time frame orders of magnitude greater than the lifetime of collisional complexes. In the univalent pathway of oxygen reduction, such spin inversion can occur during the lifetime of the intermediates produced by successive, single electron reductions of dioxygen. This is the most facile route of oxygen reduction and is most often followed in spontaneous oxidation. The spin restriction is more thoroughly discussed by Taube[1] and by Hamilton.[2] Figure 1 outlines the univalent pathway of oxygen reduction and presents some of the properties of the intermediates on this pathway.

III. O_2^- AS SUBSTRATE

A. General Properties of O_2^-

The superoxide anion, O_2^- is the conjugate base of a weak acid, the perhydroxyl radical, $HO_2\cdot$, whose pK_a is 4.8.[3] The superoxide radical is unstable in protic solvents, and will spontaneously dismute to generate hydrogen peroxide and oxygen. The rate of the spontaneous dismutation is dependent on the pH; this rate is highest at pH equal to pK_a (4.8). At a lower pH, HO_2 becomes the predominant form while at a higher pH O_2^- predominates. The dismutation reactions of these species and their rate constants[4] are

$$HO_2 + HO_2 \rightarrow H_2O_2 + O_2 \qquad K_2 \simeq 7.6 \times 10^5 \; M^{-1} \; sec^{-1} \qquad (2)$$

$$HO_2 + O_2^- + H^+ \rightarrow H_2O_2 + O_2 \qquad K_2 \simeq 8.8 \times 10^7 \; M^{-1} \; sec^{-1} \qquad (3)$$

$$O_2^- + O_2^- + 2H^+ \rightarrow H_2O_2 + O_2 \qquad K_2 < 0.3 \; M^{-1} \; sec^{-1} \qquad (4)$$

The reaction of O_2^- with itself is exceedingly slow, and it is this reaction that is catalyzed by the superoxide dismutases.

The dismutation reactions and other properties of O_2^- have been reviewed.[5]

The superoxide radical can act as a reductant (E_0' for O_2^-/O_2 is about -0.33 V) as well as an oxidant (E_0' for H_2O_2/O_2^- is $+0.87$ V). The chemical methods for the detection of O_2^- make use of these properties. Thus, indicating scavengers, at concentrations sufficient to compete effectively with the spontaneous dismutation reaction, can detect every O_2^- radical generated in the course of a reaction. There are, of course, reductants or oxidants other than O_2^- that can react with the indicating scavengers, but superoxide dismutase can be used to distinguish reactions that are mediated by O_2^- from those that are not.

HO_2 absorbs at 230 nm where $E_m = 1220\ M^{-1}\ cm^{-1}$, while O_2^- absorbs maximally at 240 nm where $E_m = 2000\ M^{-1}\ cm^{-1}$.[3] These optical properties of this radical have been used to study its production by pulse radiolysis and its decay kinetics, both spontaneous and catalyzed. The rapid freezing of aqueous solutions has permitted stabilization of O_2^- and its detection by electron paramagnetic resonance (EPR) where $g_\perp = 2.00$ and $g_\parallel = 2.08$.[6]

This method has been applied to the detection of O_2^- produced by the aerobic xanthine oxidase reaction. Lack of sensitivity of these nonintegrative methods greatly limits their application to systems of interest to biochemists.

B. Biological Generation of O_2^-

The superoxide free radical, O_2^-, is generated in many biological reactions that reduce molecular oxygen. Thus, the autoxidation of leucoflavins, hydroquinones, catecholamines, thiols, tetrahydropterins, hemoproteins, and reduced ferredoxins, have all been shown to generate O_2^- (for references, see Fridovich[7,8]). O_2^- is also produced during the action of several oxidative enzymes, i.e., xanthine oxidase, aldehyde oxidase, and dihydroorotic dehydrogenase. Subcellular organelles, such as mitochondria[9] and chloroplasts,[10,11] have been shown to produce O_2^-. Finally, O_2^- is produced by activated polymorphonuclear leukocytes (PMN) and macrophages during the respiratory burst which accompanies activation and phagocytosis.[12]

Attempts to measure the rates of O_2^- generation inside cells are hampered by the ubiquity of superoxide dismutases. A recent approach to this problem[13] used a specific inhibitory antibody to suppress the activity of superoxide dismutase and found that 17% of the oxygen consumed by crude extracts of *Streptococcus faecalis,* in the presence of NADH, was by a univalent pathway that generated O_2^-. Although the quantitatively most significant sources of O_2^-, within any given type of cell remains unknown, it is safe to conclude that superoxide radical is a commonplace product of biological oxygen reduction.

C. Cytotoxicity of O_2^-

Superoxide radical is much less reactive than the hydroxyl radical.[5,14] However, it has been demonstrated that enzymatically, photochemically, or electrochemically generated fluxes of O_2^- will decompose methional to ethylene,[15] inactivate virus,[16] kill bacteria,[17,18] lyse erythrocytes,[19] kill granulocytes,[20] damage myoblasts in culture,[21] initiate the peroxidation of linolenate,[22] depolymerize hyaluronate,[23] inactivate enzymes,[16,19] and damage DNA.[24] In most cases the damaging effects of O_2^- depended upon the simultaneous presence of H_2O_2. Furthermore, superoxide dismutase, catalase, or compounds that scavenge $OH \cdot$, protected against the damaging effects of O_2^-. These observations led to the proposal[15] that O_2^- reduces H_2O_2 to produce $OH \cdot$, a reaction originally suggested by Haber and Weiss.[25] Attempts to demonstrate the direct interaction between O_2^- and H_2O_2, have failed.[26,27] However, McCord and Day[28] have demonstrated that iron complexes catalyze the generation of $OH \cdot$ from O_2^- plus H_2O. Thus O_2^- reduces the ferric complex to the ferrous state which, in turn, reduces H_2O_2 to $OH \cdot$ as shown in Eqs. (5)–(7).

$$O_2^- + Fe^{3+}\text{-EDTA} \rightarrow O_2 + Fe^{2+}\text{-EDTA} \tag{5}$$

$$Fe^{2+}\text{-EDTA} + H_2O_2 \rightarrow Fe^{3+}\text{-EDTA} + OH^- + OH\cdot \tag{6}$$

$$\overline{O_2^- + H_2O_2 \rightarrow O_2 + OH^- + OH\cdot} \tag{7}$$

This would explain the failure to demonstrate the reaction between O_2^- and H_2O_2 in scrupulously clean systems.

Since O_2^- itself generates H_2O_2, it is clear that O_2^- production could lead to cell damage via the ultimate production of $OH \cdot$. Superoxide dismutases minimize the production of hydroxyl radical and possibly other reactive species, as well, by keeping the steady-state concentration of superoxide radical vanishingly low.

IV. SUPEROXIDE DISMUTASE—THE ENZYME

A. Assays for Superoxide Dismutase

Reliable and convenient assays are one of the cornerstones of enzymology. Superoxide dismutases are unique among enzymes in that their substrate (O_2^-) is an unstable free radical. This led to the necessity of special assays: some are direct and others are indirect. Since these methods have been reviewed,[29,30] this presentation will be brief.

1. Direct Assays

These assays usually require special equipment which is not widely available. However, they are very useful for studying the kinetics and mechanisms of the purified enzymes.

Direct measurements of the catalytic action of superoxide dismutase was made possible by the use of pulse radiolysis to generate O_2^- in aqueous solutions, combined with photometric monitoring of its concentration. With this technique, a rate constant of $1.9 \times 10^9\ M^{-1}\ sec^{-1}$ at pH between 5 and 9.5 was reported for the bovine erythrocyte superoxide dismutase.[31,32]

Rapid-freeze EPR has also been used to directly detect the activity of superoxide dismutase. Thus, the enzyme diminished the EPR signal of O_2^- generated in aerobic media by the reoxidation of redox dyes.[33,34]

McClune and Fee[26] used potassium superoxide as a source of O_2^- and followed changes in its concentration at 275 nm using stopped flow spectrophotometry. A similar technique was used[35] for assaying the enzyme at alkaline pH (8.9 to 12.7), where the spontaneous dismutation rate is suppressed.

A polarographic procedure in which a dropping mercury electrode, coated with triphenylphosphine oxide, acted both as a source of O_2^- and as a detector of its concentration, was used to assay superoxide dismutase.[36] The method gave linear response at superoxide dismutase concentrations between 10^{-10} and $1.5 \times 10^{-8}\ M$, and was useful at pH values between 9 and 10.

Fluorine-19 nuclear magnetic relaxation (^{19}F NMR) rate of CuZn- and Mn-superoxide dismutase has been used to assay the enzyme.[37] The method is reported to be dependent on the native state of the enzyme, although it is independent of its catalytic activity. It depends on the ability of halides to bind to the active center of the enzyme. The ^{19}F relaxation rate was linearly dependent on the enzyme concentration up to $10^{-5}\ M$. The method is very much less sensitive when compared with others that depend on the catalytic activity of the enzyme.

2. Indirect Assays

Sensitive and convenient assays are based upon combining a continuing source of O_2^- with an indicating scavenger of O_2^-. Superoxide dismutase then competes with the indicating scavenger for the available flux of O_2^- and thus signals its presence by its effect on the reaction of the indicating scavenger with O_2^-. It should be understood that indirect assays are done at very low steady-state levels of O_2^- (about 10^{-12} M) and respond to superoxide dismutase in the range 5 to 500 ng/ml, i.e., 0.1 to 10 nM enzyme. All of the indirect assays are conceptually identical, although they may be divided into two categories based upon a superficial difference. In one group the superoxide dismutase decreases the rate being followed, while in the second group it augments that rate. These may be called negative and positive assays, respectively.

a. Negative Type Assays. The discovery of superoxide dismutase was based, in part, on its ability to inhibit the reduction of cytochrome c by xanthine oxidase.[38,39] Thus, xanthine oxidase, acting on xanthine in the presence of oxygen, generates O_2^- which reduces cytochrome c. In turn, superoxide dismutase inhibits this O_2^--mediated cytochrome c reduction, without interfering with the catalytic turnover of xanthine oxidase. One unit of superoxide dismutase is defined as that amount of the enzyme which yields 50% inhibition of the rate of cytochrome c reduction, under specified conditions, i.e., pH, temperature, cytochrome c concentration, and rate of generation of O_2^-. The assay is reliable and remains widely used.

Other indirect assays were developed in which other sources of O_2^- and other indicating scavengers of O_2^- were used. Thus, O_2^- can be generated photochemically during the spontaneous reoxidation of photo-reduced flavins,[40] or chemically via the autoxidation of epinephrine to adrenochrome at elevated pH,[41] or via the autoxidation of pyrogallol at pH 8.2,[42] or by the oxidation of NADH with phenazine methosulfate.[43]

Several indicating scavengers of O_2^- have also been used. Thus O_2^- can reduce Nitro Blue Tetrazolium to the blue formazan which can be measured at 560 nm.[44] This was the basis for a very useful activity stain for the localization of superoxide dismutase activity bands on polyacrylamide gels, and for the dismutase assay in solutions as well, in which photochemically generated O_2^- was used to reduce Nitro Blue Tetrazolium.[44] Also O_2^- can initiate sulfite oxidation[45,46] and nitrite formation from hydroxyl ammonium chloride.[47]

b. Positive Type Assays. Recently, Misra and Fridovich[48,49] have developed a "positive assay" for the enzyme which is simple and sensitive. It eliminated the use of expensive substances (such as xanthine oxidase, cytochrome c, or Nitro Blue Tetrazolium) and is applicable to both liquid assays and to polyacrylamide gel electropherograms. In this procedure the photooxidation of dianisidine, as sensitized by riboflavin, is augmented by the dismutase; the rate can be measured at 460 nm. In this assay, the dismutase also acts by catalytically scavenging O_2^-. It appears to increase the photooxidation of dianisidine because O_2^-, generated photochemically, reduces an intermediate dianisidine radical and thereby decreases the rate of net dianisidine oxidation. The enzyme eliminates this back reaction and appears to accelerate the oxidation of dianisidine.

Marshall and Worsfold[50] used a Clark oxygen electrode, to measure the rate of oxygen consumption during photochemical generation of O_2^- in a reaction mixture containing both Nitro Blue Tetrazolium and superoxide dismutase. In the absence of the enzyme, the tetrazolium dye will oxidize the photochemically generated O_2^- back to oxygen without net oxygen uptake. However, in the presence of superoxide dismutase, which competes with the dye for available O_2^-, a net increase in oxygen uptake is observed.

B.   Classes and Distribution of Superoxide Dismutases

Superoxide dismutase is found in all oxygen-consuming organisms,[51] in some aerotolerant anaerobes,[52] and in some obligate anaerobes.[53] All superoxide dismutases are metalloproteins. The enzymes that have been isolated, from a wide range of organisms, fall into three classes depending on the metal found in their active centers; these in turn fall into two phylogenetic classes depending on amino acid sequence homology. Thus, there are superoxide dismutases (SOD) that contain both copper and zinc (CuZn-SOD), iron (Fe-SOD), *or* manganese (Mn-SOD). The iron and manganese-containing superoxide dismutases are characteristic of prokaryotes and show a high degree of sequence homology.[54,55] The copper–zinc-containing superoxide dismutases are characteristic of eukaryotic cytosols and share no sequence homology with Fe- or Mn-containing class of enzymes.[55] Recently, the complete amino acid sequence of the Mn-superoxide dismutase of *Escherichia coli* B[56] confirmed the absence of any degree of homology with the erythrocyte CuZn-superoxide dismutase. Therefore, it seems that these two classes of superoxide dismutases are of independent evolutionary origins, and that they have evolved in response to a common selection pressure, i.e., the appearance of molecular oxygen in the biosphere.

Mitochondria contain a Mn-superoxide dismutase that has high degree of sequence homology with the prokaryotic enzyme.[54,55] This supports the theory that mitochondria may have evolved from a prokaryote via an endocellular symbiosis.[57]

The finding of a CuZn-superoxide dismutase in the symbiotic prokaryote, *Photobacterium leiognathi*,[58] represents the only known exception to the rule that this class of enzymes are characteristic of eukaryotic cytosols. It appears possible that *P. leiognathi* might have acquired its CuZn enzyme from its eukaryotic host, the pony fish, via gene transfer.[59] This proposal is currently under test. The Mn-superoxide dismutase is not restricted to the mitochondrial matrix in all species: in human and baboon livers the enzyme has been found both in the mitochondria and in the cytosol.[60]

The presence of superoxide dismutase activity in some "anaerobic" bacteria may provide insight into the evolution of these microorganisms and of the enzymes themselves. It is of interest that most of the primitive anaerobes, e.g., the sulfur-reducing *Desulfovibrio desulfuricans*[61] and the photosynthetic *Chromatium*,[11] contain an iron-containing superoxide dismutase similar to that found in *E. coli* when grown anaerobically.[62] This raises two possibilities: (1) Fe-SOD is the most primitive form of superoxide dismutase and was acquired by the early strict anaerobes to protect themselves against the selective pressure of oxygen that was produced in the early oceans[63] as a result of the Urey effect, i.e., photolysis of water by UV; or (2) Fe-SOD in the anaerobes is a recent acquisition, via a plasmid or other means of gene transfer, in order to allow survival during transient exposure to oxygen. Although the erratic appearance of superoxide dismutase in anaerobic organisms[52,53] seems to favor the latter hypothesis, it is clear that the evolutionary history of superoxide dismutase retains many unanswered and fascinating questions.

C. Physicochemical Properties and Purification

The copper-zinc-superoxide dismutases (originally known as cupreins) have been isolated from a wide range of eukaryotes including bovine erythrocytes, bovine brain, bovine heart, horse liver, human liver, baboon liver, human erythrocytes, human brain, yeast, *Photobacter leiognathi*, *Neurospora*, spinach leaves, garden peas, wheat germ, and chicken liver, etc. (for complete list of references, see Fridovich[64]). The cupreins (CuZn-SOD) from the different human organs were found to be immunologically identical.[65,66] Furthermore, immunological cross-reactivity was seen between the human and monkey cupreins but no cross-reactivity was found between the human and pig, chicken, or cow cupreins.[65] Cell hybridization techniques have allowed the assignment of the human CuZn-SOD gene to chromosome 21[67] and the human Mn-SOD gene to chromosome 6.[68,69]

The CuZn-superoxide dismutases from vertebrates, fungi, and plants have a very similar amino acid composition, with a high degree of structural homology. The enzyme is a homodimer whose molecular weight is

32,000, and which contains one copper and one zinc per subunit. The complete amino acid sequence of the bovine erythrocyte enzyme[70-72] and X-ray diffraction analysis[73,74] have demonstrated the structural uniqueness of this group of enzymes. The subunit resembles a cylinder composed of eight strands of the peptide chain arranged in an antiparallel β-structure. This β-barrel structure involves residues 2-11, 13-23, 26-35, 38-47, 80-88, 91-100, 112-118, and 142-149. Two nonhelical coils, involving residues 48-79 and 119-141, protrude from one side of the β-barrel and together enclose and constitute the active center. Each subunit is stabilized by an intrachain disulfide bond between Cys-55 and Cys-144. The protein is devoid of tryptophan.[70] The enzyme is remarkably stable and remains active in the presence of 9.0 M urea or 4% sodium dodecyl sulfate.[75] Dialysis at low pH against EDTA caused loss of the copper and the zinc, with concomitant loss of activity. Only the addition of copper to the apoenzyme restored full activity.[39,76,77] On the other hand, zinc may be replaced by Co^{2+}, Hg^{2+} or Cd^{2+} with some changes in thermal stability, but without gross change in activity.[75] Thus, it appears that the copper plays a catalytic function while the zinc plays a structural role. The copper is ligated to histidines-44, -46, and -118 and is joined to the zinc by histidine-61; zinc is also ligated to histidines-67 and -78 and to aspartate-81. The copper is exposed to solvent[73]; cyanide binds to this copper and causes a reversible inhibition of the enzyme.[32] Hydrogen peroxide irreversibly inactivates the enzyme[78-80] via its interaction with the copper.[81] The sensitivity of CuZn-SOD to cyanide has been used to identify this enzyme in eukaryotes, since the Mn-superoxide dismutase is cyanide insensitive.[82]

The enzyme is very stable, and its activity is not affected by relatively harsh treatments, such as heating to 70°, exposure to chloroform plus ethanol, and precipitation with acetone. Actually, most of the isolation procedures have made use of these unusual properties. The enzyme can be isolated from lysed red cells after treatment with chloroform plus ethanol to remove hemoglobin (Tsuchihashi procedure), by adding dibasic potassium phosphate until an organic-rich phase is salted out.[39] The enzyme may then be precipitated from the organic phase with cold acetone. Ion exchange chromatography is usually used as a last step to remove residual impurities.

The manganese- and iron-containing superoxide dismutases[83,84] are totally unrelated to the CuZn-containing enzymes, except for their catalytic activity. They are resistant to cyanide and are destroyed by the "Tsuchihashi" treatment. The iron-containing enzyme is irreversibly inactivated with hydrogen peroxide[85]; this property has been used to distinguish Fe-SOD from Mn-SOD[13] since Mn-SOD is unaffected by such treatment.

Azide reversibly inhibited the activities of all forms of superoxide dismutase.[86] The degree of inhibition was dependent on the form of the enzyme, with the iron-superoxide dismutase as the most sensitive and CuZn-SOD as the least sensitive to azide treatment. Fifty percent inhibition of the Fe-, Mn-, and CuZn-SOD enzymes was achieved by 4, 20, and 32 mM azide, respectively.

The amino acid sequence of the Mn-SOD from *E. coli* has recently been completed.[56] The structural analysis of this class of enzymes is not yet completed. The bacterial enzymes are most often dimeric with subunit molecular weight of about 20,000; the manganese-containing enzymes from *Thermus thermophilus* and *Thermus aquaticus* are tetrameric.[87,88] The Mn-SOD's from chicken liver[82] and from yeast mitochondria[89] were found to be tetramers. The number of metal binding sites per subunit of Fe-SOD or Mn-SOD remains to be clarified, since published data show a range from 0.5 to 1 metal atom per subunit.

Metal-free apoproteins have been prepared from the Mn-SOD of *E. coli*[90] and of *Bacillus stearothermophilus*.[91] Removal of manganese resulted in complete loss of catalytic activity and reconstitution of a fully active enzyme was only possible when manganese was restored. These results demonstrated that the manganese is essential for the activity of Mn-superoxide dismutases.

The purification procedure for Mn-SOD and Fe-SOD from *E. coli*[83,84] consists of (a) treatment of cell-free extracts with streptomycin sulfate in order to remove the nucleic acids, although it was found recently[92] that streptomycin treatment generates electrophoretically modified, but active, forms of the enzyme in extracts of *E. coli* and of *Streptococcus faecalis;* (b) the supernatant liquid was subjected to ammonium sulfate fractionation and the precipitate was dissolved in and dialyzed against in potassium acetate at pH 5.5; (c) the clarified dialysate was passed onto a carboxymethyl cellulose (CM-cellulose) column which adsorbs the Mn-SOD; (d) Fe-SOD was not adsorbed and the dialyzed effluent was adsorbed onto a column of DEAE-cellulose from which Fe-SOD was eluted as a single symmetrical peak by a linear phosphate gradient (e) Mn-SOD was eluted from the CM-cellulose with a gradient of potassium acetate of pH 5.5; (f) the fractions containing the Mn-SOD were concentrated by ultrafiltration and rechromatographed on a DEAE-cellulose.

D. Catalytic Mechanism

Studies of the CuZn-SOD, by pulse radiolysis, revealed cyclical changes in the valence of the copper brought about by O_2^-.[31,32,93-95] Partial bleaching of the enzyme was noticed during catalysis and was interpreted as being due to the reduction of cupric to cuprous at the active center. When the enzyme was first reduced to the cuprous state and then exposed to a pulse of O_2^-, a partial oxidation of the enzyme was observed. Similar findings have been obtained with the manganese- and iron-containing superoxide dismutases.[96-99] A general mechanism for catalysis[59] is presented byEqs. (8) and (9) where E is the enzyme and Me is the metal. The copper in CuZn-SOD is represented as oscillating between cupric and cuprous states, whereas iron and manganese in Fe-SOD and Mn-SOD oscillate between the trivalent and the divalent states.

V. THE PHYSIOLOGICAL ROLE OF SUPEROXIDE DISMUTASES

All evidence available to date indicates that the dismutation of superoxide radical is the true and the only known biological function of the superoxide dismutases. *In vitro* as well as *in vivo* studies have demonstrated that O_2^- is cytotoxic, either by direct reaction with cellular components or as the result of its ability to interact with H_2O_2, under the influence of metal catalysis, to generate OH· and possibly singlet oxygen. Superoxide dismutases protects against this toxicity. Although the physiology of this detoxifying enzyme has received most attention in prokaryotes, many of the insights gained from this work may be applicable to higher organisms.

A. Regulation and Synthesis

Exposure to elevated pO_2 elicits increased synthesis of superoxide dismutase. This has been seen both in prokaryotes and in eukaryotes: exposure of *Streptococcus faecalis*,[100] *Escherichia coli*,[101] *Saccharomyces cerevisiae*,[102] *Euglena gracilis*,[11] adult rats,[103] and numerous other organisms to oxygen resulted in increased intracellular level of superoxide dismutase.

$$E\text{-}Me^n + O_2^- \rightarrow E\text{-}Me^{n-1} + O_2 \tag{8}$$

$$E\text{-}Me^{n-1} + O_2^- + 2H^+ \rightarrow E\text{-}Me^n + H_2O_2 \tag{9}$$

In *E. coli,* which contains two superoxide dismutases, Fe-SOD appears to be constitutive and is made in anaerobically grown cells, while Mn-SOD is inducible by oxygen and is absent from anaerobically grown cells. Transfer of anaerobically grown *E. coli* to air caused rapid synthesis of Mn-SOD.[62]

The synthesis of Mn-SOD in *E. coli* seems to be under rigorous control. The cells are thus able to modulate efficiently the level of this enzyme to meet their defense requirements. When $E.$ $coli$ was maintained in a glucose-limited chemostat, under constant and abundant aeration, the rate of respiration and the concentration of superoxide dismutase varied in proportion to the specific growth rate.[104] These results provided the first clue that the inducer is not molecular oxygen per se, but rather a product of its metabolism, since oxygenation of the culture was constant and abundant at all the specific growth rates. The subsequent finding that the level of superoxide dismutase in the cell was dependent on the metabolic state of the cell, as elicited by dependence upon different carbon sources, supported the view that O_2 is not the true inducer. Thus, cells grown on glucose, as the sole source of carbon, contained lower amounts of superoxide dismutase than cells grown on succinate, lactate, or trypticase–peptone medium.[105] When cells were shifted from a glucose-minimal medium to a rich medium (trypticase soy plus yeast extract), which contained limited amount of glucose (0.25%), the cells' content of the enzyme decreased in the early phase, while glucose was being utilized, and then increased abruptly as the glucose was exhausted and the cells began to utilize other components of the medium by a more oxidative type of metabolism. The rate of O_2^- production appeared to be low during glucose utilization, but increased during the oxidative phase so that cells modulated the level of dismutase to meet their changing needs for the scavenging of O_2^-. The effect of glucose was not due to a classic type catabolite repression, since adding cyclic AMP had no effect on the synthesis of the enzyme.[105] Recent findings[105,106] that increasing the intracellular production of O_2^-, via the cyclic oxidation-reduction of paraquat at constant pO_2, caused increased synthesis of Mn-SOD, lent strong support to the thesis that the level of this enzyme is regulated by the rate of O_2^- generated intracellularly. Several redox-active compounds were found to behave like paraquat.[107] Pyocyanine, phenazine methosulfate, streptonigrin, juglone, menadione, plumbagin, methylene blue, and azure c were all effective in elevating intracellular production of O_2^- and all increased the synthesis of Mn-SOD in $E.$ $coli.$ These redox-active compounds were found to increase the rate of cyanide-resistant respiration in the cells, which was taken as a measure of the rate of O_2^- being generated.[106-108] The increase of biosynthesis of the Mn-SOD by paraquat was prevented by inhibitors of transcription or of translation, but not by an inhibitor of replication.[107] Thus, the increase in Mn-SOD biosynthesis caused by paraquat, and presumably by the other compounds capable of the intracellular generation of O_2^-, was due to de $novo$ enzyme synthesis activated or derepressed at the level of transcription; O_2^-, or a unique product of it, was the effector. Extracellularly generated O_2^- did not cause induction of the enzyme, and evidence was presented showing that O_2^- does not cross the $E.$ $coli$ envelope.[108]

Our understanding of the regulation of Mn-SOD in prokaryotes is far from complete, but steady progress is being made towards this goal.

B. Protection against Oxygen Toxicity

Oxygen is toxic[7,8,29,59,64,109-111] and mutagenic.[112-115] The degree of toxicity is dependent on the concentration of oxygen and on the specific organism. Thus, mammals cannot survive long exposure to pure oxygen, and they quickly die from central nervous system damage when exposed to high pressures of oxygen. Some prokaryotes can survive such exposure, while others die when exposed to normal or even to reduced partial pressures of oxygen.

The superoxide theory of oxygen toxicity states that superoxide radical is an important agent of the toxicity of oxygen, and that superoxide dismutases provide the primary defense against such toxicity.[8,51,64] There are several lines of evidence which support this theory. Only strict anaerobes, that are oxygen sensitive, or organisms that do not utilize oxygen, were found to lack superoxide dismutase.[51] Positive correlation was found between the levels of superoxide dismutase, in those anaerobes that have it, and their tolerance to oxygen.[52] The presence of superoxide dismutase in some anaerobes, that do not consume oxygen in their normal metabolism, is an indication of their potential vulnerability to superoxide radicals. They have apparently acquired the dismutase to ensure their survival during transient exposures to oxygen.

Mutants of *E. coli* deficient in superoxide dismutase and/or catalase were found to be ultrasensitive to oxygen toxicity. One mutant was temperature sensitive with respect to oxygen tolerance and was found to be temperature sensitive with respect to superoxide dismutase biosynthesis.[116] Recently, oxygen-sensitive mutants of *E. coli* were found to fall into two groups.[117] The first had lost the ability to induce both catalase and peroxidase when exposed to oxygen but retained ability to induce Mn-SOD. The second group had lost the ability to induce all three enzymes. Oxygen was lethal to both groups. However, when a member of the first group of mutants was prevented from inducing Mn-SOD by the presence of puromycin, susceptibility to the lethal effect of oxygen was greatly increased. These results indicated that enzymatic scavenging of both O_2^- and H_2O_2 provides the ultimate defense against oxygen toxicity. The mutants were found to revert to wild-type phenotype (oxygen tolerance) at a very high frequency, possibly as the result of endogenous mutagenesis caused by the reactivities of the unscavenged O_2^- and H_2O_2. The revertants were found to have regained the ability to induce the activity of the missing enzymes, except for one revertant from the first group that had become oxygen tolerant without regaining the activities of catalase and peroxidase. Further characterization of this revertant revealed that it had lost the ability to respire, and therefore could dispense with catalase and peroxidase activities, since it could no longer generate O_2^- or H_2O_2. This revertant mutant is physiologically similar to the strain of *Lactobacillus plantarum* that was found to be oxygen indifferent and was devoid of catalase and superoxide dismutase activities.[17,51] Transfer of anaerobically grown wild-type *E. coli* K-12 to air, in the presence of puromycin which prevented it from inducing the Mn-SOD, resulted in loss of viability and sufficient structural damage to the cells that it was evident in electron micrographs.[62] Similar results were obtained with the obligate anaerobic mutants.[117]

Increased levels of Mn-SOD in *E. coli* as induced by (1) oxygenation, (2) increasing the specific growth rate or (3) by growth in presence of redox-active compounds imparted increased resistance toward the toxicity of hyperbaric oxygen or toward the oxygen enhancement of streptonigrin lethality.[62,104,106,118]

Exposure of rats to 85% oxygen caused increases in the level of their lung superoxide dismutase which showed a temporal correlation with the acquisition of tolerance toward 100% oxygen.[103] Recently, it was found that exposure of mammalian lungs to 95% O_2 for 24 hr, or of lung slices and cells in culture to 100% O_2 for 5 hr, increased the level of Mn-SOD, but not of CuZn-SOD.[119]

C. Protection against Oxygen-Dependent Lethalities

There are several redox-active compounds that can easily be reduced within cells and whose reduced forms can in turn react rapidly with oxygen to generate O_2^- and H_2O_2. These compounds tend to increase the rate of production of O_2^- and H_2O_2 in the cells and thereby exacerbate the toxicity of normal concentrations of oxygen. Paraquat, streptonigrin, pyocyanine, juglone, menadione, plumbagin, mitomycin, daunomycin, etc., belong to this class of compounds. The lethality of paraquat,[118] of streptonigrin,[120] and of pyocyanine[121] all required the presence of an electron source and of oxygen. *Escherichia coli* routinely responded to this threat by increasing the rate of synthesis of the Mn-SOD, and variably (depending on the compound used) also of catalase.[106,107,118] Interference with the rate of synthesis of Mn-SOD, either by adding inhibitors of protein biosynthesis or by the use of minimal media, during exposure to these redox compounds, augmented the susceptibility of *E. coli* to their lethality. In minimal medium, paraquat was lethal to *E. coli* under aerobic conditions, but had no effect anaerobically. However, in rich media, that can support faster growth rates and more rapid protein synthesis, *E. coli* was much more tolerant to aerobic paraquat, and this tolerance correlated with the content of Mn-SOD.[106,118] An oxygen enhancement of paraquat lethality has been seen in rats,[122] and the involvement of O_2^- and of lipid peroxidation in its lethality has been documented.[123]

Oxygen is also known to enhance the cytotoxic effects of ionizing radiation, and there are several reports that indicated the involvement of oxygen radicals. Superoxide dismutase was seen to protect against radiation damage to DNA,[24] viruses and mammalian cells in culture,[21] dilute suspensions of *E. coli*,[124] bacteriophage,[16] and mice.[125] In the case of *E. coli*,[124] catalase and scavengers of OH· also reduced the oxygen enhancement ratio (OER), and it was concluded that both O_2^- and H_2O_2, generated by X irradiation in the surrounding medium, were important agents of oxygen enhancement.

The weight of evidence makes it vividly clear that the generation of O_2^- in biological systems is frequently unavoidable and poses a threat to the integrity of the cell. Superoxide dismutases are indispensable for protection against this cytotoxic byproduct of aerobic existence.

REFERENCES

1. Taube, H. (1965). Mechanisms of oxidation with oxygen. *J. Gen. Physiol.* **49**, Suppl., 29–52.
2. Hamilton, G. A. (1974). Chemical models and mechanisms of oxygenases. *In* "Molecular Mechanisms of Oxygen Activation" (O. Hayaishi, ed.), pp. 405–451. Academic Press, New York.
3. Behar, D., Czapski, G., Rabani, J., Dorfman, L. M., and Schwarz, H. A. (1970). Acid dissociation-constant and decay kinetics of perhydroxyl radical. *J. Phys. Chem.* **74**, 3209–3213.
4. Bielski, B. H. J., and Allen, A. O. (1977). Mechanism of the disproportionation of superoxide radicals. *J. Phys. Chem.* **81**, 1048–1050.
5. Czapski, G. (1971). Radiation chemistry of oxygenated aqueous solutions. *Annu. Rev. Phys. Chem.* **22**, 171–208.
6. Nilsson, R., Pick, F. M., Bray, R. C., and Fielden, M. (1969). ESR evidence for O_2^- as a long-lived transient in irradiated oxygenated alkaline aqueous solutions. *Acta Chem. Scand.* **23**, 2554–2556.
7. Fridovich, I. (1975). Superoxide dismutases. *Annu. Rev. Biochem.* **44**, 147–159.
8. Fridovich, I. (1979). Superoxide and superoxide dismutases. *In* "Advances in Inorganic Biochemistry" (G. L. Eichhorn and D. L. Marzilli, eds.), pp. 67–90. Am. Elsevier, New York.
9. Cadenas, E., Boveris, A., Ragan, C. I., and Stoppani, A. O. M. (1977). Production of superoxide radical and hydrogen peroxide by NADH-ubiquinone reductase and ubiquinol-cytochrome *c* reductase from beef-heart mitochondria. *Arch. Biochem. Biophys.* **180**, 248–257.

10. Halliwell, B. (1975). Hydroxylation of p-coumaric acid by illuminated chloroplasts: The role of superoxide. *Eur. J. Biochem.* **55**, 355–360.
11. Asada, K., Kanematsu, S., Takehashi, M., and Kona, Y. (1976). Superoxide dismutases in photosynthetic organisms. *Adv. Exp. Med. Biol.* **74**, 551–564.
12. Babior, B. M. (1978). Oxygen-dependent microbial killing by phagocytes. *N. Engl. J. Med.* **298**, 659–668.
13. Britton, L., Malinowski, D. P., and Fridovich, I. (1978). Superoxide dismutase and oxygen metabolism in *Streptococcus faecalis* and comparison with other organisms. *J. Bacteriol.* **134**, 229–236.
14. Bielski, B. H. J., and Richter, H. W. (1977). A study of the superoxide radical chemistry by stopped-flow radiolysis and radiation induced oxygen consumption. *J. Am. Chem. Soc.* **99**, 3019–3023.
15. Beauchamp, C., and Fridovich, I. (1970). A mechanism for the production of ethylene from methional: The generation of the hydroxyl radical by xanthine oxidase. *J. Biol. Chem.* **245**, 4641–4646.
16. Lavelle, F., Michelson, A. M., and Dimitrejevic, L. (1973). Biological protection by superoxide dismutase. *Biochem. Biophys. Res. Commun.* **55**, 350–357.
17. Gregory, E. M., and Fridovich, I. (1974). Oxygen metabolism in *Lactobacillus plantarum. J. Bacteriol.* **117**, 166–169.
18. Babior, B., Curnutte, J. T., and Kipnes, R. S. (1975). Biological defense mechanisms: Evidence for the participation of superoxide in bacterial killing by xanthine oxidase. *J. Lab. Clin. Med.* **85**, 235–244.
19. Kellogg, E. W., III, and Fridovich, I. (1977). Liposome oxidation and erythrocyte lysis by enzymically-generated superoxide and hydrogen peroxide. *J. Biol. Chem.* **252**, 6721–6728.
20. Salin, M. L., and McCord, J. M. (1975). Free radicals and inflammation. Protection of phagocytosing leukocytes by superoxide dismutase. *J. Clin. Invest.* **56**, 1319–1323.
21. Michelson, A. M., and Buckingham, M. E. (1974). Effects of superoxide radicals on myoblast growth and differentiation. *Biochem. Biophys. Res. Commun.* **58**, 1079–1086.
22. Kellogg, E. W., III, and Fridovich, I. (1975). Superoxide, hydrogen peroxide, and singlet oxygen in lipid peroxidation by a xanthine oxidase system. *J. Biol. Chem.* **250**, 8812–8817.
23. McCord, J. M. (1974). Free radicals and inflammation: Protection of synovial fluid by superoxide dismutase. *Science* **185**, 529–531.
24. Van Hemmen, J. J., and Meuling, W. J. A. (1975). Inactivation of biologically active DNA by γ-ray-induced superoxide radicals and their dismutation products singlet molecular oxygen and hydrogen peroxide. *Biochim. Biophys. Acta* **402**, 133–141.
25. Haber, F., and Weiss, J. (1934). The catalytic decomposition of hydrogen peroxide by iron salts. *Proc. R. Soc. London, Ser. A* **147**, 332–351.
26. McClune, G. J., and Fee, J. A. (1976). Stopped flow spectrophotometric observation of superoxide dismutation in aqueous solution. *FEBS Lett.* **67**, 294–298.
27. Halliwell, B. (1976). An attempt to demonstrate a reaction between superoxide and hydrogen peroxide. *FEBS Lett.* **72**, 8–10.
28. McCord, J. M., and Day, E. D., Jr. (1978). Superoxide-dependent production of hydroxyl radical catalyzed by iron-EDTA complex. *FEBS Lett.* **86**, 139–142.
29. Fridovich, I. (1974). Superoxide dismutases. *Adv. Enzymol.* **41**, 35–97.
30. McCord, J. M., Crapo, J. D., and Fridovich, I. (1977). Superoxide dismutase assays: A review of methodology. *In* "Superoxide and Superoxide dismutases" (A. M. Michelson, J. M. McCord, and I. Fridovich, eds.), pp. 11–17. Academic Press, New York.
31. Klug, D., Rabani, J., and Fridovich, I. (1972). A direct demonstration of the catalytic

action of superoxide dismutase through the use of pulse radiolysis. *J. Biol. Chem.* **247,** 4839–4842.

32. Rotilio, G., Bray, R. C., and Fielden, E. M. (1972). A pulse radiolysis study of superoxide dismutase. *Biochim. Biophys. Acta* **268,** 605–609.

33. Ballou, D., Palmer, G., and Massey, V. (1969). Direct demonstration of superoxide anion production during the oxidation of reduced flavin and of its catalytic decomposition by erythrocuprein. *Biochem. Biophys. Res. Commun.* **36,** 898–904.

34. Orme-Johnson, W. H., and Beinert, H. (1969). On the formation of the superoxide anion radical during the reaction of reduced iron–sulfur proteins with oxygen. *Biochem. Biophys. Res. Commun.* **36,** 905–911.

35. Marklund, S. (1976). Spectrophotometric study of spontaneous disproportionation of superoxide anion radical and sensitive direct assay for superoxide dismutase. *J. Biol. Chem.* **251,** 7504–7507.

36. Rigo, A., Viglino, P., and Rotilio, G. (1975). Polarographic determination of superoxide dismutase. *Anal. Biochem.* **68,** 1–8.

37. Rigo, A., Viglino, P., Argese, E., Terenzi, M., and Rotilio, G. (1979). Nuclear magnetic relaxation of ^{19}F as a novel assay method of superoxide dismutase. *J. Biol. Chem.* **254,** 1759–1760.

38. McCord, J. M., and Fridovich, I. (1968). The reduction of cytochrome *c* by milk xanthine oxidase. *J. Biol. Chem.* **243,** 5753–5760.

39. McCord, J. M., and Fridovich, I. (1969). Superoxide dismutase: An enzymic function of erythrocuprein. *J. Biol. Chem.* **244,** 6049–6055.

40. Massey, V., Strickland, S., Mayhew, S. G., Howell, L. G., Engel, P. C., Matthews, R. G., Schuman, M., and Sullivan, P. A. (1969). The production of superoxide anion radicals in the reaction of reduced flavins and flavoproteins with molecular oxygen. *Biochem. Biophys. Res. Commun.* **36,** 891–897.

41. Misra, H. P., and Fridovich, I. (1972). The role of superoxide anion in the autoxidation of epinephrine and a simple assay for superoxide dismutase. *J. Biol. Chem.* **247,** 3170–3175.

42. Marklund, S., and Marklund, G. (1974). Involvement of the superoxide anion radical in the autoxidation of pyrogallol and a convenient assay for superoxide dismutase. *Eur. J. Biochem.* **47,** 469–474.

43. Nishikimi, M., Rao, N. A., and Yagi, K. (1972). The occurrence of superoxide anion in the reaction of reduced phenazine methosulfate and molecular oxygen. *Biochem. Biophys. Res. Commun.* **46,** 847–853.

44. Beauchamp, C., and Fridovich, I. (1971). Superoxide dismutase: Improved assays and an assay applicable to polyacrylamide gels. *Anal. Biochem.* **44,** 276–287.

45. McCord, J. M., and Fridovich, I. (1969). The utility of superoxide dismutase in studying free radical reactions. I. Radicals generated by the interaction of sulfite, dimethyl sulfoxide, and oxygen. *J. Biol. Chem.* **244,** 6056–6063.

46. Tyler, D. D. (1975). Polarographic assay and intracellular distribution of superoxide dismutase in rat liver. *Biochem. J.* **147,** 493–504.

47. Elstner, E., and Heupel, A. (1976). Inhibition of nitrite formation from hydroxylammonium chloride: A simple assay for superoxide dismutase. *Anal. Biochem.* **70,** 616–620.

48. Misra, H. P., and Fridovich, I. (1977). Superoxide dismutase: "Positive" spectrophotometric assays. *Anal. Biochem.* **79,** 553–560.

49. Misra, H. P., and Fridovich, I. (1977). Superoxide dismutase and peroxidase: A positive activity stain applicable to polyacrylamide gel electropherograms. *Arch. Biochem. Biophys.* **183,** 511–515.

50. Marshall, M. J., and Worsfold, M. (1978). Superoxide dismutase: A direct, continuous linear assay using the oxygen electrode. *Anal. Biochem.* **86**, 561–573.

51. McCord, J. M., Keele, B. B., Jr., and Fridovich, I. (1971). An enzyme-based theory of obligate anaerobiosis: The physiological function of superoxide dismutase. *Proc. Natl. Acad. Sci., U.S.A.* **68**, 1024–1027.

52. Tally, F. P., Goldin, H. R., Jacobus, N. V., and Gorbach, S. L. (1977). Superoxide dismutase in anaerobic bacteria of clinical significance. *Infect. Immun.* **16**, 20–25.

53. Hewitt, J., and Morris, J. G. (1975). Superoxide dismutase in some obligately anaerobic bacteria. *FEBS Lett.* **50**, 315–318.

54. Steinman, H. M., and Hill, R. L. (1973). Sequence homologies among bacterial and mitochondrial superoxide dismutases. *Proc. Natl. Acad. Sci., U.S.A.* **70**, 3725–3729.

55. Harris, J. I., and Steinman, H. M. (1977). Amino acid sequence homologies among superoxide dismutases. In "Superoxide and Superoxide Dismutases" (A. M. Michelson, J. M. McCord, and I. Fridovich, eds.), pp. 225–230. Academic Press, New York.

56. Steinman, H. M. (1978). The amino acid sequence of mangano superoxide dismutase from *Escherichia coli* B. *J. Biol. Chem.* **253**, 8708–8720.

57. Fridovich, I. (1974). Evidence for the symbiotic origin of mitochondria. *Life Sci.* **14**, 819–826.

58. Puget, K., and Michelson, A. M. (1974). Isolation of a new copper-containing superoxide dismutase, bacteriocuprein. *Biochem. Biophys. Res. Commun.* **58**, 830–838.

59. Fridovich, I. (1978). The biology of oxygen radicals. *Science* **201**, 875–880.

60. McCord, J. M., Boyle, J. A., Day, E., Jr., Rizzolo, L. J., and Salin, M. L. (1977). A manganese-containing superoxide dismutase from human liver. In "Superoxide and Superoxide Dismutases" (A. M. Michelson, J. M. McCord, and I. Fridovich, eds.), pp. 129–138. Academic Press, New York.

61. Hatchikian, C. E., LeGall, J., and Bell, G. R. (1977). Significance of superoxide dismutase and catalase activities in the strict anaerobes, sulfate reducing bacteria. In "Superoxide and Superoxide Dismutases" (A. M. Michelson, J. M. McCord, and I. Fridovich, eds.), pp. 159–172. Academic Press, New York.

62. Hassan, H. M., and Fridovich, I. (1977). Enzymatic defenses against the toxicity of oxygen and of streptonigrin in *Escherichia coli* K12. *J. Bacteriol.* **129**, 1574–1583.

63. Lumsden, J., and Hall, D. O. (1975). Superoxide dismutase in photosynthetic organisms provides an evolutionary hypothesis. *Nature (London)* **257**, 670–672.

64. Fridovich, I. (1976). Oxygen radicals, hydrogen peroxide, and oxygen toxicity. In "Free Radicals in Biology" (W. A. Pryor, ed.), Vol. 1, pp. 239–277. Academic Press, New York.

65. Shields, G. S., Markowitz, H., Klassen, W. H., Cartwright, G. E., and Wintrobe, M. M. (1961). Studies on copper metabolism. XXXI. Erythrocyte copper. *J. Clin. Invest.* **40**, 2007–2015.

66. Hartz, J. W., Funakoshi, S., and Deutsch, H. F. (1973). The levels of superoxide dismutase and catalase in human tissues as determined immunochemically. *Clin. Chim. Acta* **46**, 125–132.

67. Tan, Y. H., Tischfield, J., and Ruddle, F. H. (1973). The linkage of genes for the human interferon-induced antiviral protein and indophenol oxidase-B traits to chromosome G-21. *J. Exp. Med.* **137**, 317–330.

68. Creagan, R., Tischfield, J., Ricciuti, F., and Ruddle, F. H. (1973). Chromosome assignments of genes in man using mouse-human somatic cell hybrids: Mitochondrial superoxide dismutase [indophenol oxidase-B, tetrameric] to chromosome 6. *Humangenetik* **20**, 203–209.

69. Sinet, P. M. (1977). SOD genes in humans: Chromosome location and electrophoretic

variants. *In* "Superoxide and Superoxide Dismutases" (A. M. Michelson, J. M. McCord, and I. Fridovich, eds.), pp. 459–465. Academic Press, New York.

70. Steinman, H. M., Naik, V. R., Abernathy, J. L., and Hill, R. L. (1974). Bovine erythrocyte superoxide dismutase. Complete amino acid sequence. *J. Biol. Chem.* **249,** 7326–7338.

71. Evans, H. M., Steinman, H. M., and Hill, R. L. (1974). Bovine erythrocyte superoxide dismutase. Isolation and characterization of tryptic, cyanogen bromide, and maleylated tryptic peptides. *J. Biol. Chem.* **249,** 7315–7325.

72. Abernethy, J. L., Steinman, H. M., and Hill, R. L. (1974). Bovine erythrocyte superoxide dismutase. Subunit structure and sequence location of the intrasubunit disulfide bond. *J. Biol. Chem.* **249,** 7339–7347.

73. Richardson, J. S., Thomas, K. A., Rubin, B. H., and Richardson, D. C. (1975). Crystal structure of bovine Cu,Zn superoxide dismutase at 3 Å resolution: Chain tracing and metal ligands. *Proc. Natl. Acad. Sci. U.S.A.* **72,** 1349–1353.

74. Richardson, J. S., Thomas, K. A., and Richardson, D. C. (1975). Alpha-carbon coordinates for bovine Cu,Zn superoxide dismutase. *Biochem. Biophys. Res. Commun.* **63,** 986–992.

75. Forman, H. J., and Fridovich, I. (1973). On the stability of bovine superoxide dismutase. The effect of metals. *J. Biol. Chem.* **248,** 2645–2649.

76. Beem, K. M., Rich, W. E., and Rajagopalan, K. V. (1974). Total reconstitution of cooper-zinc superoxide dismutase. *J. Biol. Chem.* **249,** 7298–7305.

77. Fee, J. A., and Briggs, R. G. (1975). Studies on the reconstitution of bovine erythrocyte superoxide dismutase. V. Preparation and properties of derivatives in which both zinc and copper sites contain copper. *Biochim. Biophys. Acta* **400,** 439–450.

78. Simonyan, M. A., and Nalbandyan, R. M. (1972). Interaction of hydrogen peroxide with superoxide dismutase from erythrocytes. *FEBS Lett.* **28,** 22–24.

79. Rotilio, G., Morpurgo, L., Calabrese, L., and Mondovi, B. (1973). On the mechanism of superoxide dismutase. Reaction of the bovine enzyme with hydrogen peroxide and ferrocyanide. *Biochim. Biophys. Acta* **302,** 229–235.

80. Fee, J. A., and DiCorleto, P. E. (1973). Observations on the oxidation-reduction properties of bovine erythrocyte superoxide dismutase. *Biochemistry* **12,** 4893–4899.

81. Hodgson, E. K., and Fridovich, I. (1975). The interaction of bovine erythrocyte superoxide dismutase with hydrogen peroxide: Inactivation of the enzyme. *Biochemistry* **14,** 5294–5299.

82. Weisiger, R. A., and Fridovich, I. (1973). Superoxide dismutase. Organelle specificity. *J. Biol. Chem.* **248,** 3582–3592.

83. Keele, B. B., Jr., McCord, J. M., and Fridovich, I. (1970). Superoxide dismutase from *Escherichia coli* B: A new manganese-containing enzyme. *J. Biol. Chem.* **245,** 6176–6181.

84. Yost, F. J., Jr., and Fridovich, I. (1973). An iron-containing superoxide dismutase from *Escherichia coli. J. Biol. Chem.* **248,** 4905–4908.

85. Asada, K., Yoshikawa, K., Takahashi, M., Maeda, Y., and Enmanji, K. (1975). Superoxide dismutases from a blue-green alga, *Plectonema boryanum. J. Biol. Chem.* **250,** 2801–2807.

86. Misra, H. P., and Fridovich, I. (1978). Inhibition of superoxide dismutases by azide. *Arch. Biochem. Biophys.* **189,** 317–322.

87. Sato, S., and Harris, J. I. (1977). Superoxide dismutase from *Thermus aquaticus:* Isolation and characterisation of manganese and apoenzymes. *Eur. J. Biochem.* **73,** 373–381.

88. Sato, S., and Nakazawa, K. (1978). Purification and properties of superoxide dismutase from *Thermus thermophilus* HB8. *J. Biochem. (Tokyo)* **83**, 1165–1171.
89. Ravindranath, S. D., and Fridovich, I. (1975). Isolation and characterization of a manganese-containing superoxide dismutase from yeast. *J. Biol. Chem.* **250**, 6107–6112.
90. Ose, D. E., and Fridovich, I. (1976). Superoxide dismutase. Reversible removal of manganese and its substitution by cobalt, nickel, or zinc. *J. Biol. Chem.* **251**, 1217–1218.
91. Brock, C. J., Harris, J. I., and Sato, S. (1976). Superoxide dismutase from *Bacillus stearothermophilus*. Preparation of stable apoprotein and reconstitution of fully active Mn enzyme. *J. Mol. Biol.* **107**, 175–178.
92. Britton, L., and Fridovich, I. (1978). Streptomycin: Irreversible association with superoxide dismutases. *Arch. Biochem. Biophys.* **191**, 198–204.
93. Klug-Roth, D., Fridovich, I., and Rabani, J. (1973). Pulse radiolytic investigation of superoxide catalyzed disproportionation. Mechanism for bovine superoxide dismutase. *J. Am. Chem. Soc.* **95**, 2786–2790.
94. Fielden, E. M., Roberts, P. B., Bray, R. C., and Rotilio, G. (1973). Mechanism and inactivation of superoxide dismutase activity. *Biochem. Soc. Trans.* **1**, 52–53.
95. Fielden, E. M., Roberts, P. B., Bray, R. C., Lowe, D. J., Mautner, G. N., Rotilio, G., and Calabrese, L. (1974). The mechanism of action of superoxide dismutase from pulse radiolysis and electron paramagnetic resonance. *Biochem. J.* **139**, 49–60.
96. Pick, M., Rabani, J., Yost, F., and Fridovich, I. (1974). The catalytic mechanism of the manganese-containing superoxide dismutase of *Escherichia coli* studied by pulse radiolysis. *J. Am. Chem. Soc.* **96**, 7329–7332.
97. McAdam, M. E., Fox, R. A., Lavelle, F., and Fielden, E. M. (1977). A pulse-radiolysis study of the manganese-containing superoxide dismutase from *Bacillus stearothermophilus*. A kinetic model for the enzyme action. *Biochem. J.* **165**, 71–79.
98. McAdam, M. E., Lavelle, F., Fox, R. A., and Fielden, E. M. (1977). A pulse-radiolysis study of the manganese-containing superoxide dismutase from *Bacillus stearothermophilus*. Further studies on the properties of the enzyme. *Biochem. J.* **165**, 81–87.
99. Lavelle, F., McAdam, M. E., Fielden, E. M., Roberts, P. B., Puget, K., and Michelson, A. M. (1977). A pulse-radiolysis study of the catalytic mechanism of the iron-containing superoxide dismutase from *Photobacterium leiognathi*. *Biochem. J.* **161**, 3–11.
100. Gregory, E. M., and Fridovich, I. (1973). Induction of superoxide dismutase by molecular oxygen. *J. Bacteriol.* **114**, 543–548.
101. Gregory, E. M., Yost, F. J., Jr., and Fridovich, I. (1973). Superoxide dismutases of *Escherichia coli:* Intracellular localization and functions. *J. Bacteriol.* **115**, 987–991.
102. Gregory, E. M., Goscin, S. A., and Fridovich, I. (1974). Superoxide dismutase and oxygen toxicity in a eukaryote. *J. Bacteriol.* **117**, 456–460.
103. Crapo, J. D., and Tierney, D. L. (1974). Superoxide dismutase and pulmonary oxygen toxicity. *Am. J. Physiol.* **226**, 1401–1407.
104. Hassan, H. M., and Fridovich, I. (1977). Physiological function of superoxide dismutase in glucose-limited chemostat cultures of *Escherichia coli*. *J. Bacteriol.* **130**, 805–811.
105. Hassan, H. M., and Fridovich, I. (1977). Regulation of superoxide dismutase synthesis in *Escherichia coli:* Glucose effect. *J. Bacteriol.* **132**, 505–510.
106. Hassan, H. M., and Fridovich, I. (1977). Regulation of the synthesis of superoxide dismutase in *Escherichia coli:* Induction by methyl viologen. *J. Biol. Chem.* **252**, 7667–7672.
107. Hassan, H. M., and Fridovich, I. (1979). Intracellular production of superoxide radical

and of hydrogen peroxide by redox active compounds. *Arch. Biochem. Biophys.* **196,** 385–395.

108. Hassan, H. M., and Fridovich, I. (1979). Paraquat and *Escherichia coli:* Mechanism of production of extracellular superoxide radical. *J. Biol. Chem.* **254,** 10846–10852.
109. Haugaard, N. (1968). Cellular mechanisms of oxygen toxicity. *Physiol. Rev.* **48,** 311–373.
110. Gottleib, S. F. (1971). Effects of hyperbaric oxygen on microorganisms. *Annu. Rev. Microbiol.* **25,** 111–152.
111. Fridovich, I. (1972). Superoxide radical and superoxide dismutase. *Acc. Chem. Res.* **5,** 321–326.
112. Fenn, W. O., Gerschmam, R., Gilbert, D. L., Terwilliger, D. E., and Gothran, F. V. (1957). Mutagenic effects of high oxygen tensions on *Escherichia coli. Proc. Natl. Acad. Sci. U.S.A.* **43,** 1027–1032.
113. Grifford, G. D. (1968). Mutation of an auxotrophic strain of *Escherichia coli* by high pressure oxygen. *Biochem. Biophys. Res. Commun.* **33,** 294–298.
114. Yost, F. J., and Fridovich, I. (1976). Superoxide and hydrogen peroxide in oxygen damage. *Arch. Biochem. Biophys.* **175,** 514–519.
115. Bruyninckx, W. J., Mason, H. S., and Morse, S. A. (1978). Are physiological oxygen concentrations mutagenic? *Nature (London)* **274,** 606–607.
116. McCord, J. M., Beauchamp, C. O., Goscin, S., Misra, H. P., and Fridovich, I. (1973). Superoxide and superoxide dismutase. *In* "Oxidases and Related Redox Systems" (T. E. King, H. S. Mason, and M. Morrison, eds.), pp. 51–76. Univ. Park Press, Baltimore, Maryland.
117. Hassan, H. M., and Fridovich, I. (1979). Superoxide, hydrogen peroxide, and oxygen tolerance of oxygen-sensitive mutants of *Escherichia coli. Rev. Infect. Dis.* **1,** 357–367.
118. Hassan, H. M., and Fridovich, I. (1978). Superoxide radical and the oxygen enhancement of the toxicity of paraquat in *Escherichia coli. J. Biol. Chem.* **253,** 8143–8148.
119. Stevens, J. B., and Autor, A. P. (1977). Induction of superoxide dismutase by oxygen in neonatal rat lung. *J. Biol. Chem.* **252,** 3509–3514.
120. White, J. R., and White, H. L. (1965). Effect of intracellular redox environment on bactericidal action of mitomycin-c and streptonigrin. *Antimicrob. Agents Chemother.* pp. 495–499.
121. Hassan, H. M., and Fridovich, I. (1979). A mechanism for the oxygen-dependent antibiotic action of pyocyanine. *Abstr., Annu. Meet. Am. Soc. Microbiol.* K112.
122. Fisher, H. K. (1977). Importance of oxygen and of pulmonary alveolar surfactant in lung injury by paraquat. *In* "Biochemical Mechanisms of Paraquat Toxicity" (A. P. Autor, ed.), pp. 57–65. Academic Press, New York.
123. Bus, J. S., Aust, S. D., and Gibson, J. E. (1977). Lipid peroxidation as a proposed mechanism for paraquat toxicity. *In* "Biochemical Mechanisms of Paraquat Toxicity" (A. P. Autor, ed.), pp. 157–172. Academic Press, New York.
124. Misra, H. P., and Fridovich, I. (1976). Superoxide dismutase and the oxygen enhancement of radiation lethality. *Arch. Biochem. Biophys.* **176,** 577–581.
125. Petkau, A., Chelack, W. S., and Pleskach, S. D. (1976). Protection of post-irradiated mice by superoxide dismutase. *Int. J. Radiat. Biol.* **29,** 297–299.

Chapter 16

Glutathione Peroxidase

ALBRECHT WENDEL

I. INTRODUCTION

Glutathione peroxidase (EC 1.11.1.9), (GSH-peroxidase) was discovered in 1957 by Mills[1] as an enzyme in erythrocytes which is capable of protecting hemoglobin from oxidative breakdown by catalyzing Eq. (1). Exhibiting a quite different but nevertheless protective action, the enzyme was discovered a second time: a mitochondrial protein capable of preventing the GSH-induced, high-amplitude swelling of mitochondria

333

ENZYMATIC BASIS OF DETOXICATION, VOL. I

and hence named contraction factor was shown to be glutathione peroxidase.[2]. In the late 1960's, the enzyme was also found to catalyze the reduction of organic hydroperoxides formed from unsaturated fatty acids.[3-5] Hence Eq. (1) may be generalized to Eq. (2). This important

$$2GSH + H_2O_2 \rightarrow GSSG + H_2O \qquad (1)$$

$$2GSH + ROOH \rightarrow GSSG + ROH + H_2O \qquad (2)$$

finding introduced a far broader role for the enzyme in detoxication and in the protection of cell membranes against oxidative damage. A new dimension was added when the search for the prosthetic group of GSH-peroxidase and the trace element research on the essentiality of selenium converged in 1973 with the finding that this peroxidase from erythrocytes contained stoichiometric amounts of selenium.[6,7] Incidentally, another development, which remains unresolved, began in the same year: the involvement of the enzyme in prostaglandin biosynthesis.[8] Several reviews concerning the enzyme cover the literature to 1976.[9-12] Quite unexpectedly, in 1976, another non-selenium-dependent glutathione peroxidase activity (Non-Se GSH-peroxidase) was identified.[13] The presence of this enzyme in many tissues other than the red cell and its relation to the glutathione transferases or "ligandins" (Chapter 4, Volume II), now requires reevaluation of the role of GSH-peroxidases in hydroperoxide metabolism on the one hand and rounds out the picture of the interplay of multifunctional proteins in detoxication on the other.

II. THE ROLE OF GLUTATHIONE PEROXIDASE IN HYDROPEROXIDE METABOLISM

Cells living in the presence of oxygen inevitably encounter the problem of protecting themselves against the reactive oxygen intermediates O_2^-, singlet oxygen (O_2^*), and H_2O_2 (Chapter 15, this volume). It is now well documented that these species are normal products of aerobic life which, however, are able to initiate free radical chain reactions that lead to organic and lipid peroxide formation.[14-16] Figure 1 illustrates schematically how and where GSH-peroxidase fulfills, in conjunction with superoxide dismutase and catalase, its function in detoxification of hydroperoxides. The first line of defense seems to be governed by superoxide dismutase. In the second, catalase and Se-GSH-peroxidase efficiently share the task of reducing H_2O_2 to water, provided the enzymes are both present in the cell or organelle, e.g., in the red blood cell. In the third and last line of defense, required only if a polyunsaturated membrane fatty acid has been peroxidized, the GSH-peroxidase reaction is the sole enzymatic system capable of preventing further propagation of a

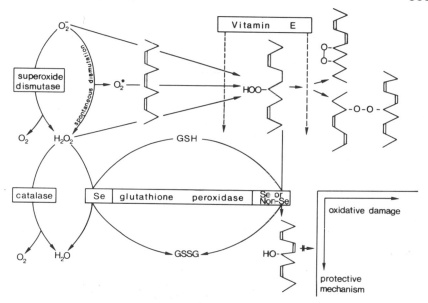

Fig. 1. General illustration of the involvement of selenium and nonselenium glutathione peroxidases in detoxication of hydroperoxides.

radical chain reaction that leads to lipid peroxidation, deterioration of membrane lipids, and severe impairment of energy-related membrane functions. This consciously simplistic picture requires refinement which would depend on the specific organ, cell type or cell organelle.

The physiological function of the selenium GSH-peroxidase in erythrocytes has been comprehensively reviewed.[9,10] Within these relatively degenerated cells, catalase and GSH-peroxidase can substitute for each other in the removal of H_2O_2 to a substantial degree. Human acatalasemics are hematologically indifferent when the red cell is not exposed to oxidative agents.[17] In contrast, all processes that cause impairment of the GSH-peroxidase reaction, (glucose-6-phosphate dehydrogenase, glutathione reductase, GSH synthetase, or GSH-peroxidase deficiencies) lead to hemolytic disorders of varying intensity. These and other findings continue to favor the importance of the GSH-peroxidase in maintaining red cell integrity.

Much new information has accumulated in the last few years, largely obtained by studying the isolated perfused rat organ, regarding the function of the enzyme in the liver and its subcellular compartments. These investigations exploited the observation that, as a consequence of the rapid GSH-peroxidase reaction and the slow rereduction of GSSG by the NADPH/GSSG-reductase system, GSSG is released from the cell. GSSG efflux was shown with erythrocytes,[18] eye lens,[19] liver,[20] and as total glutathione with lung.[21] If the catalase-resistant substrates, *tert*-butyl hydroperoxide or cumene hydroperoxide, are infused into the isolated rat liver, a strictly proportional amount of extra GSSG is released[22] that represents approximately 3% of the actual rate of the GSH-peroxidase reaction. Similar effects have been observed during intracellular drug oxidation[23], hyperoxic oxidative stress[21] or after intracellular generation of H_2O_2.[24] Recently, the interesting observation was made that, under normal conditions, a low steady-state efflux of GSH from the liver into the perfusate takes place[25] and that the greater part of extra GSSG release upon oxidative stress is released into the bile.[26] These studies show conclusively by a noninvasive technique that GSH-peroxidase is effectively detoxifying H_2O_2 and organic hydroperoxides within intact cells. An independent series of impressive studies using isolated hepatocytes led to similar conclusions with respect to the function of GSH-peroxidase.[27]

Measuring the ethane expiration of selenium- and/or vitamin E-deficient rats as an index of *in vivo* lipid peroxidation enabled the demonstration that this event occurs *in vivo,* is most probably responsible for tissue damage, and can be greatly reduced by vitamin E and by a normal Se-GSH-peroxidase activity.[28] However, this work did not take into consideration the non-selenium GSH-peroxidase that had just been discovered. A very recent study, taking advantage of the different substrate specificity of the two peroxidases, i.e., Se and non-Se, showed that isolated perfused livers from selenium-deficient rats that were virtually free of Se-GSH-peroxidase, did not release GSSG upon infusion of high levels H_2O_2, but released the usual quantities of GSSG upon infusion of the non-Se-peroxidase substrate, *tert*-butyl hydroperoxide.[29] Thus, both types of enzyme appear to be active in the removal of organic hydroperoxides, and debate as to which is of greater importance must await additional evidence. Controversy of this sort may be as needless as the dispute, lasting for a decade, about the role of catalase and the Se-GSH-peroxidase in the elimination of H_2O_2, a debate that ended with the recognition that both fulfill this task in different cell compartments.

III. DETERMINATION OF GLUTATHIONE PEROXIDASE ACTIVITY AND DEFINITION OF UNITS

A. Direct Assay by Determination of Glutathione (GSH)

Under experimental conditions to the GSH-peroxidase reaction can be saturated with respect to ROOH but not with GSH. Practically, this means that linearity of the reaction rate with time is impossible; therefore, the usual definition of an enzyme unit as μmol substrate consumed per minute is not appropriate. If the unit is defined as

$$U = \Delta\log [GSH] \min^{-1} \qquad (3)$$

a plot of $\log [GSH]_0 - \log [GSH]_t$ versus time yields a straight line with a slope that is proportional to the amount of enzyme.[30] In other words 1 U GSH-peroxidase consumes 90% of the initial GSH concentration within 1 min, irrespective of the absolute concentration of GSH.

This definition is the basis of the fixed-time assay systems which measure the remaining GSH either polarographically (procedure A, Table I), by the p-chloromercuribenzoic acid method (procedure B, Table I) or by the dithionitrobenzoid acid method (procedure C, Table I). Due to many points of interference with components of biological material, e.g., other thiols and various ions, these types of assay can only be used with purified enzyme.[30-40]

B. Coupled Test Procedure

The second method for determining GSH-peroxidase activity is the regeneration of the product enzymatically by glutathione reductase and the observation of the decrease in NADPH concentration. Although, the indicator reaction, under the appropriate conditions, follows pseudo-zero-order kinetics, the reaction rate still depends on the total glutathione concentration in the enzymatic cycle. The units of the direct assay are identical with the ones of the coupled test[30,36] if

$$U = \frac{\Delta[NADPH]}{\min} \frac{0.868}{[GSH]_0} \qquad (4)$$

For a rough comparison of the plethora of units that have been used in publications (Table I may be helpful to complete confusion), division of the values by the total glutathione concentration $[GSH]_0$ used in any specific assay, and a temperature dependence factor of $2.3/10°$, may be useful.

The coupled test does not allow kinetic studies, but is suitable for determination of GSH-peroxidase activity in biological material. In blood, hemoglobin, which exhibits a peroxidase-like activity, must be converted to cyanomethemoglobin in order to obtain reproducible blanks. The variant G is recommended with respect to saturating ROOH, low blank, no interference by excess hexacyanoferrate and no interference by the NADPH inhibition of glutathione reductase.

C. Suitable Substrates for the Assays

The broad specificity of the enzyme allows a very wide range of electron acceptor substrates. *tert*-Butyl and cumene hydroperoxide have the

TABLE I

Assay Types for Glutathione Peroxidase Activity[a]

Procedure	Substrates		Temperature	pH		Definition of unit	Principle	Reference
	GSH	ROOH						
A	1.0	1.25	H_2O_2	25°	7.0		$\Delta \log$ GSH min^{-1}	Polarographical GSH-determination	30
B	0.4	0.25	H_2O_2	37°	7.0		$\Delta \log$ GSH min^{-1}	Determination of GSH by PCMB	31
C	0.35	0.2	COOH[b]	32°	7.6		$\Delta \ln$ GSH min^{-1}	Semiautomatic DTNB method	32
D	4.0	0.1	BOOH[c]	37°	7.0		μmole GSH min^{-1}	Coupled test	33
E	1.0	0.2	H_2O_2	25°	7.0		μmole GSH min^{-1}	Coupled test	34
F	0.25	0.2	COOH[b]	37°	7.3		μmole NADPH min^{-1}	Coupled test	35
G	1.0	1.25	BOOH	37°	7.0	0.86	μmole GSH min^{-1} [GSH]$_0^{-1}$	Coupled test	36
H	5.0	0.073	H_2O_2	20°	7.0		μmole NADPH min^{-1}	Coupled test	37
I	1.0	0.2	H_2O_2	25°	7.0		μmole GSH min^{-1}	Coupled test	38
K	0.25	0.02	H_2O_2	23°	7.0		ΔA_{340} min^{-1}	Coupled test	39
L	0.5	0.1	H_2O_2	23°	7.9		μmole GSH min^{-1}	Coupled test	40

[a] All concentrations are given millimolar.
[b] COOH, cumene hydroperoxide.
[c] BOOH, *tert*-butyl hydroperoxide.

advantage for assay in that their spontaneous reaction with GSH is lower than is that with H_2O_2, and neither of the hydroperoxides are metabolized by catalase. However, with the discovery of the non-Se-GSH-peroxidase, it became clear that what was being measured with either of these organic peroxides were the sum of the Se-containing and the non-Se-enzyme. Thus it seems necessary to come back to H_2O_2 as a substrate, thereby allowing accurate measurement of the Se-enzyme if catalase is inhibited by 1 mM azide. Since the V_{max} for GSH-peroxidase is independent of the nature of the hydroperoxide (Section VI), both peroxidases can be determined when separate assays are conducted: one with H_2O_2 and the other with butyl hydroperoxide. The difference, $V_{BOOH} - V_{H_2O_2}$, corrected for the (different) spontaneous reaction rates, are equal to the concentration of non-selenium GSH-peroxidase activity, i.e., glutathione transferase.

IV. PURIFICATION

Within the cell, GSH-peroxidase is usually present in very low concentrations. It constitutes approximately 0.01% of the total protein of bovine erythrocytes[41,42] and roughly 0.1% of total extractable rat liver protein.[34,43] Since the enzyme tends to become unstable upon purification, sophisticated and rapid techniques are needed for a successful preparation. Table II is a compilation of the results of purification from various sources.[5,31,33,34,38-46] Our currently used isolation strategy includes countercurrent hollow-fiber dialysis of the hemolysate, ion exchange chromatography on DEAE-Sephadex, hydrophobic chromatography on phenyl-Sepharose, gel filtration on BioGel P-100 and hydroxylapatite chromatography. This procedure yields 200 mg of homogeneous enzyme of a specific activity[41,42] of 550 U/mg from 180 liter of bovine blood, and represents a purification of 13,500-fold and a yield of 14%.

Two partial purifications have been reported for the non-Se-GSH-peroxidase.[47,48] In its property as an GSH: organic hydroperoxide oxidoreductase little data on the purified enzyme are available. Considerable evidence supports the assumption that the enzyme is identical with GSH transferase B[47-50] (Chapter 4, Volume II).

V. PHYSICAL AND CHEMICAL PROPERTIES

A. Molecular Data

The properties of GSH-peroxidases from various sources are compiled in Table III.[7,33,34,41,45,46,51,52] The impression that the red cell enzyme

TABLE II

Purification of Glutathione Peroxidase from Various Sources

Source	Final specific activity (U/mg)[a]	Assay and definition of units[b]	Purification (-fold)	Yield (%)	Reference
Erythrocyte					
Bovine	105	A	5,000	2.5	41
Bovine	550	G	13,500	13	42
Ovine	60	B	4,000	23	31
Human	103	D	14,000	2	33
Pig	3 470	K	2,500	41	39
Liver					
Rat	70	E	400	12	34
Rat cytosol	278	F	1,000	1	43
Rat mitochondria	128	F	600	1.6	44
Rabbit	96	I	1 300	12	38
Lung, rat	173	F	2,700	4	45
Placenta, human	202	D	3,300	40	46
Lens, bovine	42	K	1,400	42	5
Aorta, pig	20	L	34		40

[a] Individual units according to the reference. For comparison cf. section III.
[b] See Table I.

exhibits small but consistently different properties than liver GSH-peroxidase with respect to molecular weight and isoelectric point, has been supported by a recent study which compared directly the enzymes from sheep.[52] This investigation revealed in addition that all selenium in the red cell is associated with GSH-peroxidase, whereas only 10% of the liver selenium is accounted for by the enzyme.[52]

The determination of the enzyme's pH optimum (cf Table III) is complicated by a large increase in nonenzymatic oxidation of GSH at alkaline pH. With the GSH analogue, mercaptoacetic acid methyl ester, which exhibits a markedly different pK_{SH} (7.8) than GSH (9.1), the same pH curve was obtained.[53]

It would appear that GSH-peroxidase is unstable at pH values approaching the isoelectric point, i.e., below pH 6.0. On the other hand, the bovine red cell enzyme is fairly stable at pH values up to 9.5. The presence of ethanol[52,54] considerably stabilizes the enzyme in solution at pH 7 as does GSH.[5,34,53,54] In the presence of peroxide and in the absence of GSH the enzyme quickly loses activity and tends to irreversibly precipitate.

TABLE III

Physical Properties of GSH-Peroxidases from Various Sources

Source	Molecular weight	Subunits	pH optimum	Isoelectric point	Multiple forms	Se content (g atoms/mole tetramer)	Reference
Erythrocyte							
Bovine	84,500	21,000	8.8	5.6–6.0	No	4	7,41,51
Ovine	89,000	22,000	—	6.0–6.5	No	4	52
Human	95,000	32,000 (A)	8.5 (A)	4.9 (A)	A + B	3.5 (A)	33
		47,000 (B)				Present (B)	
Liver							
Rat	76,000	17,000	—	—	—		34
Ovine	80,000	20,000	—	—	—	4	52
Lung, rat	80,000	20,000	8.8–9.1	—	No	Present	45
Placenta, human	85,000	22,000	—	—	No	4	46

B.   Substrate Specificity

The selenium GSH-peroxidase is extremely specific for GSH. Among 30 other thiols, only slight activity was observed with mercaptoacetic acid methyl ester.[55] In contrast, nearly any organic hydroperoxide can act as the electron acceptor substrate (reviewed in Flohé and Günzler).[56] Recently, evidence was presented that only one of the GSH-dependent reaction steps is strictly specific for GSH, whereas the other one can be substituted by other thiols, e.g., mercaptoethanol.[32] Non-Se-GSH-peroxidase is unable to reduce H_2O_2, whereas it shows a similar broad activity with organic hydroperoxide.[47,48,57]

The Se-GSH-peroxidase is inhibited by iodoacetate[56] or chloroacetate[58] if reduced by GSH or KBH_4. Curiously, iodoacetamide does not inactivate the enzyme but binds with a stoichiometry of 1 : 1 in the presence of 6 M urea.[59] The oxidized enzyme is inactivated and depleted of its selenium after prolonged exposure to high concentrations of KCN.[60] For practical reasons it is essential to note that high concentrations of phosphate and sulfate reversibly inhibit the selenium enzyme. In contrast, the non-Se-GSH-peroxidase is not inhibited by iodoacetate, but by N-ethylmaleimide. Cyanide inhibited competitively with respect to GSH ($K_i \approx 1$ mM).[50] Bromosulfophthalein, a glutathione transferase substrate, leads to competitive inhibition of the non-selenium-enzyme with respect to organic peroxide.[48]

VI.   KINETIC DATA

The kinetic properties of bovine erythrocyte Se-GSH-peroxidase have been thoroughly investigated.[61] The enzyme follows a ter uni ping-pong mechanism which may be written, in Cleland's terms, as follows

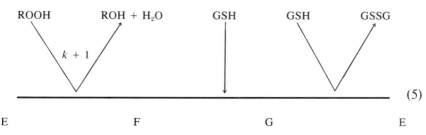

$$(5)$$

The first order rate constant, $k + 1$ of the oxidation of the enzyme by hydroperoxide equals $10^8 M^{-1}$ sec^{-1} at pH 7 and 37°.

There are several features of this type of reaction mechanism that

deserve mention: (1) No stable ternary complexes between enzyme and both substrates is found. (2) The enzyme definitively shows no saturation with respect to GSH, i.e., the extrapolated V_{max} is infinite. This implies that no true K_M values can be given, since the apparent maximum velocities for infinite peroxide concentration are a linear function of the concentration of GSH. (3) The kinetically derived reaction mechanism postulates the existence of three enzyme species (E,F,G) of different redox states (Section VII). This proposed mechanism is additionally supported by initial rate kinetics with different peroxide substrates which show, as expected, different apparent K_M values but identical V_{max}.[62] The data given in Table IV must be evaluated within these limitations.

No such detailed knowledge is available for peroxides active with the non-selenium-peroxidase. In spite of this, comparison of the data shows that the K_M values of the non-Se-enzyme for organic peroxidase seem to range consistently above those for the Se-enzyme. A sequential type mechanism has been discussed.[49,50]

VII. ACTIVE SITE, STRUCTURE, AND MECHANISM OF CATALYSIS

A. Chemical Nature of the Selenium Moiety

The first attempts to identify a low molecular weight selenium compound in hydrolysates of bovine red cell GSH-peroxidase yielded several Se-containing fractions upon amino acid analysis, including selenocysteine,[53] but the recovery of selenium was poor. After demonstration of participation of selenium in the catalytic cycle of the enzyme,[63] chloroacetate was chosen as an active-site directed reagent suitable for an inactivating, chemical modification of the protein-bound selenium.[58] From ^{14}C-carboxymethylated, homogeneous GSH-peroxidase, radioactive Se-carboxymethylselenocysteine and its oxidation product were identified after total acid and enzymatic hydrolysis.[59] By difference spectroscopy, evidence was obtained that the freshly prepared, so-called "native enzyme" may not contain the free selenol: after reduction of GSH-peroxidase by KBH$_4$ in urea, the difference spectrum exhibited a peak at 248 nm, indicative of 4 SeH per tetramer. If reduced by GSH, an increase in absorption at 237 nm is observed which is interpreted as the involvement of a sulfhydryl group.[59] At the same time, independent studies with the partially purified rat liver enzyme led to the identification of alkyl and aminoethyl derivatives of selenocysteine following chemical modification of the enzyme.[64] Thus the structure of the selenium-containing moiety of

TABLE IV

Kinetic Data of Glutathione Peroxidases[a]

Enzyme source	K_M (apparent) for ROOH		at [GSH]	K_M (apparent) for GSH	GSH	at [ROOH]	Reference
Se-GSH-peroxidase							
Bovine erythrocytes	0.027^c	(COOH)[b]	1	0.017^c	0.001	(BOOH)	61,62
	0.05^c	(BOOH)[b]	1	0.03^c	0.001	(COOH)	
	0.008^c	(H$_2$O$_2$)	2.5	0.13^c	0.001	(H$_2$O$_2$)	
	0.031^c	(H$_2$O$_2$)	10				
	0.0035^c	(H$_2$O$_2$)	1				
Human erythrocytes	0.052^e	(COOH)		4.1^e			33
Ovine erythrocytes	0.08	(COOH)	1				47
	0.12	(BOOH)	1				
Non-Se-GSH-peroxidase							
Rat liver	0.18	(COOH)	1				47
	5.3	(BOOH)	1				
	0.57^d	(COOH)		0.2^d			50
	0.55	(COOH)	1.2				48
	2.32	(BOOH)	1.2				

[a] All concentrations are given as millimolar.

[b] COOH, cumene hydroperoxide; BOOH, *tert*-butyl hydroperoxide.

[c] Calculated from the data of the cited references.

[d] Limiting extrapolated K_M.

[e] No details given.

the protein appears to be either Se-cysteine or the corresponding seleninic acid.

B. Protein Structure

Se-GSH-peroxidase from bovine red blood cells crystallizes in the presence of 1.25 M phospate at pH 7.[53] X-Ray structural analysis is being conducted,[65] and a complete model of the tetramer at 2.8 Å resolution is available.[66] The enzyme consists of four identical monomers each of which is nearly spherical with a radius of approximately 19 Å. The subunit of about 180 amino acid residues is built up from a central core of two parallel and two antiparallel strands of pleated sheet surrounded by four α-helices. One of the helices runs antiparallel to the neighboring β-strands. A similar $\beta\alpha\beta$ substructure has been found in other dehydrogenases, thioredoxin, flavodoxin, and rhodanese.[66] A selenocysteine fits into the electron density of residue 35, where the selenium atoms have been clearly identified at the surface of the subunit. Near this position, two neighboring subunits are flatly depressed and in contact with each other. The tentative amino acid sequence shows an accumulation of aromatic amino acid residues near the active site and the presence of a histidine. From a difference Fourier systhesis between oxidized and substrate-reduced enzyme, it is concluded that no other prominent changes take place than at the selenium atom, which may have bound oxygen after oxidation.

C. Reaction Mechanism

In view of the kinetic and structural data, the insight gained by X-ray photoelectron and UV difference spectroscopy, chemical modification and inhibition, a tentative reaction mechanism may be formulated. The scheme presented in Fig. 2 extends earlier proposals[53,67] to include the recent findings.

VIII. DISTRIBUTION OF GLUTATHIONE PEROXIDASE

In the animal world, GSH-peroxidase seems to be ubiquitously distributed. GSH-peroxidase-like activities have been reported in plants but not further characterized. Based on present knowledge, it is difficult to evaluate early work which could not account for non-Se-GSH-peroxidase activity. The interspecies differences between two selected tissues, blood (devoid of non-Se-GSH-peroxidase) and liver, are summarized in Table

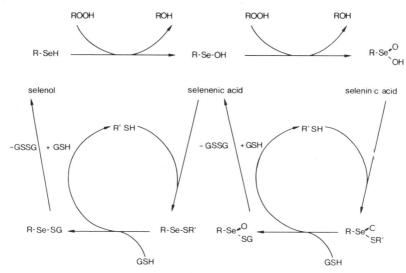

Fig. 2. Reaction mechanism proposed for the selenium glutathione peroxidase. R-SeH, protein-bound selenocysteine; R'-SH, a protein-bound cysteine. Nonenzymatic reduction is possible with KBH₄ and slow nonenzymatic oxidation with oxygen. Intracellularly, the left cycle may dominate, whereas the so-called native enzyme may exist *in vitro* in the form of the intramolecular mixed thioselenenate (left cycle) or its oxidation product (right).

V.[33,36,57,68-76] It should be kept in mind that the GSH-peroxidase activity is subject to numerous influences, including dietary selenium[67,77-79] (see review[10,12]), age and sex,[80] estrus cycle,[81] and to other environmental factors including exposure to ozone[82] or ingestion of peroxidized lipids.[83]

It is remarkable that guinea pig liver does not contain measurable amounts of Se-GSH-peroxidase, whereas mouse liver shows very high activites of the Se-dependent enzyme. In blood, an enormous interspecies variation of Se-GSH-peroxidase activity is observed. Mouse and carp exhibit by far the highest activities, whereas man shows very low concentrations. It is abundantly documented that GSH-peroxidase activity is subjected to large species differences in all organs.

The picture is clearer when organ distribution is studied in a single species. Mills[1] reported that the highest activity in the rat was in liver; moderately high values were found in red cells, heart, lung and kidneys; and low, possibly insignificant enzyme concentration were in the intestinal tract and skeletal muscle. Such data are not available for man. The lens of different animal species[5,75,83a] contains remarkably high GSH-peroxidase activity in conjunction with unusually high GSH levels.[83b] The subcellular distribution of Se-GSH-peroxidase has interesting implica-

TABLE V

Species Distribution of Glutathione Peroxidases in Blood and Liver

Enzyme source	Se-GSH-peroxidase			Non-Se-GSH-peroxidase	Reference
Liver					
Human	3.6	±	2	2.7 ± 1.2	68
	0.95	±	0.23[a]	6.1 ± 0.68[a]	57,69
Rat	14.2	±	3.2[a]	6.1 ± 1[a]	57,69
Sheep	2.7	±	0.6[a]	12.7 ± 1.9[a]	57
Guinea pig		n.d.[b]		5.3 ± 0.5[a]	57
Hamster	18.8	±	2[a]	13.8 ± 0.9[a]	57
Mouse	56	±	10	5	70
Blood					
Human	0.54	±	0.15[a]	n.d.	36,71
	0.11[a]			n.d.	33
	0.22	±	0.04	n.d.	72
	0.55	±	0.14	n.d.	73
	0.7	±	0.18	n.d.	74
Rat	8.9				75
Cow	0.06	−	0.9		70
Rabbit	1.9	±	0.3		74
Mouse	22	±	4		74
Sheep	19.3	±	9		74
Carp	20.3	±	0.86		76

[a] Data roughly calculated from the reference to yield units per gram of tissue.
[b] n.d., not detectable.

tions in relation to its function: 70% of the activity in rat liver is present in the cytosol, 30% is localized in the mitochondrial matrix,[83c] and essentially none is found in the catalase-rich peroxisomes,[83c] i.e., the major H_2O_2-producing organelle.[84] Very recently, non-Se-GSH-peroxidase activity, abundantly present in the cytosol, was demonstrated in mitochondria.[85] Thus, liver catalase and GSH-peroxidase are functionally separated in different subcellular compartments. Furthermore, the mitochondrial GSH-peroxidase has been shown to be involved in the regulation of mitochondrial substrate oxidations.[86]

IX. COMMENTS

Several lines of independent work based on studies with single cells or isolated organs within a short-term time scale, i.e., up to several hours, lead to the conclusion that the glutathione peroxidases constitute a major defense and repair system against oxidative damage to essential intracellular low-molecular weight compounds, proteins, and polyunsaturated fatty acids. It has never been shown directly, however, that within systems retaining their biological form of organization, monohydroxypolyenic fatty acids are formed from polyunsaturated membrane lipid as a consequence of the successful protective function of GSH-peroxidase. An attempt in this direction failed to yield proof,[87] possibly because of methodological difficulties. Futhermore, the mechanism by which enzymes present in the soluble fraction of cells react with membrane-bound fat components, is not clear. Within a longer time scale, there is a lack of evidence which relates the function of GSH-peroxidase to events associated with radical-induced[88] aging of cells, organs, or organisms.[16] Teleologically, however, the ability of GSH-peroxidase to reduce nearly any organic hydroperoxide, including peroxidized DNA, offers an attractive enzymatic basis for speculation.

Conclusions about the biological relevance of GSH-peroxidase have been derived from studies on the effect of selenium deficiency on various parameters related more or less directly to metabolites of the catalyzed reaction. The interpretation of such experiments is not only complicated by the presence and adaptive behavior of non-Se-GSH-peroxidase[13,29] but also by the fact that the GSH-peroxidase reaction is obviously not the only selenium-dependent enzymatic process in differentiated mammalian cells.[12,89,90] Furthermore, enzymological work is needed to characterize all cellular selenoproteins and to refine techniques for selectively analyzing reaction products in complex biological systems. Detailed knowledge of the localization of GSH-peroxidase, based on immunochemical evidence and including accessibility of substrates, is needed before long-term experiments with organisms are appropriate.

ACKNOWLEDGMENT

Substantial parts of the author's work on this subject were supported by the Deutsche Forschungsgemeinschaft (Grant We 686/5).

REFERENCES

1. Mills, G. C. (1957). Hemoglobin catabolism. I. Glutathione peroxidase, an erythrocyte enzyme which protects hemoglobin from oxidative breakdown. *J. Biol. Chem.* **229,** 189–197.

2. Neubert, D., Wojtczak, A. B., and Lehninger, A. L. (1962). Purification and enzymatic identity of mitochondrial contraction factors I and II. *Proc. Natl. Acad. Sci. U.S.A.* **48,** 1651–1658.

3. Little, C., and O'Brien, P. J. (1968). An intracellular GSH-peroxidase with a lipid peroxide substrate. *Biochem. Biophys. Res. Commun.* **31,** 145–150.

4. Christophersen, B. O. (1968). Formation of monohydroxypolyenic fatty acids from lipid peroxides by a glutathione peroxidase. *Biochim. Biophys. Acta* **164,** 35–46.

5. Holmberg, N. J. (1968). Purification and properties of glutathione peroxidase from bovine lens. *Exp. Eye Res.* **7,** 570–580.

6. Rotruck, J. T., Pope, A. L., Ganther, H. E., Swanson, A. B., Hafeman, D., and Hoekstra, W. G. (1973). Selenium: Biochemical role as a component of glutathione peroxidase. *Science* **179,** 588–590.

7. Flohé, L., Günzler, W. A., and Schock, H. H. (1973). Glutathione peroxidase: A selenoenzyme. *FEBS Lett.* **32,** 132–134.

8. Nugteren, D. H., and Hazelhof, E. (1973). Isolation and properties of intermediates in prostaglandin biosynthesis. *Biochim. Biophys. Acta* **326,** 448–461.

9. Flohé, L. (1970). Die Glutathionperoxidase: Enzymologie und biologische Aspekte. *Klin. Wochenschr.* **49,** 669–683.

10. Ganther, H. E., Hafeman, D. G., Lawrence, R. A., Serfass, R. E., and Hoekstra, W. G. (1976). Selenium and glutathione peroxidase in health and disease—A review. *In* "Trace Elements in Human Health and Disease" (A. Prasad, ed.), Vol. 2, pp. 165–234. Academic Press, New York.

11. Flohé, L., Günzler, W. A., and Ladenstein, R. (1976). Glutathione peroxidase. *In* "Glutathione: Metabolism and Function" (I. M. Arias and W. B. Jakoby, eds.), pp. 115–135. Raven, New York.

12. Diplock, A. T. (1976). Metabolic aspects of selenium action and toxicity. *Crit. Rev. Toxicol.* **4,** 271–329.

13. Lawrence, R. A., and Burk, R. F. (1976). Glutathione peroxidase activity in selenium-deficient rat liver. *Biochem. Biophys. Res. Commun.* **71,** 952–958.

14. Halliwell, B. (1974). Superoxide dismutase, catalase and glutathione peroxidase: Solutions to the problem of living with oxygen. *New Phytol.* **73,** 1075–1086.

15. Chance, B., Boveris, A., Nakase, Y., and Sies, H. (1978). Hydroperoxide metabolism: An overview. *In* "Functions of Glutathione in Liver and Kidney" (H. Sies and A. Wendel, eds.), Proc. Life Sci., pp. 95–106. Springer-Verlag, Berlin and New York.

16. Tappel, A. L. (1978). Protection against free radical lipid peroxidation reactions. *Adv. Exp. Med. Biol.* **97,** 111–131.

17. Aebi, H., Heininger, J. P., and Lauber, E. (1964). Methämoglobinbildung in Erythrocyten durch Peroxideinwirkung. Versuche zur Beurteilung der Schutzfunktion von Katalase und Glutathionperoxidase. *Helv. Chim. Acta* **47,** 1428–1440.

18. Srivastava, S. K., and Beutler, E. (1969). The transport of oxidized glutathione from human erythrocytes. *J. Biol. Chem.* **244,** 9–16.

19. Srivastava, S. K., and Beutler, E. (1969). Cataract produced by tyrosinase and tyrosin systems in rabbit lens *in vitro*. *Biochem. J.* **112,** 421–425.

20. Sies, H., Gerstenecker, C., Menzel, H., and Flohé, L. (1972). Oxidation of the NADPH system and release of GSSG from hemoglobin-free perfused rat liver during peroxidatic oxidation of glutathione by hydroperoxides. *FEBS Lett.* **27,** 171–175.

21. Nishiki, K., Jamieson, D., Oshino, N., and Chance, B. (1976). Oxygen toxicity in the perfused rat liver and lung under hyperbaric conditions. *Biochem. J.* **160,** 343–355.

22. Sies, H., and Summer, K. H. (1975). Hydroperoxide metabolizing systems in rat liver. *Eur. J. Biochem.* **57,** 503–512.

23. Oshino, N., and Chance, B. (1977). Properties of glutathione release observed during reduction of organic hydroperoxides, demethylation of aminopyrine and oxidation of some substances in perfused rat liver and their implications for the physiological function of catalase. *Biochem. J.* **162**, 509–525.
24. Sies, H., Bartoli, G. M., Burk, R. F., and Waydhas, C. (1978). Glutathione efflux from perfused rat liver after phenobarbital treatment during drug oxidations and in selenium deficiency. *Eur. J. Biochem.* **89**, 113–118.
25. Bartoli, G., and Sies, H. (1978). Reduced and oxidized glutathione efflux from liver. *FEBS Lett.* **86**, 89–91.
26. Sies, H., Wahlländer, A., and Waydhas, C. (1978). Properties of glutathione disulfide (GSSG) release and glutathione-S-conjugate release from perfused rat liver. *In* "Functions of Glutathione in Liver and Kidney" (H. Sies and A. Wendel, eds.), pp. 120–126. Springer-Verlag, Berlin and New York.
27. Jones, D. P., Thor, H., Andersson, B., and Orrenius, B. (1978). Detoxification reactions in isolated hepatocytes. Role of glutathione peroxidase, catalase and formaldehyde dehydrogenase in reactions relating to N-demethylation by the cytochrome P-450 system. *J. Biol. Chem.* **253**, 6031–6037.
28. Hafeman, D. G., and Hoekstra, W. G. (1977). Lipid peroxidation *in vivo* during vitamin E and selenium deficiency in the rat as monitored by ethane evolution. *J. Nutr.* **107**, 666–672.
29. Burk, R. F., Nishiki, K., Lawrence, R. A., and Chance, B. (1978). Peroxide removal by selenium-dependent and selenium-independent glutathione peroxidases in hemoglobin-free perfused rat liver. *J. Biol. Chem.* **253**, 43–46.
30. Flohé, L., and Brand, I. (1970). Some hints to avoid pitfalls in quantitative determination of glutathione peroxidase (EC 1.11.1.9). *Z. Klin. Chem. Klin. Biochem.* **8**, 156–161.
31. Oh, S. H., Ganther, H. E., and Hoekstra, W. G. (1974). Selenium as a component of glutathione peroxidase isolated from ovine erythrocytes. *Biochemistry* **13**, 1825–1829.
32. Zakowski, J. J., and Tappel, A. L. (1978). A semiautomated system for measurement of glutathione in the assay of glutathione peroxidase. *Anal. Biochem.* **89**, 430–436.
33. Awasthi, Y. C., Beutler, E., and Srivastava, S. K. (1975). Purification and properties of human erythrocyte glutathione peroxidase. *J. Biol. Chem.* **250**, 5144–5149.
34. Nakamura, W., Hosoda, S., and Hayashi, K. (1974). Purification and properties of rat liver glutathione peroxidase. *Biochim. Biophys. Acta* **358**, 251–261.
35. Tappel, A. L. (1978). Glutathione peroxidase and hydroperoxides. *In* "Methods of Enzymology" (S. Fleischer and L. Packer, eds.), Vol. 52, pp. 506–513. Academic Press, New York.
36. Günzler, W. A., Kremers, H., and Flohé, L. (1974). An improved coupled test procedure for glutathione peroxidase (EC 1.11.1.9) in blood. *Z. Klin. Chem. Klin. Biochem.* **12**, 444–448.
37. Paglia, D. E., and Valentine, W. N. (1967). Studies on the quantitative and qualitative characterization of erythrocyte glutathione peroxidase. *J. Lab. Clin. Med.* **70**, 158–169.
38. Santoro, L., del Boccio, G., Sacchetta, P., Polidoro, G., Canella, C., Rotunno, M., and Federici, G. (1976). Parziale purificazione e proprietá della glutathione perossidasi dal fegato di coniglio. *Boll. Soc. Ital. Biol. Sper.* **52**, 922–927.
39. Little, C., Olinescu, R., Reid, K. G., and O'Brien, P. J. (1970). Properties and regulation of glutathione peroxidase. *J. Biol. Chem.* **245**, 3632–3636.
40. Smith, A. G., Harland, A., and Brooks, C. J. W. (1973). Glutathione peroxidase in human and animal aortas. *Steroids Lipids Res.* **4**, 122–128.
41. Flohé, L., Eisele, B., and Wendel, A. (1971). Glutathionperoxidase. I. Reindarstellung und Molekulargewichtsbestimmungen. *Hoppe-Seyler's Z. Physiol. Chem.* **353**, 987–999.

42. Kerner, B. (1978). Die chemische Natur der selenhaltigen Komponente von Glutathion-peroxidase aus Rindereythrocyten. Ph.D. Thesis, University of Tübingen.
43. Stults, F. H., Forstrom, J. W., Chiu, D. T. Y., and Tappel, A. L. (1977). Rat liver glutathione peroxidase: Purification and study of multiple forms. *Arch. Biochem. Biophys.* **183**, 490–497.
44. Zakowski, J. J., and Tappel, A. L. (1978). Purification and properties of rat liver mitochondrial glutathione peroxidase. *Biochim. Biophys. Acta* **526**, 65–76.
45. Chiu, D. T. Y., Stults, F. H., and Tappel, A. L. (1976). Purification and properties of rat lung soluble glutathione peroxidase. *Biochim. Biophys. Acta* **445**, 558–566.
46. Awasthi, Y. C., Dao, D. D., Lal, A. K., and Srivastava, S. K. (1979). Purification and properties of glutathione peroxidase from human placenta. *Biochem. J.* **177**, 471–476.
47. Prohaska, J. R., and Ganther, H. E. (1977). Glutathione peroxidase activity of the glutathione-S-transferases purified from rat liver. *Biochem. Biophys. Res. Commun.* **76**, 437–445.
48. Lawrence, R. A., Parkhill, L. K., and Burk, R. F. (1978). Hepatic cytosolic non-selenium dependent glutathione peroxidase activity: Its nature and the effect of selenium deficiency. *J. Nutr.* **108**, 981–987.
49. Prohaska, J. R. (1980). The glutathione peroxidase activity of glutathione-S-transferases. *Biochim. Biophys. Acta* **611**, 87–98.
50. Pierce, S., and Tappel, A. L. (1978). Glutathione peroxidase activities from rat liver. *Biochim. Biophys. Acta* **523**, 27–36.
51. Flohé, L. (1969). Spezifischer Nachweis der Glutathionperoxidase auf Cellogel-Elektrophoresestreifen und Bestimmung des isoelektrischen Punktes. *Hoppe-Seyler's Z. Physiol. Chem.* **350**, 856–858.
52. Sunde, R. A., Ganther, H. E., and Hoekstra, W. G. (1979). A comparison of ovine liver and erythrocyte glutathione peroxidase *Fed. Proc., Fed. Am. Soc. Exp. Biol.* **37**, 757 (abstr.).
53. Günzler, W. A. (1974). Glutathionperoxidase. Kristallisation, Selengehalt, Aminosäurezusammensetzung und Modellvorstellungen zum Reaktionsmechanismus. Ph.D. Thesis, University of Tübingen.
54. Ganther, H. E., Prohaska, J. R., Oh, S. H., and Hoekstra, W. G. (1978). The labile nature of selenium in oxidized glutathione peroxidase. *Trace Elem. Metab. Man Anim., Proc. Int. Symp., 3rd, 1977* pp. 77–84.
55. Flohé, L., Günzler, W., Schaich, E., and Schneider, F. (1971). Glutathionperoxidase. II. Substratspezifität und Hemmbarkeit durch Substratanaloge. *Hoppe-Seyler's Z. Physiol. Chem.* **352**, 159–169.
56. Flohé, L., and Günzler, W. A. (1974). Glutathione peroxidase. *In* "Glutathione" (L. Flohé, H. C. Benöhr, H. Sies, H. D. Walker, and A. Wendel, eds.), pp. 132–145. Thieme, Stuttgart.
57. Burk, R. F., and Lawrence, R. A. (1978). Non-selenium-dependent glutathione peroxidase. *In* "Functions of Glutathione in Liver and Kidney" (H. Sies and A. Wendel, eds.), Proc. Life Sci., pp. 114–119. Springer-Verlag, Berlin and New York.
58. Wendel, A., and Kerner, B. (1977). Modification of red cell glutathione peroxidase by alkyl halides. *Hoppe-Seyler's Z. Physiol. Chem.* **358**, 1296 (abstr.).
59. Wendel, A., Kerner, B., and Graupe, K. (1978). The selenium moiety of glutathione peroxidase. *In* "Functions of Glutathione in Liver and Kidney" (H. Sies and A. Wendel, eds.), Proc. Life Sci., pp. 107–113. Springer-Verlag, Berlin and New York.
60. Prohaska, J. R., Oh, S. H., Hoekstra, W. G., and Ganther, H. E. (1977). Glutathione peroxidase: Inhibition by cyanide and release of selenium. *Biochem. Biophys. Res. Commun.* **74**, 64–71.

61. Flohé, L., Loschen, G., Günzler, W. A., and Eichele, E. (1972). Glutathione peroxidase. V. The kinetic mechanism. *Hoppe-Seyler's Z. Physiol. Chem.* **353**, 987–999.

62. Günzler, W. A., Vergin, H., Müller, I., and Flohé, L. (1972). Glutathionperoxidase. VI. Die Reaktion der Glutathionperoxidase mit verschiedenen Hydroperoxiden. *Hoppe-Seyler's Z. Physiol. Chem.* **353**, 1001–1004.

63. Wendel, A., Pilz, W., Ladenstein, R., Sawatzki, G., and Weser, U. (1975). Substrate-induced redox change of selenium in glutathione peroxidase studied by X-ray photoelectron spectroscopy. *Biochim. Biophys. Acta* **377**, 211–215.

64. Forstrom, J. W., Zakowski, J. J., and Tappel, A. L. (1978). Identification of the catalytic site of rat liver glutathione peroxidase as selenocysteine. *Biochemistry* **17**, 2639–2644.

65. Ladenstein, R., and Wendel, A. (1976). Crystallographic data of the selenoenzyme glutathione peroxidase. *J. Mol. Biol.* **104**, 877–882.

66. Ladenstein, R., Epp, O., Bartels, K., Jones, A., Huber, R., and Wendel, A. (1979). Structural analysis and molecular model of the selenoenzyme glutathione peroxidase at 2.8 Å resolution. *J. Mol. Biol.* **134**, 199–218.

67. Ganther, H. E. (1975). Selenoproteins. *Chem. Scr.* **8A**, 79–84.

68. Konz, K. H. (1979). Die Aktivität des Peroxid-metabolisierenden Systems in Human-Leber. *Z. Klin. Chem. Klin. Biochem.* **17**, 353–359.

69. Lawrence, R. A., and Burk, R. F. (1978). Species, tissue and subcellular distribution of Non-Se-dependent glutathione peroxidase-activity. *J. Nutr.* **108**, 211–215.

70. Unpublished observations of our laboratory.

71. Günzler, W. A. (1973). Some critical remarks on the determination of glutathione peroxidase in blood. *In* "Glutathione" (L. Flohé, H. C. Benöhr, H. Sies, H. D. Waller, and A. Wendel, eds.), pp. 180–183. Thieme, Stuttgart.

72. Perona, G., Guidi, G. C., Piga, A., Cellerino, R., Menna, R., and Zatti, M. (1978). *In vivo* and *in vitro* variations of human erythrocyte glutathione peroxidase activity as result of cell aging, selenium availability and peroxide activation. *Br. J. Haematol.* **39**, 399–408.

73. Shukla, V. K. S., Jensen, G. E., and Clausen, J. (1977). Erythrocyte glutathione peroxidase deficiency in multiple sclerosis. *Acta Neurol. Scand.* **56**, 524–550.

74. Marcel, J., Puget, K., and Michelson, A. M. (1977). Comparative study of superoxide dismutase, catalase and glutathione peroxidase levels in erythrocytes of different animals. *Biochem. Biophys. Res. Commun.* **77**, 1525–1535.

75. Lawrence, R. A., Sunde, R. A., Schwartz, G. L., and Hoekstra, W. G. (1974). Glutathione peroxidase activity in rat lens and other tissues in relation to dietary selenium intake. *Exp. Eye Res.* **18**, 563–569.

76. Mazeaud, F., Marcel, J., and Michelson, A. M. (1979). Distribution of superoxide dismutase and glutathione peroxidase in the carp: Erythrocyte manganese SOD. *Biochem. Biophys. Res. Commun.* **86**, 1161–1168.

77. Chow, C. K., and Tappel, A. L. (1974). Response of glutathione peroxidase to dietary selenium in rats. *J. Nutr.* **104**, 444–451.

78. Scott, D., Kelleher, J., and Losowsky, M. S. (1977). The influence of dietary selenium and vitamin E on glutathione peroxidase and glutathione in the rat. *Biochim. Biophys. Acta* **497**, 218–224.

79. Toy, P., Hatfield, S., Bull, R., and Couri, D. (1978). The effects of different levels of selenium administered to rats in drinking water on distribution and glutathione peroxidase. *Res. Commun. Chem. Pathol. Pharmacol.* **21**, 115–131.

80. Pinto, R. E., and Bartley, W. (1969). The effect of age and sex on glutathione reductase and glutathione peroxidase activities and on aerobic glutathione oxidation in rat liver homogenates. *Biochem. J.* **112**, 109–115.

81. Pinto, R. E., and Bartley, W. (1969). The nature of the sex-linked differences in glutathione peroxidase activity and aerobic oxidation of glutathione in male and female rat liver. *Biochem. J.* **115**, 449–455.

82. Chow, C. K., Dillard, C. J., and Tappel, A. L. (1974). Glutathione peroxidase system and lysozyme in rats exposed to ozone or nitrogen dioxide. *Environ. Res.* **7**, 31–39.

83. Reddy, K., and Tappel, A. L. (1974). Effect of dietary selenium and autoxidized lipids on the glutathione peroxidase system of the gastrointestinal tract and other tissues of the rat. *J. Nutr.* **104**, 1069–1078.

83a. Pierie, A. (1965). Glutathione peroxidase in lens and a source of hydrogen peroxide in aqueous humour. *Biochem. J.* **96**, 244–253.

83b. Beutler, E., and Srivastava, S. K. (1974). GSH metabolism of the lens. *In* "Glutathione" (L. Flohé, H. C. Benöhr, H. Sies, H. D. Waller, and A. Wendel, eds.), pp. 201–203. Thieme, Stuttgart.

83c. Flohé, L., and Schlegal, W. (1971). Glutathionperoxidase. IV. Intrazelläre Verteilung des Glutathionperoxidase-Systems in der Rattenleber. *Hoppe-Seyler's Z. Physiol. Chem.* **352**, 1401–1410.

84. Sies, H. (1974). Biochemistry of the peroxisome in the liver cell. *Angew. Chem., Int. Ed. Engl.* **13**, 706–718.

85. Wahlländer, A., Soboll, S., and Sies, H. (1979). Hepatic mitochondrial and cytosolic glutathione content and the subcellular distribution of GSH-S-transferases. *FEBS Lett.* **97**, 138–140.

86. Sies, H., and Moss, K. (1978). A role of mitochondrial glutathione peroxidase in modulating mitochondrial oxidations in liver. *Eur. J. Biochem.* **84**, 377–383.

87. McCay, P. B., Gibson, D. D., Fong, K.-L., and Hornbrook, K. R. (1976). Effect of glutathione peroxidase activity on lipid peroxidation in biological membranes. *Biochim. Biophys. Acta* **431**, 459–468.

88. Gordon, P. (1974). Free radicals and the aging process. *In* "Theoretical Aspects of Aging" (M. Rockstein, ed.), pp. 61–81. Academic Press, New York.

89. Burk, R. F., and Masters, B. S. S. (1975). Some effects of selenium deficiency on the hepatic microsomal cytochrome *P*-450 system in the rat. *Arch. Biochem. Biophys.* **170**, 124–131.

90. Black, R. S., Tripp, M. J., Whanger, P. D., and Weswig, P. H. (1978). Selenium proteins in ovine tissues. III. Distribution of selenium and glutathione peroxidases in tissue cytosol. *Bioinorg. Chem.* **8**, 161–172.

Chapter 17

Monoamine Oxidase

KEITH F. TIPTON

I. INTRODUCTION

The enzyme monoamine oxidase (amine:oxygen oxidoreductase (deaminating) flavin-containing EC 1.4.3.4) catalyses the oxidative deamination of amines according to the Eq. (1). It is widely distributed within the cells of most mammalian tissues.[1,2] It functions in the breakdown of the endogenously produced neurotransmitter amines and the hormone adrenaline as well as those taken in the diet or generated by intestinal microorganisms. In the central nervous system the role of the enzyme has been generally assumed to be mainly concerned with the modulation of neurotransmitter turnover, since the blood–brain barrier will largely exclude most circulating amines, although those with simple hydrophobic side chains such as phenethylamine may readily enter this organ.

$$RCH_2NH_2 + O_2 + H_2O \rightarrow RCHO + NH_3 + H_2O_2 \qquad (1)$$

The enzyme is present in presynaptic nerve terminals as well as in nerve cell bodies and glial cells. Transmitter amines, such as norad-

355

ENZYMATIC BASIS OF DETOXICATION, VOL. I

renaline, dopamine, and serotonin, are stored in presynaptic storage vesicles[3] where they are unavailable for degradation by monoamine oxidase. This stored amine, however, will be in equilibrium with a much lower concentration of free cytoplasmic amine (which has been estimated, by analogy with the situation in the adrenal medulla, to be about 100 μM (see Tipton[4])) that will be available for degradation. After a transmitter amine has been released into the synaptic cleft as a result of nerve depolarization, the principal way in which its activity is terminated is by an energy-dependent high-affinity transport system into presynaptic terminals,[5] where it either may be taken up into storage vesicles for reuse or degraded by monoamine oxidase. A comparison between the K_m values for vesicle uptake and oxidation suggests that the former process will predominate unless the vesicles are approaching saturation. Inhibition of presynaptic monoamine oxidase activity will result in a slow buildup of free transmitter amines in the nerve terminals which, in turn, will impair the uptake process[6].

Uptake of amines by extraneuronal cells occurs by a low-affinity transport system,[5] although the methylation-dependent uptake system for catecholamines that has a relatively high affinity has recently been demonstrated in some peripheral tissues.[7] The enzyme is present in extraneuronal cells, as well as in nerves, and in such organs as heart[8] and liver.[9] Detoxication of exogenous amines presumably occurs in a number of tissues, and the relatively high levels of activity in the intestine and liver suggest important roles for these organs in this respect. In the blood, the enzyme is present in platelets but absent from the erythrocytes.

II. ASSAY AND PURIFICATION

There is no shortage of assay methods for determining the activity of the enzyme.[2] This account will restrict itself to discussing the advantages and disadvantages of those that have found most common use. For routine work the method of Tabor et al.,[10] which involves following the increase in absorbance at 250 nm as benzylamine is converted to benzaldehyde, has been widely used. It provides a simple and convenient continuous assay for the enzyme, but it can be difficult to use with crude homogenates that contain material absorbing strongly at this wavelength. In addition, two forms of monoamine oxidase are now recognized as present in most tissues, and benzylamine is a substrate for only one of them (see Section VII). Although a number of other colorimetric substrates have been developed for the enzyme, only kynuramine[11] has found extensive use.

Of assays that are not restricted in the choice of substrates, measurement of oxygen uptake either manometrically[12] or with an oxygen electrode[13] has great flexibility but can be difficult with crude mitochondrial preparations that show a high rate of endogenous oxygen consumption. Several investigators have included cyanide in the assay mixture to inhibit such reactions,[12,13] but this compound is a reversible inhibitor of the enzyme[14] as well as potentiating the action of several irreversible inhibitors[15]; its use cannot be recommended. A convenient spectrophotometric assay which is applicable to most substrates is to follow the reduction of NAD at 340 nm when the aldehyde produced is oxidized in the presence of aldehyde dehydrogenase.[16] The generally low K_m values of aldehyde dehydrogenase make it ideal for use as a coupling enzyme and quite impure preparations may be used. Since aldehydes containing a 2-hydroxyl group are rather poor substrates, this assay is not suitable for use with noradrenaline. It is also unsuitable for use with crude extracts that contain activities capable of reoxidizing the NADH produced.

The sensitivity of this assay may be increased if the fluorescence change that accompanies NADH formation is measured. An alternative fluorescence assay is to couple the hydrogen peroxide formed to the oxidation of a compound such as homovanillic acid, by the action of peroxidase.[17] Unfortunately, a number of phenolic amines will also be oxidized by this system, and, although in the case of tyramine it is possible to minimize this effect by having homovanillic acid present in a great excess, it has not proved possible to use the method with noradrenaline, dopamine, or serotonin as substrates.

When relatively small amounts of enzyme activity are available, the most widely used assay for studies is to follow the conversion of a radioactively labeled amine to the corresponding aldehyde. The methods involved have been recently summarized by Tipton and Youdim.[18] After incubating the amine and enzyme in the appropriate buffer, the reaction is stopped by acidification. The aldehyde product, together with any acid and alcohol metabolites that have been formed, is separated by ion exchange chromatography or by extraction into an organic solvent, and its concentration is determined by liquid scintillation counting. Since this is a discontinuous assay, it is necessary to determine the amount of product formed after a series of fixed time periods in order to ensure that the reaction is proceeding linearly for the time interval chosen for routine work.

Monoamine oxidase is an integral membrane protein being tightly bound to the mitochondrial outer membrane,[19] and extremely vigorous procedures, such as detergent treatment, sonication, and extraction into organic solvents, are necessary to bring it into solution.[2,20] Although a number of solubilization and purification procedures have been devised[21,22] it appears that the process of freeing the enzyme from its normal membrane-bound environment results in changes in its inhibitor sensitivity[9,23,24] and in the details of the kinetic mechanism obeyed.[25] Thus, although such purified preparations will be of great value in studies of the chemical mechanism of the enzyme-catalyzed reaction, the relationship between their properties and those of the enzyme within the cell is far from clear.

III. SUBSTRATE SPECIFICITY

The enzyme can catalyze the oxidation of a wide variety of amines,[1,2] including secondary and tertiary amines in which the amine substituents are methyl groups. The amine group must be attached to an unsubstituted methylene group; aniline and amphetamine are not substrates. The enzyme cannot catalyze the oxidation of methylamine, but ethylamine is a poor substrate in some species. Activity increases with increase in aliphatic chain length to an optimum of C_5 or C_6.[1,2] The specificity of the enzyme from rat liver for aromatic amines is shown in Table I.[26]

A source of continued confusion is the distinction between monoamine oxidase and the other amine oxidases, including diamine oxidase (see Chapter 9, this volume) and the plasma amine oxidases. The confusion arises from overlapping specificities. For example, the plasma enzyme from some species is active toward benzylamine and some other monoamines but not to diamines, and diamine oxidase is active toward some monoamines, whereas monoamine oxidase itself will oxidize long-chain aliphatic diamines such as 1,12-dodecamethylenediamine.[1,2] A number of attempts have been made to clarify the situation[2]: although the distinctions between the other amine oxidases remains unclear, monoamine oxidase appears to differ from them in being a flavoprotein, in its insensitivity to inhibition by carbonyl reagents such as semicarbazide, and in its mitochondrial localization. The sensitivity to carbonyl reagents of the other amine oxidases suggests that such a group plays an essential role in their catalytic mechanisms. Many of the amine oxidases have been shown to contain bound copper, whereas this is not the case with monoamine oxidase.[27]

IV. PROPERTIES OF THE ENZYME

A number of purified preparations of the enzyme have been shown to be capable of aggregating in solution, and molecular weight estimations have

TABLE I

Substrate Specificity of Rat Liver Monoamine Oxidase[a]

Substrate	K_m (μM)	Relative maximum velocity
Benzylamine	245	(100)
2-Phenethylamine	21	118
Tyramine	282	200
Dopamine	405	112
4-Methoxy-2-phenethylamine	24	96
3,4-Dimethoxy-2-phenethylamine	1360	108
4-Methoxy-3-hydroxy-2-phenethylamine	475	113
p-Hydroxybenzylamine	2500	114
p-Chlorobenzylamine	300	254
Vanillylamine	201	68
N-Methylbenzylamine	714	91
Octopamine	625	149
Adrenaline	400	70
Noradrenaline	416	71
m-O-Methylnoradrenaline	200	54
m-O-Methyladrenaline	267	65
Tryptamine	19	81
5-Hydroxytryptamine	187	124
5-Methoxytryptamine	16	66
Oxygen[b]	156	—

[a] Rat liver mitochondrial outer membrane preparations were used as the enzyme source. Data are taken from Houslay and Tipton.[26] Maximum velocities are expressed as percentages of that obtained with benzylamine.

[b] Determined with benzylamine as substrate.[14]

yielded values ranging from about 10^5 to more than 10^6.[28,29] It appears, however, that disaggregation can be achieved by the use of detergents and values of about 100,000[28,30] and even smaller[31] have been obtained by these means. The enzyme is a flavoprotein that has been reported to contain one molecule of FAD per 110,000 g,[21,22] covalently linked to a cysteine residue in the protein in a thioether bond to the 8α-position.[32] Electrophoresis of the enzyme in sodium dodecyl sulfate has revealed the subunit molecular weight of the enzyme to be about 55,000.[21,22,33,34] Thus, it appears that an enzyme molecule with a molecular weight of about 100,000 will be made up of two subunits of apparently identical size, only one of which contains bound FAD. The minimum active size of about 100,000 is also supported by kinetic studies that indicated that a maximum of 1 mole each of aldehyde and ammonia could be released by this weight of enzyme anaerobically.[35,36]

The roles of other functional groups in the activity of the enzyme are

not clearly defined. The preparation from ox liver contains eight sulfhydryl groups per 100,000 g, but it is unclear whether any of these are
essential for activity[37,38] (see also McEwen *et al.*[39] and White and
Glassman[40]). Gorkin,[41] however, has reported that oxidation of some of
these groups can result in profound alterations in the specificity of the
enzyme. As previously mentioned, copper is not involved in the activity
of the enzyme. Although some preparations have been reported to contain
iron,[42] there appeared to be no correlation between the content of this
metal and the activity of the enzyme from pig liver.[28] A purified preparation from ox liver was not found to contain appreciable amounts of iron,
manganese, cobalt, or molybdenum.[27]

V. INHIBITORS

A number of reversible and irreversible inhibitors of monoamine
oxidase have been found to be pharmacologically useful and the structures of some of these are shown in Table II. The most commonly used
irreversible inhibitors fall into three classes: propargylamines (such as
pargyline, clorgyline, and deprenyl), cyclopropylamine derivatives such
as Lilly 51641 and tranylcypromine, and substituted hydrazines such as
phenethylhydrazine. Although iproniazid, the first monoamine oxidase
inhibitor to be discovered, is a hydrazide, it appears that the actual
inhibitor is in fact isopropylhydrazine which is formed following hydrolysis.[43]

The mechanism of inhibition by a number of these compounds has
recently been elucidated. In the case of N,N-dimethylpropargylamine,
kinetic studies indicate that an initial noncovalent enzyme–inhibitor complex is formed before covalent bond formation, and protection by substrate indicates that binding occurs at or near the active site. Inhibition is
accompanied by reduction of the enzyme-bound flavin, and it is suggested
that the adduct shown in Fig. 1a is formed by reaction of oxidized or
partially oxidized inhibitor with the nitrogen atom in position 5 of the
reduced or partially reduced flavin.[44] The reaction of hydrazine derivatives is more complicated. Kenny *et al.*[45] have shown that phenylhydrazine reacts with the flavin at position 4a to give the compound shown in
Fig. 1b. They also observed reduction of the flavin accompanied inactivation and suggested that the hydrazine was first oxidized to the corresponding azine ($C_6H_5N{:}NH$) which then reacted with the reduced flavin. In
these studies, however, only partial reduction of the flavin was observed
and it was suggested that, in addition to reaction with the flavin, an
alternative mode of inhibition by reaction with other groups on the

TABLE II

Some Monoamine Oxidase Inhibitors

Inhibitor	Structure	Nomenclature
Irreversible Deprenyl	(phenyl)—CH$_2$CHNCH$_2$C≡CH with CH$_3$ on the CH and CH$_3$ on the N	Phenylisopropylmethylpropinylamine
Clorgyline	(2,4-dichlorophenyl)—O(CH$_2$)$_3$NCH$_2$C≡CH with CH$_3$ on N	N-Methyl-N-propargyl-3-(2,4-dichlorophenoxyl)propylamine
Pargyline	(phenyl)—CH$_2$NCH$_2$C≡CH with CH$_3$ on N	N-Methyl-N-2-propynylbenzylamine
Phenelzine (Nardil)	(phenyl)—CH$_2$CH$_2$NHNH$_2$	2-Phenethylhydrazine
Iproniazid	(pyridyl)—CNHNHCH(CH$_3$)$_2$ with =O on carbonyl	N'-Isonicotinoyl-N²-isopropylhydrazine
Tranylcypromine	(phenyl)—(cyclopropyl)—NH$_2$	2-Phenylcyclopropylamine
Lilly 51641	(2-chlorophenyl)—O(CH$_2$)$_2$NH—(cyclopropyl)	N-[2-(O-chlorophenoxy)ethyl]cyclopropylamine
PCO	(phenyl)-substituted 1,2,4-oxadiazole—CH$_2$CH$_2$NH—(cyclopropyl)	5-Phenyl-3-(N-cyclopropyl)ethylamine-1,2,4-oxadiazole
Reversible Harmine	CH$_3$O—(β-carboline ring) with CH$_3$	
Harmaline	CH$_3$O—(dihydro-β-carboline ring) with CH$_3$	
Amphetamine	(phenyl)—CH$_2$CHNH$_2$ with CH$_3$	
Amitriptyline	(dibenzocycloheptene) =CHCH$_2$CH$_2$N(CH$_3$)$_2$	
Imipramine	(dibenzazepine)—N—CH$_2$CH$_2$CH$_2$N(CH$_3$)$_2$	

enzyme also occurred. The inhibition of the enzyme by tranylcypromine also involves reduction of the enzyme, but, in this case, the reduction is not irreversible, suggesting that this compound is oxidized to the corresponding imine which then reacts with a sulfhydryl group in the enzyme.[46]

VI. KINETICS AND MECHANISM

The pH optimum of the enzyme-catalyzed reaction has been shown to vary with the substrate used and its concentration as well as that of oxygen.[47,48] A number of studies have indicated that the true substrate for the enzyme is probably the un-ionized form of the amine,[49,50] and it is believed that the reaction proceeds by way of an imine [Eq. (2)] which is subsequently hydrolysed [Eq. (3)].

$$FAD \qquad FADH_2$$

$$RCH_2NH_2 \longrightarrow RCH{=\!=}NH \qquad (2)$$

$$RCH{=\!=}NH + H_2O \longrightarrow RCHO + NH_3 \qquad (3)$$

Kinetic studies with the enzyme from rat liver suggest that hydrolysis of the imine occurs while it is still bound to the enzyme rather than nonenzymically after it has been released.[14,25] Without any firm knowledge of the essential residues that are present at the active site, it has not yet been possible to do more than speculate on the details of the chemical mechanism involved. Two possible mechanisms for the oxidation of tertiary amines are shown in Fig. 2.[50-52]

The kinetic mechanism followed by the enzyme from a number of sources has been shown to be of the double-displacement (or ping-pong) type[14,25,35,53,54] which can be represented by the system

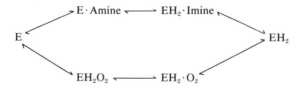

where EH_2 represents the form of the enzyme with its flavin component reduced. An enzyme following such a mechanism will obey a kinetic equation of the general form of Eq. (4).

$$v = \frac{V}{1 + (K_m^{\text{Amine}}/[\text{Amine}]) + K_m^O/[O_2]} \qquad (4)$$

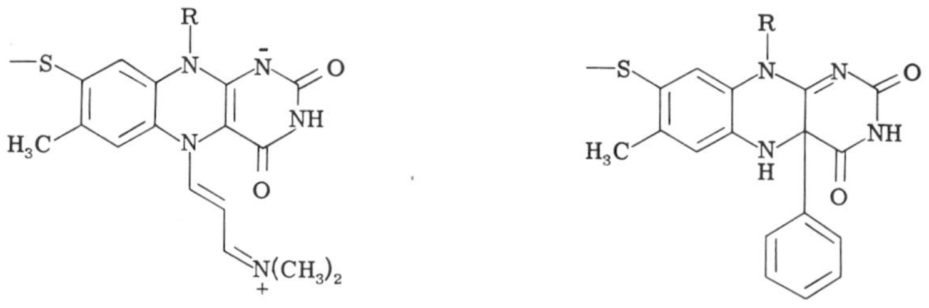

Fig. 1. Proposed structures of the flavin adducts formed by reaction of irreversible inhibitors with monoamine oxidase.

where V is the maximum velocity and K_m^{Amine} and K_m^0 are, respectively, the amine and oxygen concentrations that give half-maximum velocity at a saturating concentration of the other substrate.

A number of kinetic studies of the enzyme have been complicated by the appearance of inhibition by high concentrations of the amine substrate. In many cases this is due to contamination of the substrate by the corresponding aldehyde,[14] although with some substrates it has been possible to detect genuine inhibition at high concentrations due to the amine forming an abortive complex with the reduced form of the enzyme.[48]

Although the K_m values shown by the enzyme toward oxygen will, to some extent, depend upon the specific amine employed,[48] they are relatively high (see Table I), being close to the oxygen concentration in air-saturated water at 37°. This might imply that the enzyme would be only half-saturated with this substrate under normal conditions and that

(a) By removal of a hydride ion[51]

$$RCH_2NR'_2 \xrightarrow{O_2} RCH{=}\overset{+}{N}R'_2 \xrightarrow{H_2O} RCHO + H_2\overset{+}{N}R'_2$$

(b) By a transfer reaction involving a lysine residue in the enzyme[50] (see also Hellerman and Erwin[52])

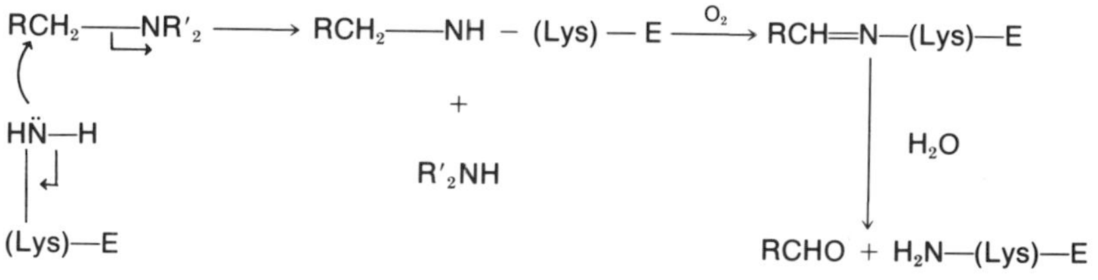

Fig. 2. Possible mechanism of amine oxidation.

the activity might be affected by fluctuations in the local oxygen concentration due to competition with other oxygen-using enzymes. It is, however, a feature of the specific type of kinetic mechanism obeyed by monoamine oxidase that it will tend to make the activity insensitive to fluctuations in oxygen concentration at low concentrations of its amine substrate.[55] The possible selective advantages of such a mechanism have recently been discussed.[56] A practical consequence of the high K_m value for oxygen is that, in assay methods that employ small volumes, it may be necessary to shake the vessel containing the reaction mixture continuously, or to bubble air or oxygen into it, to ensure that the rate of the reaction does not decline because of a fall in the dissolved oxygen concentration.

VII. MULTIPLE FORMS

A number of irreversible inhibitors of the enzyme have been shown to affect the activity toward some substrates at considerably lower concentrations than are necessary to inhibit that towards others. This has led to the suggestion that many tissues contain two forms of the enzyme which have been termed the A and B forms. Some of the specificities and inhibitor sensitivities of the two forms in rat liver are summarized in Table III. Generally, it appears that the two enzyme forms from rat and human brain have similar properties to those shown in Table III, and, based on the sensitivities toward clorgyline and deprenyl, the relative proportions of the two forms have been shown to vary in a number of organs from the

TABLE III

Some Properties of the Two Forms of Monoamine Oxidase from Rat Liver

Characterization	A form		B form
Substrates oxidized	Serotonin Noradrenaline Adrenaline		2-Phenethylamine Benzylamine
Common substrates		Tyramine Dopamine	
Selective irreversible inhibitors	Clorgyline Lilly 51641 PCO		Deprenyl Pargyline
Selective reversible inhibitors	Amphetamine Harmaline Harmine		Amitriptyline Imipramine Benzylcyanide

rat.[9] There are, however, a number of organs in which the specificities of the two species appear to differ. For example, serotonin is a substrate for both forms in ox heart and benzylamine is a substrate for both forms in rat heart.[57] An additional complication is that the specificities of the two forms are not absolute, and substrates for the A form have been shown to be oxidized by the B form at high concentrations and *vice versa*.[58] Such an effect may explain different results that have been reported for the specificities of the two forms in the same organ, e.g., different groups have reported dopamine to be a substrate for both forms[40,59] or only the B form[60] in human brain.

The nature of the two forms has been the subject of considerable speculation. It has been proposed that they represent two distinct proteins,[61] a single enzyme that they may be modified in different ways by its membrane-bound environment[16,62] or a single enzyme with two different active sites with different specificities.[63,64] If there are two distinct enzymes their properties must be very similar, since they show similar isoelectric points and molecular weights[40] and, in rat liver and brain, they are immunologically identical.[65] The suggestion that the difference arose from differences in the membrane environment of a single enzyme stem from studies indicating that agents resulting in weakening of lipid–protein interactions, lead to abolition of the clear distinction between inhibitor sensitivities of the two forms of the enzyme.[16,62] A number of other studies have shown that the properties[9,47,66-68] and inhibitor sensitivities[23,24] of the two forms of the enzyme are affected by their lipid environments, but it has not yet proved possible to demonstrate interconversion of the two forms by lipid substitution. Recently Russell et al.[69] have found that the two forms in human liver are situated on opposite faces of the mitochondrial outer membrane. If this is the case, they might be expected to be in different lipid environments. However, it is not clear what directs the enzyme, which is believed to be synthesized extramitochondrially,[70] to partition between these two locations. Small differences in protein structure or differences in the degree or type of glycosylation of the protein could be involved. If the two forms arose from a single enzyme with two active sites, membrane localization may have allowed one site to be on the inner surface of the outer membrane and the other to be on the outer. White and Glassman[40] have discussed the possibility that the properties of two sites may be modified in different ways by their lipid environments. Solubilized preparations of the enzyme have been shown to be separable into a number of different bands of activity by polyacrylamide gel electrophoresis,[61,71] but these have since been shown to be artifacts of the solubilization procedure releasing enzyme with different amounts of lipid attached.[16,72]

Since the A form of the enzyme has been shown to be active toward the amines that are generally believed to be neurotransmitters, it has been suggested that this form would be largely neuronal, whereas the B form would be present in extraneuronal cells. Denervation experiments indicate that the A form is largely confined to the nerves in pineal gland.[73] However, the situation is not that simple in other organs, where both forms may be present neuronally as well as extraneuronally.[74]

The activity of the enzyme is not constant with age. In rat brain, but not liver, little activity of the B form can be detected 6 days prepartum; activity increases with development until it reaches about 40% of the total activity (determined with tyramine) in the adult.[75] Increase in the total monoamine oxidase activity have been reported to occur with age in a number of other organs.[60] In the rat heart this increase has been shown to be due to a decrease in the rate of enzyme degradation, with no charge in its rate of enzyme synthesis,[76] and it has been suggested that the age-related increase in human brain may explain the increased incidence of affective disorders in middle age.[77]

REFERENCES

1. Blaschko, H. (1952). Amine oxidase and amine metabolism. *Pharmacol. Rev.* **4**, 415–453.
2. Tipton, K. F. (1975). Monoamine oxidase. *In* "Handbook of Physiology" (H. Blaschko and A. D. Smith, eds.), Sect. 7, Vol. 6, pp. 677–697. Am. Physiol. Soc., Washington, D.C.
3. Langercrantz, H. (1976). On the composition and function of large dense cored vesicles in sympathetic nerves. *Neuroscience* **1**, 81–92.
4. Tipton, K. F. (1973). Biochemical aspects of monoamine oxidase. *Br. Med. Bull.* **29**, 116–119.
5. Iversen, L. L. (1973). Catecholamine uptake processes. *Br. Med. Bull.* **29**, 130–135.
6. Trendelenburg, U., Draskoczy, P. R., and Graefe, K. H. (1972). The influence of intraneuronal MAO on neuronal net uptake of noradrenaline and sensitivity to noradrenaline. *Adv. Biochem. Psychopharmacol.* **5**, 371–377.
7. Trendelenburg, U. (1978). Extraneuronal uptake and metabolism of catecholamines as a site of loss. *Life Sci.* **22**, 1217–1222.
8. Horita, A., and Lowe, M. C. (1972). On the extraneuronal nature of cardiac monoamine oxidase in the rat. *Adv. Biochem. Psychopharmacol.* **5**, 227–242.
9. Tipton, K. F., Houslay, M. D., and Mantle, T. J. (1976). The nature and locations of the multiple forms of monoamine oxidase. *Ciba Found. Symp.* **39**, (new ser.), 5–16.
10. Tabor, C. W., Tabor, H., and Rosenthal, S. M. (1954). Purification of amine oxidase from beef plasma. *J. Biol. Chem.* **208**, 645–661.
11. Weissbach, H., Smith, T. E., Daly, J. W., Witkop, B., and Udenfriend, S. (1960). A rapid spectrophotometric assay of monoamine oxidase based on the rate of disappearance of kynuramine. *J. Biol. Chem.* **235**, 1160–1163.

12. Creasey, N. H., (1956). Factors which interfere in the manometric assay of monoamine oxidase. *Biochem. J.* **64**, 178–183.

13. Tipton, K. F., and Dawson, A. P. (1968). The distribution of monoamine oxidase and α-glycerophosphate dehydrogenase in pig brain. *Biochem. J.* **108**, 95–99.

14. Houslay, M. D., and Tipton, K. F. (1973). The reaction pathway of membrane bound rat liver mitochondrial monoamine oxidase. *Biochem. J.* **135**, 735–750.

15. Green, A. L. (1962). The inhibition of monoamine oxidase by arylalkylhydrazines. *Biochem. J.* **84**, 217–223.

16. Houslay, M. D., and Tipton, K. F. (1973). The nature of the electrophoretically separable multiple forms of rat liver monoamine oxidase. *Biochem. J.* **135**, 173–186.

17. Snyder, S. H., and Hendley, E. D. (1968). A simple and sensitive fluorescence assay for monoamine oxidase and diamine oxidase. *J. Pharmacol. Exp. Ther.* **163**, 386–392.

18. Tipton, K. F., and Youdim, M. B. H. (1976). Assay of monoamine oxidase. *Ciba Found. Symp.* **39**, (new ser.), 393–403.

19. Greenawalt, J. W. (1972). Localization of monoamine oxidase in rat liver mitochondria. *Adv. Biochem. Psychopharmacol.* **5**, 207–226.

20. Youdim, M. B. H. (1975). Monoamine oxidase. *Int. Rev. Biochem. Ser. One* **12**, 189–209.

21. Minamiura, N., and Yasunobu, K. T. (1978). Bovine liver monoamine oxidase. A modified purification procedure and preliminary evidence for two subunits and one FAD. *Arch. Biochem. Biophys.* **189**, 481–489.

22. Salach, J. I. (1979). Monoamine oxidase from beef liver mitochondria: Simplified isolation procedure, properties and determination of its cysteinyl flavin content. *Arch. Biochem. Biophys.* **192**, 128–137.

23. Houslay, M. D. (1977). A model for the selective mode of action of the irreversible monoamine oxidase inhibitors clorgyline and deprenyl, based on studies of their ability to activate a $Ca^{2+} - Mg^{2+}$ ATPase in defined lipid environments. *J. Pharm. Pharmacol.* **29**, 664–669.

24. Schurr, A., Porath, O., Krup, M., and Livne, A. (1978). The effects of hashish components and their mode of action on monoamine oxidase from the brain. *Biochem. Pharmacol.* **27**, 2513–2517.

25. Houslay, M. D., and Tipton, K. F. (1975). Rat liver mitochondrial monoamine oxidase: A change in the reaction mechanism on solubilization. *Biochem. J.* **145**, 311–321.

26. Houslay, M. D., and Tipton, K. F. (1974). A kinetic evaluation of monoamine oxidase activity in rat liver mitochondrial outer membranes. *Biochem. J.* **139**, 645–652.

27. Yasunobu, K. T., Igaue, I., and Gomes, B. (1968). The purification and properties of beef liver mitochondrial monoamine oxidase. *Adv. Pharmacol.* **6a**, 43–59.

28. Oreland, L. (1971). Purification and properties of pig liver mitochondrial monoamine oxidase. *Arch. Biochem. Biophys.* **139**, 410–421.

29. Gomes, B., Igaue, I., Kloepfer, H., and Yasunobu, K. T. (1969). Amine oxidase. XIV. Isolation and characterisation of the multiple beef liver amine oxidase components. *Arch. Biochem. Biophys.* **132**, 16–27.

30. Yasunobu, K. T., Wantabe, K., and Zeidan, H. (1979). Monoamine oxidase: Some new findings. *In* "Monoamine Oxidase: Structure, Function and Altered Functions" (R. W. von Korff, T. P. Singer, and D. Murphy, eds.) Academic Press. New York pp. 251–263.

31. Aohi, S., Manabe, T., and Okuyama, S. (1977). Molecular weight estimation of bovine brain monoamine oxidase. *J. Biochem. (Tokyo)* **82**, 1533–1539.

32. Walker, W. H., Kearney, E. B., Seng, R. L., and Singer, T. P. (1971). The covalently bound flavin of hepatic monoamine oxidase. 2. Identification and properties of cysteinyl flavin. *Eur. J. Biochem.* **24**, 328–331.

33. Youdim, M. B. H., and Collins, G. G. S. (1971). The dissociation and reassociation of rat liver mitochondrial monoamine oxidase. *Eur. J. Biochem.* **18**, 73–78.
34. Oreland, L., Kinemuchi, H., and Stigbrand, T. (1973). Pig liver monoamine oxidase: Studies on the subunit structure. *Arch. Biochen. Biophys.* **159**, 854–860.
35. Tipton, K. F. (1968). The reaction pathway of pig brain mitochondrial monoamine oxidase. *Eur. J. Biochem.* **5**, 316–320.
36. Oi, S., and Yasunobu, K. T. (1973). Mechanistic aspects of the oxidation of amines by monoamine oxidase. *Biochem. Biophys. Res. Commun.* **53**, 631–637.
37. Gomes, B., Naguwa, G., Kloepfer, H. G., and Yasunobu, K. T. (1969). Amine oxidase. XV. The sulfydryl groups of beef liver mitochrondrial monoamine oxidase. *Arch. Biochem. Biophys.* **132**, 28–33.
38. Gomes, B., Kloepfer, H. G., Oi, S., and Yasunobu, K. T. (1976). The reaction of sulfydryl reagents with bovine hepatic monoamine oxidase: Evidence for the presence of two cysteine residues essential for activity. *Biochim. Biophys. Acta* **483**, 347–357.
39. McEwen, C. M., Sasaki, G., and Jones, D. C. (1970). Human liver mitochondrial monoamine oxidase. III. Kinetic studies concerning time-dependent inhibitions. *Biochemistry* **8**, 3963–3972.
40. White, H. L., and Glassman, A. T. (1977). Multiple binding sites of human brain and liver monoamine oxidase: Substrate specificities, selective inhibitions, and attempts to separate enzyme forms. *J. Neurochem.* **29**, 987–997.
41. Gorkin, V. Z. (1976). Monoamine oxidase inhibitors and the transformation of monoamine oxidases. *Ciba Found. Symp.* **39**, (new ser.), 61–79.
42. Youdim, M. B. H., and Sourkes, T. L. (1966). Properties of purified, soluble monoamine oxidase. *Can. J. Biochem.* **44**, 1397–1400.
43. Smith, T. E., Weissbach, H., and Udenfriend, S. (1963). Studies on monoamine oxidase: The mechanism of inhibition of monoamine oxidase by iproniazid. *Biochemistry* **2**, 746–751.
44. Maycock, A., Abeles, R. H., Salach, J. I., and Singer, T. P. (1976). Structure of the flavin inhibitor adduct of monoamine oxidase. *Biochemistry* **15**, 114–125.
45. Kenny, W. C., Nagy, J., Salach, J., and Singer, T. P. (1979). Structure of the covalent phenylhydrazine adduct of monoamine oxidase. In "Monoamine Oxidase: Structure, Function and Altered Functions" (T. P. Singer, R. W. von Korff, and D. Murphy, eds.). Academic Press, New York pp. 25–37.
46. Paech, C., Salach, J. I., and Singer, T. P. (1980). Mechanism of the suicide inactivation of monoamine oxidase by transphenylcyclopropylamine. In "Monoamine Oxidase: Structure Function and Altered Functions" (T. P. Singer, R. W. von Korff, and D. Murphy, eds.). Academic Press, New York pp. 39–50.
47. Gabay, S., Achee, F. M., and Mentes, G. (1976). Some parameters affecting the activity of monoamine oxidase in purified bovine brain mitochondria. *J. Neurochem.* **27**, 415–424.
48. Roth, J. A. (1979). Effects of drugs on inhibition of oxidized and reduced form of MAO. In "Monoamine Oxidase: Structure, Function and Altered Functions" (T. P. Singer, R. W. von Korff, and D. Murphy, eds.). Academic Press, New York pp. 153–168.
49. McEwen, C. M., Sasaki, G., and Jones, D. C. (1969). Human liver mitochondrial monamine oxidase. II. Determinants substrate and inhibitor specificites. *Biochemistry* **8**, 3952–3962.
50. Williams, C. H. (1974). Monoamine oxidase. I. Specificity of some substrates and inhibitors. *Biochem. Pharmacol.* **23**, 615–628.
51. Smith, T. E., Weissbach, H., and Udenfriend, S. (1962). Studies on the mechanism of action of monamine oxidase: Metabolism of N,N-dimethyltryptamine and N,N-dimethyltryptamine N-oxide. *Biochemistry* **1**, 137–143.

52. Hellerman, L., and Erwin, V. G. (1968). Mitochondrial monoamine oxidase. II. Action of various inhibitors for the bovine kidney enzyme. Catalytic mechanism. *J. Biol. Chem.* **243,** 5234–5243.

53. Fischer, A. G., Schulz, A. R., and Oliner, L. (1968). Thyroidal biosynthesis of iodothyronines. II. General characteristics and purification of mitochondrial monoamine oxidase. *Biochim. Biophys. Acta* **159,** 460–471.

54. Oi, S., Shimada, K., Inamasu, M., and Yasunobu, K. T. (1970). Mechanistic studies of beef liver mitochondrial amine oxidase. XVII. Amine oxidase. *Arch. Biochem. Biophys.* **139,** 28–37.

55. Tipton, K. F. (1972). Some properties of monoamine oxidase. *Adv. Biochem. Psychopharmacol.* **5,** 11–24.

56. Tipton, K. F., and Mantle, T. J. (1977). Dynamic properties of monoamine oxidase. *In* "Structure and Function of Monoamine Enzymes" (E. Usdin, N. Weiner, and M. B. H. Youdim, eds.), pp. 559–584. Dekker, New York.

57. Fowler, C. J., Callingham, B. A., Mantle, T. J., and Tipton, K. F. (1978). Monoamine oxidase A and B: A useful concept? *Biochem. Pharmacol.* **27,** 97–101.

58. Suzuki, O., Katsumata, Y., Oya, M., and Matsumoto, T. (1979). Effect of β-phenylethylamine concentration on its substrate specificity for type A and type B monoamine oxidase. *Biochem. Pharmacol.* **28,** 953–956.

59. Roth, J. A., and Feor, K. (1978). Deamination of dopamine and its 3-O-methylated derivative by human brain monoamine oxidase. *Biochem. Pharmacol.* **27,** 1606–1608.

60. Glover, V., Sandler, M., Owen, F., and Riley, G. J. (1977). Dopamine is a monoamine oxidase B substrate in man. *Nature (London)* **265,** 80–81.

61. Youdim, M. B. H. (1972). Multiple forms of monoamine oxidase and their properties. *Adv. Biochem. Psychopharmacol.* **5,** 67–77.

62. Tipton, K. F., Houslay, M. D., and Garrett, N. J. (1973). Allotopic properties of human brain monoamine oxidase. *Nature (London)* **246,** 213–214.

63. White, H. L., and Wu, J. C. (1975). Multiple binding sites of human brain monoamine oxidase as indicated by substrate competition. *J. Neurochem.* **25,** 21–26.

64. Mantle, T. J., Wilson, K., and Long, R. F. (1975). Kinetic studies of membrane bound rat liver monoamine oxidase. *Biochem. Pharmacol.* **24,** 2039–2046.

65. Dennick, R. G., and Mayer, R. J. (1977). Purification and immunochemical characterisation of monoamine oxidase from rat and human liver. *Biochem. J.* **167,** 167–174.

66. Huang, R. H., and Eiduson, S. (1977). Significance of multiple forms of brain monoamine oxidase *in situ* as probed by electron spin resonance. *J. Biol. Chem.* **252,** 284–290.

67. Sawyer, S. T., and Greenawalt, J. W. (1979). Association of monamine oxidase with lipid. A comparative study of mitochondria from Novikoff hepatoma and rat liver. *Biochem. Pharmacol.* **28,** 1735–1744.

68. Naoi, M., and Yagi, K. (1979). Beef heart mitochondrial monoamine oxidase: Effect of phospholipids on enzyme activity. *Abstr., Int. Congr. Biochem. 11th, 1979* p. 266.

69. Russell, S. M., Davey, J., and Mayer, R. J. (1979). Vectorial orientation of monoamine in the mitochondrial outer membrane. *Biochem. J.* **181,** 7–14.

70. Neupert, W., Brdiczka, D., and Bucher, T. (1967). Incorporation of amino acids into the outer and inner membrane of isolated rat liver mitochondria. *Biochem. Biophys. Res. Commun.* **27,** 488–493.

71. Diaz Borges, J. M., and D'Iorio, A. (1972). Multiple forms of mitochondrial monoamine oxidase. *Adv. Biochem. Psychopharmacol.* **5,** 79–89.

72. Kandaswami, C., Diaz Borges, J. M., and D'Iorio, A. (1977). Studies on the fractionation of monoamine oxidase from rat liver mitochondria. *Arch. Biochem. Biophys.* **183,** 273–280.

73. Goridis, C., and Neff, N. H. (1971). Evidence for a specific monoamine oxidase associated with sympathetic nerves. *Neuropharmacology* **10,** 557–564.
74. Achee, F. M., Gabay, S., and Tipton, K. F. (1977). Some aspects of monoamine oxidase activity in brain. *Prog. Neurobiol.* **18,** 325–348.
75. Mantle, T. J., Garrett, N. J., and Tipton, K. F. (1976). The development of monoamine oxidase in rat liver and brain. *FEBS Lett.* **64,** 227–230.
76. Della Corte, L., and Callingham, B. A. (1977). The influence of age and adrenalectomy on rat heart monoamine oxidase. *Biochem. Pharmacol.* **26,** 407–415.
77. Robinson, D. S., Nies, A., Davis, J. N., Bunney, W. E., Coppen, A. J., Davis, J. M., Colburn, R. W., Bowne, H. R., and Shaw, D. M. (1972). Ageing, monoamines and monoamine oxidase levels. *Lancet* **1,** 290–291.

Cumulative Index for Volumes I and II

The roman numerals preceeding page numbers indicate volume number.